T0419843

127
Advances in Biochemical Engineering/Biotechnology

Series Editor: T. Scheper

Advances in Biochemical Engineering/Biotechnology

Series Editor: T. Scheper

Recently Published and Forthcoming Volumes

Genomics and Systems Biology of Mammalian Cell Culture
Volume Editors: Hu, W.-S., Zeng, A.-P.
Vol. 127, 2012

Tissue Engineering III: Cell - Surface Interactions for Tissue Culture
Volume Editors: Kasper, C., Witte, F., Pörtner, R.
Vol. 126, 2012

Biofunctionalization of Polymers and their Applications
Volume Editors: Nyanhongo G.S., Steiner W., Gübitz, G.M.
Vol. 125, 2011

High Resolution Microbial Single Cell Analytics
Volume Editors: Müller S., Bley, T.
Vol. 124, 2011

Bioreactor Systems for Tissue Engineering II
Strategies for the Expansion and Directed Differentiation of Stem Cells
Volume Editors: Kasper C., van Griensven M., Pörtner, R.
Vol. 123, 2010

Biotechnology in China II
Chemicals, Energy and Environment
Volume Editors: Tsao, G.T., Ouyang, P., Chen, J.
Vol. 122, 2010

Biosystems Engineering II
Linking Cellular Networks and Bioprocesses
Volume Editors: Wittmann, C., Krull, R.
Vol. 121, 2010

Biosystems Engineering I
Creating Superior Biocatalysts
Volume Editors: Wittmann, C., Krull, R.
Vol. 120, 2010

Nano/Micro Biotechnology
Volume Editors: Endo, I., Nagamune, T.
Vol. 119, 2010

Whole Cell Sensing Systems II
Volume Editors: Belkin, S., Gu, M.B.
Vol. 118, 2010

Whole Cell Sensing Systems I
Volume Editors: Belkin, S., Gu, M.B.
Vol. 117, 2010

Optical Sensor Systems in Biotechnology
Volume Editor: Rao, G.
Vol. 116, 2009

Disposable Bioreactors
Volume Editor: Eibl, R., Eibl, D.
Vol. 115, 2009

Engineering of Stem Cells
Volume Editor: Martin, U.
Vol. 114, 2009

Biotechnology in China I
From Bioreaction to Bioseparation and Bioremediation
Volume Editors: Zhong, J.J., Bai, F.-W., Zhang, W.
Vol. 113, 2009

Bioreactor Systems for Tissue Engineering
Volume Editors: Kasper, C., van Griensven, M., Poertner, R.
Vol. 112, 2008

Food Biotechnology
Volume Editors: Stahl, U., Donalies, U. E. B., Nevoigt, E.
Vol. 111, 2008

Protein-Protein Interaction
Volume Editors: Seitz, H., Werther, M.
Vol. 110, 2008

Biosensing for the 21st Century
Volume Editors: Renneberg, R., Lisdat, F.
Vol. 109, 2007

Genomics and Systems Biology of Mammalian Cell Culture

Volume Editors:
Wei-Shou Hu · An-Ping Zeng

With contributions by

M. Berger · V. Blanchard · L. Botezatu · M. Castro-Melchor
C. Clemens · L. Gama-Norton · Z. P. Gerdtzen · H. Hauser
E. Heinzle · B. Hitzmann · P. Hossler · W.-S. Hu · U. Jandt
C. Kasper · H. Kaufmann · M. Kaup · D. Landgrebe
H. Le · M. Lübbecke · K. Mutz · J. Niklas · J. Schaub
T. Scheper · R. Schucht · T. W. Schulz · S. Sievers
F. Stahl · J. Walter · D. Wirth · A.-P. Zeng

Editors
Prof. Dr. Wei-Shou Hu
Department of Chemical Engineering
and Material
University of Minnesota
Washington Ave. SE 421
Minneapolis MN 55455
USA

Prof. Dr. An-Ping Zeng
Technische Universität Hamburg-Harburg
Institut für Bioprozess- und
Biosystemtechnik
Denickestr. 15
21073 Hamburg
Germany

ISSN 0724-6145
e-ISSN 1616-8542
ISBN 978-3-642-28349-9
e-ISBN 978-3-642-28350-5
DOI 10.1007/978-3-642-28350-5
Springer Heidelberg New York Dordrecht London

Library of Congress Control Number: 2012931409

Printed on acid-free paper

Springer is part of Springer Science+Business Media (www.springer.com)

Series Editor

Prof. Dr. T. Scheper

Institute of Technical Chemistry
University of Hannover
Callinstraße 3
30167 Hannover, Germany
scheper@iftc.uni-hannover.de

Volume Editors

Prof. Dr. Wei-Shou Hu

Department of Chemical Engineering
and Material
University of Minnesota
Washington Ave. SE 421
Minneapolis MN 55455
USA

Prof. Dr. An-Ping Zeng

Technische Universität Hamburg-Harburg
Institut für Bioprozess- und
Biosystemtechnik
Denickestr. 15
21073 Hamburg
Germany

Editorial Board

Honorary Editor

Founding Editors

Advances in Biochemical Engineering/Biotechnology

Advances in Biochemical Engineering/Biotechnology is included in Springer's ebook package *Chemistry and Materials Science*. If a library does not opt for the whole package the book series may be bought on a subscription basis. Also, all back volumes are available electronically.

For all customers who have a standing order to the print version of *Advances in Biochemical Engineering/Biotechnology*, we offer free access to the electronic volumes of the Series published in the current year via SpringerLink.

If you do not have access, you can still view the table of contents of each volume and the abstract of each article by going to the SpringerLink homepage, clicking on "Chemistry and Materials Science," under Subject Collection, then "Book Series," under Content Type and finally by selecting *Advances in Biochemical Bioengineering/Biotechnology*

You will find information about the

- Editorial Board
- Aims and Scope
- Instructions for Authors
- Sample Contribution

at springer.com using the search function by typing in *Advances in Biochemical Engineering/Biotechnology*.

Color figures are published in full color in the electronic version on SpringerLink.

Aims and Scope

Advances in *Biochemical Engineering/Biotechnology* reviews actual trends in modern biotechnology.

Its aim is to cover all aspects of this interdisciplinary technology where knowledge, methods and expertise are required for chemistry, biochemistry, microbiology, genetics, chemical engineering and computer science.

Special volumes are dedicated to selected topics which focus on new biotechnological products and new processes for their synthesis and purification. They give the state-of-the-art of a topic in a comprehensive way thus being a valuable source for the next 3–5 years. It also discusses new discoveries and applications.

In general, special volumes are edited by well-known guest editors. The series editor and publisher will however always be pleased to receive suggestions and supplementary information. Manuscripts are accepted in English.

In references *Advances in Biochemical Engineering/Biotechnology* is abbreviated as *Adv. Biochem. Engin./Biotechnol.* and is cited as a journal.

Special volumes are edited by well-known guest editors who invite reputed authors for the review articles in their volumes.

Impact Factor in 2010: 2.139; Section "Biotechnology and Applied Microbiology": Rank 70 of 160

Attention all Users of the "Springer Handbook of Enzymes"

Information on this handbook can be found on the internet at springeronline.com

A complete list of all enzyme entries either as an alphabetical Name Index or as the EC-Number Index is available at the above-mentioned URL. You can download and print them free of charge.

A complete list of all synonyms (more than 57,000 entries) used for the enzymes is available in print form (ISBN 978-3-642-14015-0) and electronic form (ISBN 978-3-642-14016-7).

Save 15%

We recommend a standing order for the series to ensure you automatically receive all volumes and all supplements and save 15% on the list price.

A Special Preface: Laudatio for Professor Wei-Shou Hu

This is a special preface for the topical volume "Genomics and Systems Biology of Mammalian Cell Culture" in the Series Advances in Biochemical Engineering/ Biotechnology. The Series Editor, Professor Thomas Scheper, suggested dedicating this volume as a "surprise gift" to the co-editor of the volume, Professor Wei-Shou Hu, on the occasion of his 60th birthday in November 2011. It is our distinct honor and pleasure to follow this suggestion and provide this laudatio. Dr. Wei-Shou Hu is currently a Distinguished McKnight University Professor at the Department of Chemical Engineering and Materials Science, University of Minnesota, USA. As a former research associate (Weichang Zhou) and a visiting scientist (An-Ping Zeng) in his laboratory, we have interacted with Dr. Wei-Shou Hu for the last twenty years in various aspects, and witnessed many of his remarkable achievements as a researcher, an educator and a leader in our community. We have great admiration for his leadership, boundless energy, and outstanding contributions in broad biochemical engineering fields.

Dr. Wei-Shou Hu is one of the internationally most respected biochemical engineering educators, researchers, and leaders. He is especially recognized for his landmark achievements spanning multiple biochemical engineering fields including cell culture engineering, metabolic engineering, genomics and systems biology, tissue and stem cell engineering, and his tireless service to students, colleagues, and the profession. He is truly a "Founding Father" of modern mammalian cell culture engineering as well as a visionary leader in applied genomics as exemplified by his forming of an academic-industrial consortium to decipher the Chinese hamster genome.

Wei-Shou Hu studied Agricultural Chemistry at the National Taiwan University. After obtaining his B.S. degree, he went to the US and studied Biochemical Engineering at the Massachusetts Institute of Technology. There he obtained his Ph.D. degree under the supervision of Professor Daniel Wang in 1983. Thereafter, he joined the Department of Chemical Engineering and Materials Science, University of Minnesota, as an Assistant Professor. He was promoted to an Associate Professor in 1989, became a full Professor in 1994, and was named a

Distinguished McKnight University Professor in 1998. Prof. Hu has had an outstandingly productive professional career that includes over 270 publications and 8 issued patents. In addition to 47 Ph.D. and 24 M.S. students who have graduated from his laboratory, a large number of postdocs, visiting scientists and international students have worked in his laboratory. Furthermore, over two thousand biotechnology professionals have passed through his well-known training course on cell culture engineering. In the following sections, we briefly summarize his key research accomplishments, contributions to the training of the next generation of biochemical engineers, leadership, and services to the profession.

I. Research Excellence

Dr. Wei-Shou Hu's research has focused on applying engineering fundamentals to the analysis of biological systems and to the advancement of biotechnological innovations. In addition to his core area of cell culture engineering, he has cultivated diverse research interests in microbial and plant systems as well as in tissue engineering and stem cell culture engineering. He always thrives on combining his intimate understanding of cell physiology and classical chemical reaction engineering to break new ground in biochemical engineering. An experimentalist by nature, Dr. Hu brings a characteristic research approach to every area he touches. He is never hesitant to apply unfamiliar new tools for systematic quantification to his research. He always reaches out to colleagues in theoretical and computational arenas to expand intellectual boundaries. His key research contributions include:

A. Cell Culture Engineering

Cell culture processes were largely an art form until the mid 1980s. They are now the workhorses of the biotechnology industry and account for over 50 billion dollars of annual product sales. Dr. Hu's pioneering injection of biochemical engineering fundamentals into cell culture processes has contributed in no small measure to the rapid advancement of this field of endeavor. His key accomplishments in cell culture engineering include:

1. Dr. Hu was amongst the first to introduce engineering quantification to cell culture technology. In his early work, he introduced the concept of distributed properties of cells and microcarrier size for optimization of the growth of anchorage-dependent cells in scalable bioreactors. He demonstrated that despite the complexity of the animal cell's structure and growth requirements, achieving steady states with one and only one growth rate limiting substrate is possible. He also was the first to demonstrate irrevocably the applicability and limitations of the Monod type model, and he correctly argued that it would be

possible to have steady state multiplicity when different metabolisms are imposed on the cell culture system, leading to the use of cybernetic models to describe that phenomenon. Since each of these steady states represents varying levels of productivity, the implications of these findings on commercial process optimization are enormous. Recently he again applied reaction engineering analysis to glycosylation pathways and presented ways that uniform glycans can be synthesized through cell engineering.

2. Dr. Hu has been a leader in introducing novel research tools to address fundamental issues in cell culture engineering. Through a combination of process insights and awareness of modern analytical techniques, he has introduced new methodologies to advance cell culture engineering. He first applied flow cytometry to measure population heterogeneity and linked that to understanding and optimizing culture productivity. He also first demonstrated the power of confocal imaging and optical sectioning techniques in revealing the structural characteristics of 3D cell supports and the behavior and functional activity of cells' interiors. Dr. Hu again was among the first to apply the tools of transcriptome and proteome analysis in cell culture engineering. Through these analyses, his research group became the first to discover the role of epigenetic regulations, specifically gene silencing, in cells employed for bioprocessing. The insights gained from this global analysis also prompted Dr. Hu to articulate that the complex trait of hyperproductivity can be attributed to global and subtle alterations in gene expression. Knowing that the vast process data in modern production plants are untapped resources for knowledge discovery, Dr. Hu's group has introduced modern data mining techniques to bioprocess data. The analysis of Genentech's Vacaville plant will lead the way for a new phase of knowledge discovery in cell culture processing.
3. Dr. Hu has been the leading advocate for the incorporation of process engineering fundamentals into cell culture processes. As the biotechnology industry matures, it is increasingly moving towards high cell density, high viability fed-batch and simple continuous culture production systems. The optimal performance of those systems requires metabolic state estimation and dynamic feeding of nutrients. Dr. Hu, as usual, foresaw this need and initiated research in this area before the need became apparent in the wider community. His research illustrates many guiding principles that have led to better performance in industrial processing. Over the years, his research has involved many industrial partners including American Home Products, Sandoz, Boehringer Mannheim, Protein Design Labs, Eli Lilly, Pfizer, Roche, Schering-Plough, Sankyo, Genentech, Centocor, Novartis, and others. More recently, he organized the Consortium for Chinese Hamster Ovary Cell Genomics under the auspices of the Society for Biological Engineers and together with the Bioprocessing Technology Institute of Singapore. Once again, with the genomic tools developed through the consortium, Dr. Hu and his colleagues are expanding the horizon of cell culture engineering by injecting biology fundamentals into the research of this economically important cell line.

B. Microbial Physiology and Regulation

Dr. Hu's research on natural product biosynthesis was amongst the first to identify the rate-limiting step in the production of cephalosporin by using theoretical prediction. This finding was elegantly exploited nearly two decades ago by using genetic modifications to alleviate the bottleneck in production, one rare example of utilized predictions from a kinetic model leading to successful metabolic engineering of the producing organism. His work on demonstrating the temporal and spatial distribution of secondary metabolism in a mycelial system is another example of elegant application of new research tools (confocal microscopy and visualization with green fluorescent protein) to bioengineering research. Recently, he again demonstrated his intellectual leadership by applying genomic tools to decipher the regulatory network of secondary metabolism using Boolean algorithms. He took a bold move to construct a whole genome DNA microarray of *Streptomyces coelicolor* and, using support vector machine, exploited time-series dynamics of transcriptome profiles to reveal regulatory structure and transcriptional unit organization. He and his colleagues combined quantitative transcriptome and proteomic analyses and demonstrated that nearly 15% of genes have different mRNA and protein expression dynamics, an important insight that will bear much importance in gene expression analysis. Finally, his mastery of classical reaction engineering is best illustrated by the demonstration of bistable behavior in convergent gene pair regulatory switches in two different microbial species of *Streptomycetes* and *Enterococcus*, one involved in antibiotic production, and the other in antibiotic resistance. His ability to bridge traditional microbiology and chemical engineering is amply illuminated.

C. Tissue and Stem Cell Culture Engineering

In this field, Dr. Hu is best known for his development of a bioartificial liver device for sustaining the life of liver failure patients. He is among the few biochemical engineers who have successfully applied their reactor engineering skills to a medical device and led a multidisciplinary team to progress to clinical trials under FDA approved IND. His invention has spurred many similar efforts around the world. He has been a major force in fostering the employment of self-assembled tissue-like 3D structures for tissue engineering applications. His demonstration of bile acid transport and cell polarization in such self-assembled liver tissue-like structures is a masterpiece of engineering 3D tissue *in vitro*. His work on bioreactor culture of human embryonic stem cells has attracted much attention in a short while. The successful differentiation of rodent and human stem cells into liver cells has led to over 500 citations. Through transcriptome analysis, he and colleagues have defined the signature of the adult pluripotent stem cells. Once again, this accomplishment demonstrates Dr. Hu's ability to operate at the frontiers of biochemical engineering. In plant tissue engineering, he introduced quantification of morphological features into the development of plant tissue cultures of somatic embryos of carrots and the economically important Douglas fir. He was one of the few biochemical engineers engaged in multi-dimensional complex data analysis long before its importance was recently recognized in genomics and DNA microarray applications. Together with his computer science colleagues, Dr. Hu recently combined protein-protein-interaction network and transcriptome data from human and mouse in a meta-analysis to shed light on the pluripotency of stem cells. Again Dr. Hu is leading the way to open a new avenue for biochemical engineering in stem cell culture. His conviction that a robust stem cell process hinges on engineers' ability to control cell fate through a better understanding of cellular regulatory network will ring true.

II. Education

A. *Textbook and Short Courses*

Dr. Hu was instrumental in attracting Dr. Paul Belter (after his retirement from the Upjohn Company) to the University of Minnesota. They collaborated with Prof. Ed Cussler to develop a course which resulted in the first textbook on "Bioseparations". The publication of that book was timed perfectly to meet the burgeoning educational needs of a rapidly expanding biotechnology industry. He also authored a laboratory manual and developed a videotape entitled "Microcarrier Culture Techniques", both of which are widely used as teaching aids. He created the international short course on Cell Culture Reactor Engineering in 1986 at the infancy of utilizing cell culture for recombinant protein production. This course has facilitated the career transition of many biotechnologists previously engaged in microbial fermentation to new opportunities in protein production by mammalian cell culture. This course, which has been offered annually for the last 25 years in the United States and periodically in other regions of the world as well as on-site in industrial institutions (including Amgen, Merck, MedImmune, Centocor, Genentech, Roche (Germany) and Novozymes (Australia)), has trained well over 2000 professionals. It has also catalyzed the injection of biochemical engineering fundamentals into industrial practice, which was generally regarded as an art. The monograph (*Cell Culture Technology for Pharmaceutical and Cellular Therapy* (*2006*)) he edited with Dr. Sadettin Ozturk is widely used by cell culture professionals around the world.

B. Students and Mentees

Dr. Hu has graduated 47 Ph.D. and 24 M.S. students. In addition, he has hosted over 30 postdoctoral fellows or visiting scientists from abroad in his labs. All have developed successful careers in various industrial and academic organizations. Those devoted to education and research are all very proactive members of the community, including Prof. Jorge Gonzalez (Argentina), Prof. Qin Meng (Zhejiang University, China), Prof. Jae Hoe Kim (KAIST, Korea), Prof. Jonghan Park (KAIST, Korea), Prof. Sarika Mehra (IIT-Bombay, India), Prof. Ziomara Gerdtzen (University of Chile, Chile), Prof. Chang-Chun Hsiao (Chang Gung University, Taiwan), Prof. Chetan Gadgil and Prof. Mughda Gadgil (NCL, Pune, India), Prof. David Umulis (Purdue University), Prof. Manolis Tzanakakis (SUNY Buffalo), and Prof. An-Ping Zeng (Hamburg University of Technology, Germany). Many are now in leadership positions in the US and international pharmaceutical and biotechnology companies including Abbott, Bayer, Biogen-IDEC, Bristol-Myers Squibb, Centocor, Dupont, Eli Lilly, Genencor, Genentech, Genzyme, Glaxo Smith Kline, Invitrogen, Merck, and Pfizer. Some of the industry notables who have trained with Prof. Hu include Mr. Tim Dodge (Staff Scientist, Genencor), Ms. Emily TY Tao (Senior Director, Genzyme), Mr. Matthew Scholz (Corporate Scientist at 3M), Dr. Vaughn Himes (former VP of Operations, Targeted Genetics, VP of Worldwide Manufacturing, Corixa; currently Executive VP, Seattle Genetics), Dr. Hugo Vits (Director General, Shell Research Center, Bangalore, India), Dr. M. V. Peshwa (former VP of Process Sciences, Dendreon; currently Executive VP, Maxcyte, Inc.), Dr. Weichang Zhou (former Associate Director of Fermentation and Cell Culture at Merck, currently Senior Director of Commercial Cell Culture Development at Genzyme), Dr. Derek Adams (Executive Director, Alexion Pharmaceuticals) and Mr. Stephane Bancel (CEO, BioMerieux). The breadth of contributions of Prof. Hu's former students is certainly noteworthy: from design and construction of modern biotech manufacturing facilities to development of large-scale manufacturing processes for production of antibodies, viral vaccines and vectors, engineered cell and tissue products, as well as biologically engineered materials. Some of his students are also actively working as physicians and surgeons at hospitals including Dr. Scott Nyberg (Mayo Clinic), Dr. Wen-Je Ko (National Taiwan University Hospital), and Dr. Bryce Pierson (Fairview Hospital).

III. Leadership and Service

A. Organization of Conferences

Dr. Hu was a co-founder of the Engineering Foundation Conference on Cell Culture Engineering in 1988. Since its inception, the Cell Culture Engineering

Conference series has served as the leading forum for the dissemination of cell culture engineering knowledge and technology. Effectively using that forum, Dr. Hu and his colleagues have taught the biotechnology community that cell culture is no more difficult than microbial fermentation if fundamental biochemical engineering principles are applied to design, operate and control cell culture processes.

Dr. Hu also helped to organize numerous national and international conferences. One meriting particular attention is the Engineering Foundation Biochemical Engineering Conference X, Kananaskis, Alberta, Canada in 1997. This Conference Series had been dwindling in attendance since the late 1980s. The late Professor Jay Bailey took the helm of the Conference Series in 1995 to rejuvenate it. To ensure that it would subsequently retain its vigor, Prof. Bailey and the advisory committee recruited Dr. Hu to organize the 10th conference in 1997. The 1997 offering was a resounding success in fundraising, in attracting international attendance, in balancing academics and industrial participation, and in bringing biochemical engineering fundamentals back into the program. This is another example of Dr. Hu's dedication to serving the community and his ability to lead with vision and unequaled determination.

B. Professional Community

Dr. Hu served as the vice chair and chair of Division 15 (Food, Pharmaceutical, Biotechnology and Bioengineering) of American Institute of Chemical Engineers (AIChE). As in his other endeavors, we witnessed Dr. Hu leading the transformation of an organization in crisis with energy and vision. To streamline the organization of the technical program for the national meeting, Dr. Hu has promoted website information dissemination to allow for better coordination and discussion. He devoted tremendous energy to revert the decline of division membership. He helped steer the formulation of the Division's relationship with and the creation of the Society for Biological Engineers (SBE), and coordinated the San Francisco AIChE annual meeting in 2003, then its most successful annual meeting in many years. His devotion and energy was evident in the vitality of the division during his stewardship. More recently, he, Prof. Miranda Yap and June Wispelwey organized the Consortium for Chinese Hamster Ovary Cells Genomics. The industrial members of the consortium also joined SBE as corporate sponsors, thus financially stabilizing SBE in its critical early years. His contribution was recognized by the distinguished service awards from both Division 15 and SBE.

C. International Support

One of Dr. Hu's passions is the promotion of international cooperation and understanding. He is one of the most active biochemical engineers from the US in the international arena. As one of the principal organizers of the Cell Culture Engineering and Biochemical Engineering Conferences, he promoted conference participation by scientists from emerging Pacific Rim countries. He has also helped many academic institutions in Asian countries to establish or expand research programs (e.g., Osaka University, National University of Singapore, King Mongkut's Institute of Technology, Thailand, Vietnam National University at Ho Chi Minh City, and University of Sydney, Australia). Recently he organized the cell culture training course at IIT Bombay, which attracted all key players in India's biotechnology industry.

Dr. Hu has distinguished himself by his research accomplishments, his leadership and services to the profession. His un-rivaled contributions in the biochemical engineering profession, in particular cell culture engineering, have had far-reaching impact on academic research and the biotechnology industry. We had the good fortune to be former members of his laboratory. Like his former and present students and co-workers, we are grateful to him for his guidance and friendship, not only during the time we spent in his laboratory, but also throughout our entire careers. We have all benefited from his sharp scientific instinct, sound judgment, his enthusiasm for applying engineering principles in solving modern biology problems, and of course his boundless energy. As members of the biotechnology industry and academic research institutions, we wish to thank Wei-Shou sincerely for his seminal contributions and honor his outstanding achievements in our profession. We wish him all the best and continued success in his pursuit of applying engineering principles and fundamental approaches at the frontiers of biochemical engineering and biotechnology.

Framingham, Massachusetts, USA, September 2011 Weichang Zhou
Hamburg, Germany, September 2011 An-Ping Zeng

Preface

Mammalian cell cultures are the dominating production system of today's biopharmaceutic products. The increasing emphasis on product quality and the demand for greater cost-efficiency and process robustness call for better fundamental understanding on cell, process and product. Whereas physiological manipulation and bioprocess renovation have played a pivotal role to the success of cell culture biomanufacturing in the past, genomic and systems approaches will drive the next innovations in mammalian cell culture technology. Genome-wide analysis of the genetic information, gene expression, signaling network and metabolism, along with the integration of vast experimental data and mathematical tools will afford a holistic and quantitative understanding of cellular processes and lead to quality products and robust processes. Although somewhat lagging behind life sciences and biomedicine, systems biology application to cell culture technology is gaining its momentum.

In this volume, we have gathered nine chapters written by authors from both academia and industry which present recent development and different facets of genomic and systems-biological approaches applied to mammalian cell cultures. The first two chapters by Stahl et al. and by Castro-Melchor et al. deal with transcriptomic analysis and methodologies of systems biology analysis of static and dynamic gene expression data, respectively. Examples of their uses in mammalian cell cultures are presented. The next two chapters by Gerdtzen and by Niklas and Heinzle describe strategies and available tools for the modeling of metabolic pathways and networks of mammalian cells, especially the method of metabolic flux analysis. The following chapter by Schaub et al. presents an interesting framework of advanced data analysis consisting multivariate data analysis, metabolic flux analysis and pathway analysis for mapping of large-scale gene expression data. This integrated data analysis approach was successfully applied to the analysis and improvement of cultures of Chinese hamster ovary cells in an industrial bioproduction process. For the production of recombinant pharmaceuticals in mammalian cells the glycosylation of the recombinant proteins is of paramount importance for their functions and efficacy. The next two chapters by Berger et al. and by Hossler review the recent developments of analytic tools for the characterization of protein glycolylation, strategies for protein glycosylation control in production processes and perspectives of genomics and systems biology for understanding and enhancing protein glycolsylation control. For glycosylation and many other key cellular processes in mammalian cells intracellular transport and compartmentation are two key aspects which are however still not well understood, at least not quantitatively and in mathematical models. The chapter by Jandt and Zeng summarizes recent developments in simulation methods and frameworks for describing intracellular transport processes. In the last chapter of this volume Botezatu et al. describes methods for targeted genetic perturbation of mammalian cells which is an essential part of an utmost systems biology approach, namely the iterative process of perturbation, data generation, modeling, hypothesis generation and verification by perturbation again. The authors go a step beyond this by combining a synthetic biology approach for targeted and predictable modification of cellular networks. Indeed, a combined systems analysis and synthesis approach will lead to the next level of advances of mammalian cell cultures.

We thank the authors for their excellent contributions and hope that this volume will give some impulse and inspiration to the genomics and systems biology of mammalian cell cultures.

Winter 2011

Wei-Shou Hu
An-Ping Zeng

Contents

Adv Biochem Engin/Biotechnol (2012) 127: 1–25
DOI: 10.1007/10_2011_102

Published Online: 28 September 2011

Transcriptome Analysis

Frank Stahl, Bernd Hitzmann, Kai Mutz, Daniel Landgrebe, Miriam Lübbecke, Cornelia Kasper, Johanna Walter and Thomas Scheper

Abstract Transcriptome analysis technologies are important systems-biology methods for the investigation and optimization of mammalian cell cultures concerning with regard to growth rates and productivity. For the production of recombinant proteins, knowledge of the expression conditions of the influencing genes is a major issue in the improvement of cell lines by means of genome engineering. This chapter presents two main techniques for transcriptome analysis: microarray technology and next-generation sequencing. Protein-based methods are also briefly outlined. Furthermore, the impact of these technologies on mammalian cell culture improvement is discussed.

Keywords Cell culture · Transcriptome · Systems biology · Microarray · Next generation sequencing

Contents

F. Stahl (✉) · K. Mutz · D. Landgrebe · M. Lübbecke · C. Kasper · J. Walter · T. Scheper
Institute for Technical Chemistry, Leibniz University, Hannover Callinstr. 5, 30167 Hannover, Germany
e-mail: stahl@iftc.uni-hannover.de

B. Hitzmann
Fg Prozessanalytik und Getreidetechnologie, Universität Hohenheim, Garbenstr. 23, 70599 Stuttgart, Germany

1 Introduction

Several sophisticated technologies for genome-wide expression monitoring are available and widely used since the development of microarray technology in the 1990s and the complete sequencing of the human genome. The technical and experimental possibilities for studying gene expression offer a snapshot of the entire genome with a resolution that would have been inconceivable some years ago. The transcriptome displays a complete collection of messenger RNAs present under defined conditions. RNA synthesis is a central process in the flow of genetic information in eukaryotic and prokaryotic cells, and isolated RNA has therefore become the target for many analytical and diagnostic techniques. The availability of simple and efficient systems for the production of synthetic RNA has likewise led to the development of techniques for studying the interactions of biomolecules with defined RNA sequences. There is also an increasing interest in the clinical use of nucleic acids. The in vitro generation of randomized RNA transcripts has led to breakthroughs in the generation of high-throughput screens to identify sequences that may have enzymatic or therapeutic applications.

Traditional gene expression analysis involves techniques such as:

- RT-PCR
- Northern blotting
- Nuclease protection assay.

More progressive techniques are:

- Differential display
- Substractive hybridization
- Representational difference analysis (RDA)
- Expressed sequence tags
- cDNA fragment fingerprinting
- Serial analysis of gene expression (SAGE)

These methods enable the discovery of unknown differentially expressed genes. The bottleneck of the traditional methods is the limitation in the number of genes which can be analyzed in parallel. While conventional methods focus on the examination of single genes, a DNA chip experiment delivers a complete gene expressions pattern of the cell. This high degree of parallelization of biochips displays a great advantage over classic molecular biological methods. During the last decade, more and more microarrays containing probes for each annotated gene in the genome have become commercially available for completely sequenced organisms. For metagenomes and for unsequenced genomes, transcriptome analysis by new-generation sequencing (e.g. 454 or Solexa) is state of the art. Therefore, this chapter describes microarray technology and next-generation sequencing in detail. The impact of transcriptome analysis on cell culture techniques and improvement of productivity will be discussed from the user's point of view.

2 Transcriptome Analysis Using Microarray Technology

DNA microarrays, developed to determine gene expression levels in living cells, have revolutionized the way scientists study gene expression [1–3]. Since they enable the analysis of the mRNA levels of a large number of genes in a single assay, DNA microarrays have become standard tools for gene expression profiling. For the understanding of biological systems with up to 30,000 genes, the measurement of the complete set of transcripts of an organism is necessary. Thus, an ideal tool for such measurements is the DNA microarray technology, a large-scale and high-throughput application utilizing amino-modified oligonucleotides or PCR products arrayed on silylated microscope slides with high-speed robotics. These microscope slides containing many immobilized DNA samples—so-called targets—are typically hybridized with fluorescently labelled cDNA probes. This results in a highly parallel, addressable, and miniaturized format, in contrast to traditional molecular-biology methods.

Applications for this technology include, amongst others, the monitoring of gene expression [4, 5], mutation detection [6, 7], clone mapping, drug development [8], tailored therapeutics, single nucleotide polymorphism (SNP) research, detection of genetically modified organisms (GMO) [9], and high-throughput screening in general. DNA microarrays can considerably simplify and accelerate a number of expensive diagnostic methods and have a profound impact on biological research [4], industrial production [4], medicine [10], diagnostics [11], environmental research [12, 13], bioprocess optimization [14, 15], and pharmacology [8], and will likely be used as the biosensors of the future. Microarray technology represents a powerful tool that allows researchers to link hypothesis testing and data. The data generated by microarray experiments can provide a large amount of information about important cellular pathways and processes.

The complete DNA sequences of various microorganisms which have been determined in the past few years can be applied to optimize strains as well as recombinant protein production. Strain optimization involves measurement of genome-wide mRNA levels in wild-type and mutant strains using DNA microarrays. Furthermore, microarray analysis can help to identify unknown genes required for recombinant protein production.

Bioprocess optimization using microarrays facilitates metabolic control analysis, modeling, and molecular biology methods to create new mutants and strains, e.g. with an optimized protein production rate [16]. Microarrays allow the investigation on a genomic scale enabling the qualitative and quantitative characterization of the cell metabolism. A better understanding of the impact of recombinant protein production can thereby be achieved using the generated "snapshot" of the actual cellular composition and activity. Additionally, the knowledge of the interaction of host cell metabolism with recombinant protein production is improved and contributes to process optimization.

By applying high-throughput screening technologies, it is possible to screen large numbers of strains that are produced by random mutagenesis and to determine the valuable ones containing beneficial mutations. By going through several rounds of random mutagenesis on the one hand and screening for the desired phenotype on the other, this process will thus identify those strains with considerably improved properties for production.

Microarrays play a pivotal role in modern biological sciences. They enable the utilization and analysis of a great amount of genetic information, for example derived from the human genome project. As a result, microarrays facilitate the understanding of gene regulation and gene function. The microarray technology enables the simultaneous analysis of complex genetic changes (the so-called "differential gene regulation") by its high degree of parallelization. This can be achieved by parallel measurements of thousands of interactions between mRNA-derived molecules and genome-derived target molecules, thereby producing large amounts of raw data.

2.1 Fabrication of DNA Microarrays

Microarrays use modified glass slides as substrate enabling high spot density (spot diameter < 200 microns). Various methods for the automated production of DNA microarrays are used at present. The oligonucleotides can either be generated directly in situ on the microarray surface in a so-called on-chip synthesis or can be synthesized separately followed by an immobilization to the surface using a so-called DNA arrayer. There are three primary technologies: photolithography, ink jetting, and contact printing, and variants thereof. Each of these manufacturing technologies has specific advantages and disadvantages.

The photolithographic approach relies on the in situ synthesis of 20–30mer oligonucleotides using photomasks (Fig. 1.1) [17–19]. By utilization of photolabile protection groups, each probe is individually synthesized on the surface of the microarray at a high density. Photolithography was developed by Fodor et al. [17] and commercialized by Affymetrix.

Affymetrix uses several 25mer oligonucleotides per gene in a perfect match and a mismatch manner, whereas Agilent uses one 60mer oligonucleotide per gene. In contrast, the ink jetting and contact printing methods attach pre-synthesized

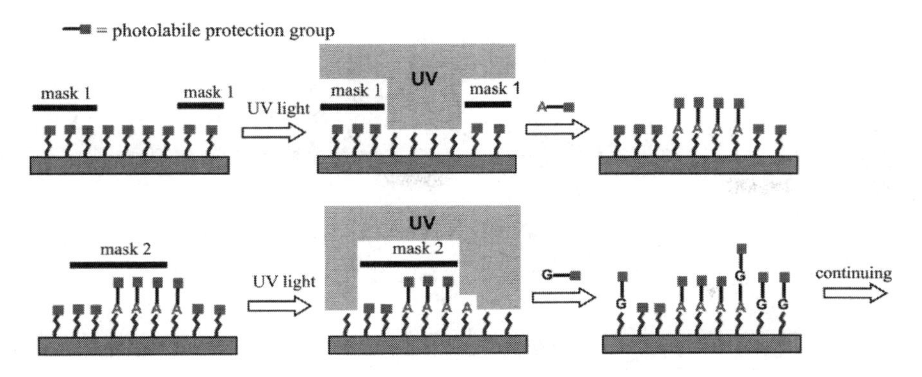

Fig. 1.1 Photolithographic process for the in-situ synthesis of oligonucleotides on microarrays

DNA probes to the chip surface. While the in situ probe synthesis requires sophisticated and expensive equipment, the contact and non-contact dispensing methods have made DNA microarrays affordable for academic research laboratories. In addition, the direct synthesis on the chip is less precise and the products cannot be sufficiently validated. In contrast, pre-synthesized oligonucleotides can be validated and thus produced at a high quality.

Since 1996, many DNA arrayers have become commercially available. Currently, the glass slide DNA microarrays represent the most popular format for gene expression profiling experiments.

2.2 *Design of Microarray Experiments*

The design of scientific experiments is an art of balancing several considerations including cost, equipment, and accuracy. For a given experiment, there would not be one 'right' design. Instead, different designs for the same scientific question may be chosen. Nevertheless, some commonsense principles are broadly accepted [20].

First of all, depending on the nature of the starting material, both, biological and technical replicates are to be selected. The biological variability of a given population needs to be calculated to enable conclusions from the investigation of a single measured effect on the entire target population of interest. Concretely, this means that, for example, investigating a nutritional supplement in a given animal model requires the measurement of several different animals (biological replicates), whereas the same effect in HepG2 cells can be tested on technical replicates.

Secondly, microarray experiments can be performed as single-component (colour) or two-component (colour) assays. Experimental standard designs of two colour assays are the so-called dye swap design (Fig. 1.2a), where the

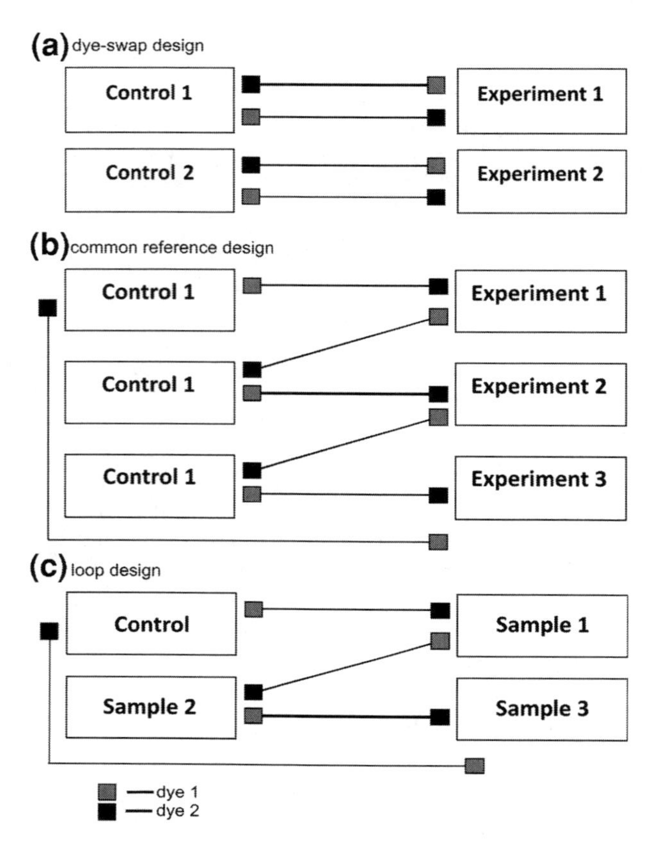

Fig. 1.2 Design of microarray experiments in two-component systems. Competitive hybridization of two labelled cDNA samples to the same microarray. The two mRNA targets are reverse transcribed into cDNA, labelled with different fluorescent dyes (usually *green* fluorescent dye, Cyanine 3 (Cy3, *dye 1*) and a *red* fluorescent dye, Cyanine 5 (Cy5, *dye 2*) mixed in equal proportions and hybridized to the spotted DNA probe molecules on the microarray surface. **a** Dye-swap design. **b** Common reference design. **c** Loop design

hybridization is repeated with a reverse labelling and the dye effect can therefore be minimized, and the so-called common reference design (Fig. 1.2b), where for every hybridization the same reference is always labelled with one dye and the samples of interest (e.g. different patients, cell lines or different time points) are labelled with the other dye. The most economic design, because a minimum number of chips is needed, is the so-called loop design (Fig. 1.2c). Here each sample is hybridized to each of two various different samples with different dye combinations [21–23].

The advantage of the two-component system is its independence of the absolute amount of the fixed DNA, as only the relative ratios of the Cy3 versus Cy5 signal intensities are analyzed for each spot separately. In contrast, one-component systems (e.g. Affymetrix GeneChips) require larger numbers of hybridizations because of their higher array-to-array variance.

2.3 Principle of Microarray Technology

On a single microarray chip, a large set of genes is arrayed in a compact and regular manner. Due to the small size of the spots (<200 μm) several thousands of different oligonucleotides can be immobilized on one single slide. Each of these so-called probes binds to the complementary nucleic acid ("target") isolated from the test and/or reference sample. The comparison of the binding efficiencies between two samples provides an easy and efficient survey of gene transcript level changes for numerous genes in a single experiment. Total or messenger RNA is used as starting material for target preparation. RNA isolation is one important step in the array experiment. Low quality RNA leads to poor hybridization results. In order to prepare the target for hybridization, first-strand cDNA is synthesized enzymatically from total RNA using oligo-d(T) primer or random primer. To exclude interference within the labelling reaction and during hybridization, DNAse I digestion of the isolated RNA is strongly recommended after RNA extraction. During reverse transcription a fluorescently labelled nucleotide (Cy3-dC/UTP or Cy5-dC/UTP) is incorporated into the nascent first-strand cDNA. Subsequently, the template RNA is degraded by chemical treatment and the first-strand cDNA is separated from primers, unincorporated nucleotides, and RNA debris. Two sets of differently labelled cDNAs can be further combined and co-hybridized to the same array under stringent conditions. After hybridization, the unbound and non-specific bound cDNA is removed from the array by thorough washing. After subsequent scanning of the array with a confocal array scanner, the fluorescence intensity of each individual spot is determined and converted to grayscale values. Following normalization of individual grayscale values for each spot, the expression ratio of each gene on the array can be calculated semi-quantitatively. Data normalization is performed by using non-linear regression procedures.

A simple array experiment consists of five basic steps (Fig. 1.3):

1. The oligonucleotides are designed and spotted onto a substrate.
2. The sample RNA is isolated.
3. The cDNA is synthesized, a procedure that also involves fluorescent labelling for later detection.
4. The labelled cDNA target molecules are hybridized to the probe oligonucleotides on the substrate.
5. The hybridization results are imaged and analyzed.

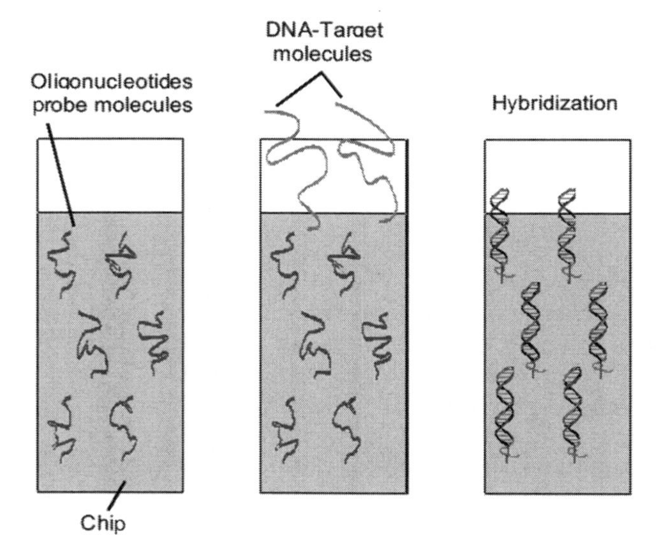

Fig. 1.3 Interaction of labelled cDNA target molecules with molecules immobilized on a glass array

3 Data Analysis

3.1 Microarray Data Analysis

Microarrays promise dynamic snapshots of cell activity, but microarray results are unfortunately not straightforward to interpret. The generation of complicated data sets and the difficulty of interpreting them requires a sound experimental design and particularly a coordinated and appropriate use of statistical methods [24, 25]. Tools for the efficient integration and interpretation of large datasets are needed. Despite the vast amount of literature available on microarray analysis, there is still a lack of standards for comparing and exchanging such data. Therefore, Minimum Information About a Microarray Experiment (MIAME) standards [26] have been established as a prerequisite for the worldwide comparability of gene expression data, and there are several URLs where these standards are available (e.g.http://www.ncbi.nlm.nih.gov/geo/).

For the planning and evaluation of microarray experiments, bioinformatics supply various procedures and algorithms [27]. One reason for the diversity is that the microarray experiments themselves are not performed in a consistent way but in different ways, depending on the objective of a project. In Fig. 1.4 the different contributions of bioinformatics are presented for the planning and evaluation of microarray experiments. They will be discussed below in detail. During all these steps, a fault detection and treatment is performed, which will not be discussed

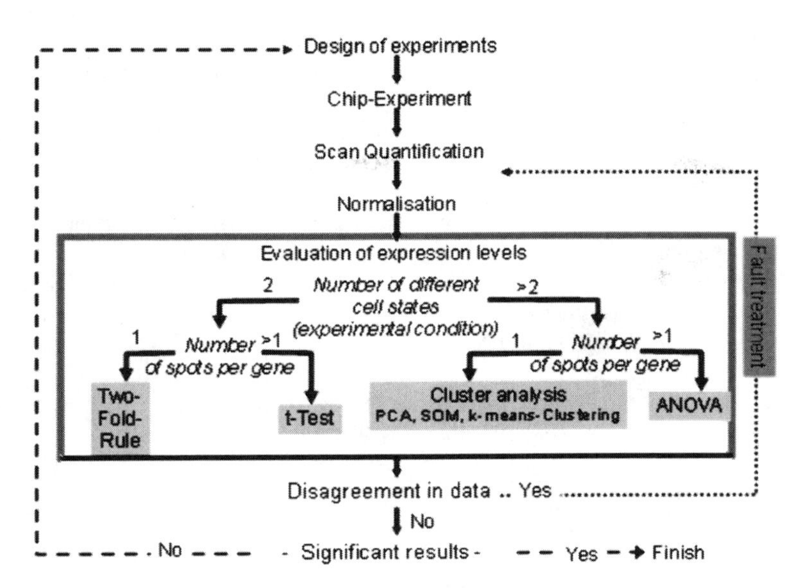

Fig. 1.4 Typical tasks and techniques of bioinformatics for the evaluation of microarray experiments

here. The first step is the design of the DNA microarray experiment. Here, for example, bioinformatics algorithms are used to select the sequence of oligonucleotides, to ensure that they exhibit high specificity for one single mRNA of the whole transcriptome. Software is used to specify the positioning of spots on the chips, in order to indentify the corresponding gene in the evaluation procedure. Sometimes just the effect of an active pharmaceutical ingredient on a cell is investigated. In this case, a single chip experiment can be performed, where the transcriptomes of untreated and treated cells are compared to each other. However, to eliminate the influence of the dye, the dye-swap experiment is carried out, as discussed above. If replicates of spots are used, the positions of these replicates on the chip have to be considered carefully, to get the maximum information. All these issues are specified during the design of the microarray experiment.

After the microarray has been hybridized it must be scanned to acquire the raw data for further evaluation. The scanning is performed for the two dyes separately, measuring the fluorescence signal for each dye and obtaining two grayscale images for evaluation. The first step in analysis is the detection of the spots. Here, different segmentation procedures such as fixed circle, adaptive circle and adaptive shape segmentation can be applied. In the first and second procedures, a circle with a fixed or a variable radius is located optimally around a spot and the pixel intensity in the inner circle area is used to quantify it. The adaptive segmentation procedure does not postulate a circle shape but an arbitrary shape to locate the spot area for quantification [28]. To take into account changes of in the background over the chip, this can be usually determined for each spot separately by using a kind of *o*-ring around a spot whose inner radius is clearly larger than the radius of the spot area. The outer radius is chosen, so that the ring will not interfere with other spots. The intensity in the ring will represent the background of the corresponding spot. For the quantification of the background and spot intensity different quantities can again be used, such as median, modal or mean values of the intensity. The difference between spot intensity and background intensity is a measure of the expression degree of the corresponding gene. In this way systematic variations in a DNA microarray experiment such as slide heterogeneity, spotting variation, changing background signals etc. can be compensated for. Image analysis software such as imaGene™ and GenePix™ software can be used for this evaluation. However, due to the fact, that there are many variabilities from the harvesting of the mRNA to this quantified value, the degree of expression degree obtained cannot be regarded as an absolute measure. Therefore, at least two states of cells are investigated simultaneously (e.g. treated and untreated) whose mRNAs are labelled differently. The values obtained are evaluated relative to each other. Because the mRNAs obtained are treated almost in the same way (except for the dye used for labelling), the ratio of the values gives a relative measure to each other, i.e. the expression change.

However, the labelling as well as the fluorescence intensities of the two dyes are not exactly the same. A normalization procedure is therefore required, which is one of the most important steps during the evaluation. Applying normalization procedures, results from different experiments are made comparable and technical imperfections are compensated for. If housekeeping genes are available, they can be used for normalization. Housekeeping genes are genes whose products are necessary for fundamental cell maintenance and which are transcribed at an almost constant level. Therefore, it can be assumed that they will not change their expression grade under the situations investigated. For these genes the expression as well as the quantified expression grades must be the same. Thus, all expression values of one evaluated image can be transformed by using a multiplication factor in such a way that after normalization the expression values of the housekeeping genes of the two images (representing the transcriptomes of treated and untreated cells) are the same. Then the ratio of the corresponding grade should give the correct change in expression.

If a whole genome chip is under investigation, another normalization procedure can be carried out. Under the assumption that the overall expression of the mRNA is the same for the treated and untreated cells, the sum of the expression grade of all genes must be the same for both cases. Therefore, each expression grade is divided by the sum of the expression grades of all corresponding genes (as a consequence, the sum of the transformed grades will then be 1). After the transformation, the individual ratios are calculated as mentioned before and will give the change in expression.

If the expression change is calculated in this way, it has the disadvantage that a twofold upregulated (ratio = 2) and a twofold downregulated (ratio = 0.5) gene will be characterized by different numerical data. If the logarithm with respect to the base 2 is calculated, then for a twofold upregulated gene the ratio is 1 and for a twofold downregulated gene the ratio is −1. The absolute values are the same for both cases. A symmetric distribution for up-and downregulated genes is therefore obtained. The $\log_2$(ratio) obtained with the normalized expression grades are used as expression levels.

A further normalization procedure based on logarithms is frequently used. Here a special *xy* plot is considered, called an MA plot. In this plot the ordinate values represent the logarithm of the ratio of the corresponding expression grades and the abscissa values represent the logarithm of the multiplied corresponding expression grades. Then a linear regression or alternatively a locally linear regression (LOWESS regression) is carried out with the data. The theoretical values of the regression curve are subtracted from the ordinate values, so that afterwards the MA plot is more symmetric to the abscissa. Further normalization procedures are described in the literature [29].

After all normalization procedures are performed, the $\log_2$(ratio) is evaluated to give expression levels. Here, further analysis depends on whether replicates of spots have been considered or not. If the expression of a gene is represented by a single spot, then the twofold rule is applied, which should be considered just as an auxiliary release. Using the twofold rule an expression level greater than 1 is considered as upregulated, and an expression level less than −1 is considered as downregulated. However, this is difficult to interpret if the expression grades of both states (from treated and untreated cells) are small. Replicates should be therefore be performed. This offers the possibility of applying the *t*-test for the decision of a differently expressed gene.

If the response of cells to different conditions (for example whether they respond to different concentrations of an active pharmaceutical ingredient) is under consideration, then, as described above, replicates should be performed and multivariate evaluation techniques such as cluster analysis, principal component analysis (PCA) or self-organized maps (SOM) should be applied to elucidate the information gained from the chips. If replicates are available, then analysis of variance (ANOVA) is the best choice.

All these multivariate data evaluation techniques do not require explicit knowledge of transcriptome data analysis and will not be discussed here in detail, but can be found in general statistical textbooks.

3.2 Exploratory Analysis

The next step in the analytical pipeline is usually gene expression clustering: a preliminary examination of data to confirm that groups are homogeneous. Many studies aim to find unknown co-regulated genes. Such studies using multi-dimensional scaling and clustering can be summarized as exploratory analysis.

Various clustering techniques can be applied for the identification of patterns in gene expression data [30]. Several software tools for cluster analysis have been developed, such as GeneSight^{TM} from Biodiscovery or Eisen-cluster. Most cluster analysis techniques are hierarchical, where the classification results in nested classes resembling a phylogenetic tree. Non-hierarchical clustering techniques involve partitioning of objects into different groups, such as *k*-means clustering.

The concluding section of analysis deals with methods and problems in determining differentially expressed genes between groups of samples, covering for example *t*-tests or different multiple-testing corrections as well as analysis of variance. The purpose of finding differently expressed genes can be achieved by statistical tests rather than cluster analysis. Finally, the results as well as an additional indication of the statistical reliability are given in order to allow further, more precise studies of gene expression (e.g. qRT-PCR) or the publication of these microarray results as evidence for changes in gene abundance. In general, it is difficult to analyze data from low-density microarray experiments with commercially available programs. Low-density microarrays consist of a choice of a few relevant genes, thus offering the possibility of a very individual chip design in order to investigate specific experimental questions. One disadvantage of low-density microarrays is that the data analysis cannot be performed by commercial software.

Following the data analysis, *k*-means clustering could be conducted depending on the experimental design. Furthermore, the resulting files include not only information about the microarray analysis but also about the pathways of known genes on the microarray coming from the KEGG (Kyoto Encyclopedia of Genes and Genomes) database.

4 Transcriptome Analysis Using Next-Generation Sequencing

Microarray technology has begun to reach its limitations [31]. It shows a relatively small dynamic range due to background, and a limited spot density. In addition, mismatches and cross-hybridization significantly affect the results [32]. Furthermore, the comparison between different experiments usually needs complex bioinformatical normalization algorithms [33]. In contrast to microarrays, which provide only relative mRNA levels, sequencing methods have major advantages with regard to true quantification and they offer statistically complex data analysis [34, 35].

In the past, genetics was only an observational science. But the invention of the DNA sequencing technique was a revolutionary point for this biological science and represents the starting point for modern genomics which can enable mechanistic understandings. DNA sequencing means the assignment of the sequence of nucleotides in a DNA molecule. To date, the genomes of over 1,000 different organisms have been analyzed[1] in total and enhancements in DNA sequencing methods are increasingly used for transcriptome analysis to replace DNA

[1] http://www.ncbi.nlm.nih.gov/guide/genomes/.

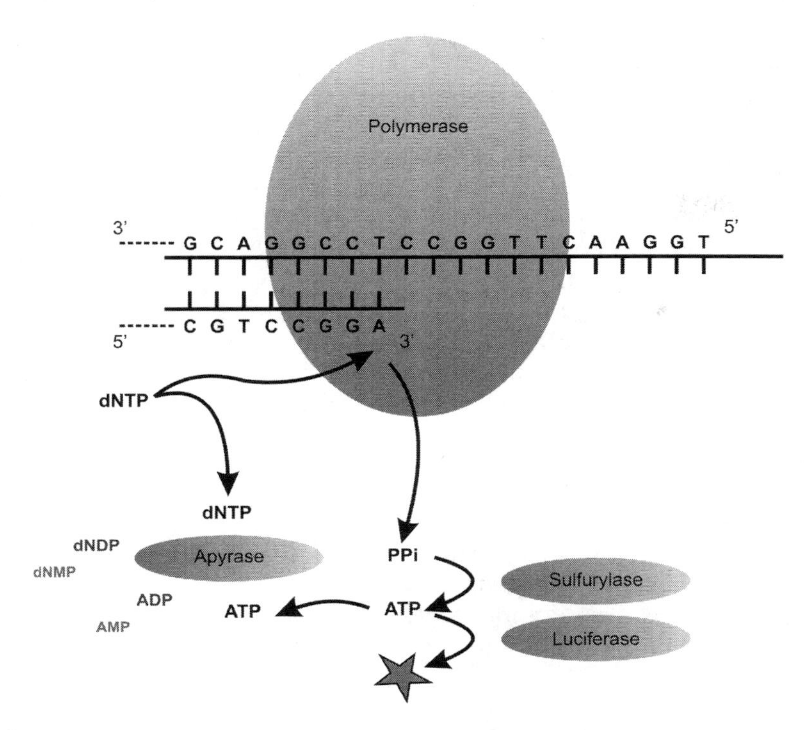

Fig. 1.5 Pyrosequencing is a genetic analysis based on sequencing by synthesis. It delivers explicit sequence data within a few minutes

microarray technology. Today, various methods for the achievement of sequence information are well established. Most of them are based on the dideoxy method drafted by Friedrich Sanger in the 1970s [36], which uses an enzymatic reaction. Starting from a short known primer, DNA polymerase elongates the complementary strand. The use of four differently labelled dideoxy-nucleotides leads to the identification of the unknown DNA sequence.

Recent developments in sequencing technology have led to a new key tool for transcriptomics: the pyrosequencing technology [37], which offers opportunities for accelerated sequencing by highly parallel approaches (Fig. 1.5). In analogy to Sanger et al., pyrosequencing-based next-generation sequencing methods use a DNA polymerase for synthesis of the complementary DNA strand. The incorporation of nucleotides during DNA sequencing is monitored by bioluminescence in real time. A luciferase-based multi-enzyme system generates visible light after nucleotide binding, which can be detected. The four different nucleotides are added successively, so that only the binding of the compatible nucleoside triphosphate generates a light signal. This shows whether a known nucleotide is incorporated or not.

The pyrosequencing technology was invented in 1986 based on the idea of following nucleotide incorporation by using released pyrophosphate (PPi) to generate a signal.

The release of an equimolar amount of PPi is a natural process during the binding of a nucleotide to the 3′ end of the primer, which is used as the starting point of sequencing. The multi-enzyme system mentioned above consists of DNA polymerase, ATP sulfurylase, luciferase and apyrase. After incorporation and release of PPi, ATP sulfurylase catalyses the conversion of the PPi to ATP in a quantitative reaction.

The luciferase uses the ATP for catalyzing the conversion of luciferin to oxyluciferin. Within this reaction visible light is emitted, which can be detected by a CCD camera [37]. A special computer program displays the recorded data as peaks in a diagram. Free ATP and nucleotides are degraded by the apyrase. This disables light emission and regenerates the reaction solution. The complete enzymatic process can be performed in a single well, offering a fast reaction time of approximately 20 min per 96-well plate [37]. It is possible to determine the DNA sequence by computer-assisted analysis of the detected light signals, because of the consecutive addition of the four different dNTPs comparable to the well-known Sanger protocol [38]. Besides Sanger sequencing, pyrosequencing is the only method which is currently commercially available [39]. Pyrosequencing is applied in the analysis of single-nucleotide polymorphisms (SNPs), as well as tag sequencing and whole genome sequencing [37]. Beyond microarrays, next-generation sequencing is pushing transcriptomics further into the digital age [40].

Although those next-generation technologies generate many short DNA fragments with reduced time and costs [41], sequencing of a whole vertebrate genome is still an extensive task. Transcriptome sequencing of cDNA has the advantage that the templates are of relative small size. This enables high-throughput applications of gene expression profiling, genome annotation or discovery of non-coding RNA. Miscellaneous data offer a parallel analysis of gene expression, genomic loci structures and, e.g. SNPs [42]. At present, three popular next-generation platforms support gene expression profiling: SOLiD, Solexa and 454 Sequencing System [40]. In the following section these platforms are presented in detail.

4.1 SOLiD System

SOLiD is the acronym for "Sequencing by Oligonucleotide Ligation and Detection" and is available from Applied Biosystems, Inc. (Foster City, CA, USA). The underlying principle is sequencing by ligation. In this method, DNA fragments of the sample to be sequenced, modified with internal and external adapters, are coupled to magnetic microparticles. The adapters are cleaved and DNA rings are formed by ligation of the adapter ends. The rings are then split again at defined positions at the left and right domain of the adapter. The first few bases of the adherent end domains are sequenced afterwards and new corresponding adapters are ligated to the ends. The new adapters allow the subsequent attachment of the fragments to the microparticles and amplification by emulsion PCR. After accumulation of the particles, octamer degenerated oligonucleotides are hybridized to the particles. These oligonucleotides are each labelled with a different fluorescent dye after the 5th base. Detection then takes place, followed by elimination of the last three bases after the analyzed nucleotide. By repeating this procedure in further cycles, the 10th and 15th bases will be identified. Other steps with shorter primers lead to the detection of the positions 4, 9, 14 and so on.[2]

4.2 454 Sequencing System

The 454 sequencing technology is the primary next-generation method in transcriptomics. The 454 Sequencing System by Roche Diagnostics Corporation (Branford, CT, USA) is an ultra-high-throughput system. It is the first next-generation sequencing technology released to the market [42] and can be described as pyrosequencing in high-density picoliter reactors. DNA fragments received by shearing are attached to streptavidin beads captured into separate droplets in an emulsion PCR. The droplets form small amplification reactors [43]. Then, any bead is transferred into a picoliter plate and analyzed by pyrosequencing. The instrument can sequence up to 120 million bases in a time of about 10 h. Limited by the pyrosequencing chemistry used, the single reading frames (250 nt) are considerably shorter than with Sanger technology (600 nt), but up to 400,000 reactions can be performed in parallel. With more than 100 research publications [42], it is the most widely published next-generation platform.[3]

4.3 Illumina/Solexa Sequencing Technology

This technique is based on a two-step mechanism, where amplification takes place first. Shear-stressed DNA fragments are tagged with different so-called "dense lawn" primers as adapters at both ends of their chain. Together with both complementary primers, all molecules are immobilized randomly to the surface of flow cell channels. DNA fragments hybridize with the complementary primers in a bridging way to start solid-phase bridge amplification immediately and the fragments become double-stranded. With further steps consisting of denaturation, renaturation and synthesis, a high density of equal DNA fragments is generated in an extremely small area. Several million of these dense clusters of double-stranded DNA are synthesized.

The second step is the real sequencing reaction. All four dNTPs labelled with different dyes and primers are added and successively incorporated by a DNA polymerase. After washing steps, a high-definition image is generated by laser

[2] www.appliedbiosystems.com.

[3] www.454.com.

excitation from each cluster. The identity of the first base is recordable. The elimination of the 3′ blocked terminus and the dyes follows. Within each new cycle, the DNA chain is elongated and more images are recorded for analysis. Here the reading frame is tenfold smaller (30 nt) than with common pyrosequencing. The whole system has is closely related to the method from Helicos BioScience.

Applications of this method are sold by both Illumina, Inc. (San Diego, CA, USA) as well as Solexa, Inc. (Hayward, CA, USA). Today the Genome Analyzer IIx is available on the market. It can be used for common DNA sequencing as well as for transcriptome analysis such as RNA-Seq, tag profiling or microRNA discovery.[4]

4.4 RNA-Seq

The transcriptomics alternative to pyrosequencing technology is called short-read high-throughput sequencing or RNA-Seq [41]. In recent years, RNA-Seq has rapidly emerged as the major quantitative transcriptome profiling system [13, 44]. RNA-Seq has been used, for example, for the global profiling of expression levels in human embryonic kidney and B cells [45], or for identification of differently expressed genes in mouse embryonic stem cells [46] and also for quantification of the whole mouse transcriptome [47]. Moreover, structural information or alternative splicing forms can be detected with this method [41].

Unlike microarrays, RNA-Seq can evaluate absolute transcript levels and detect novel transcripts and isoforms. As a consequence, it can be used to determine expression levels more precisely than microarrays [48]. Microarrays, on the other hand, have the power to measure the expression of thousands of genes in parallel, but they are not able to display the coding sequences of the transcripts. The results are calculated from indirect hybridization data, which is gives rise to reproducibility and comparability problems [42]. One great advantage of RNA-Seq compared to microarrays is the possibility of capturing transcriptome dynamics across different cell culture conditions without normalization of data sets. Therefore, RNA-Seq is the method of choice in projects for transcript discovery, especially the analysis of metagenomes.

4.5 Applications

The major application area for next-generation sequencing technologies is biomedical research associated with key goals like “the USD1000 genome” [49]. Next-generation sequencing is used for the detection of sequence variations within

---

[4] www.illumina.com.

an individual genome such as SNPs, deletions, insertions, or structural changes [50]. Here, RNA-Seq is typically used for the analysis of non-coding RNAs as crucial regulators [51].

Secondary next-generation sequencing was adopted for high-throughput research performed mostly by microarrays. To date, it has been possible to generate transcriptome sequencing libraries for important cell cultures for recombinant protein production, e.g. *Arabidopsis thaliana*, *Caenorhabditis elegans* and human cell line transcriptomes were successfully interrogated with 454 technology [52]. For HeLa cells Illumina technology was used [53]. The understanding of whole transcriptional networks was also enabled by RNA-Seq. Examination of small non-coding miRNAs allows a global view of the transcriptome [48].

A major focus of systems-biotechnology work is the quantitative understanding of molecular principles behind protein synthesis, modification and secretion derived from basic production strains as well as mutants and rationally engineered strains. Next-generation sequencing provides the tools for rationalize inverse metabolic engineering approaches so that they can be implemented in future into rational system-wide modeling and optimization strategies [54]. The functional complexity of a transcriptome cannot be fully elucidated with expressed sequence tags and microarrays. RNA-Seq can reveal more precisely the boundaries of untranslated regions at single nucleotide resolution and is useful for analyzing complex transcriptome and sequence variations, e.g. alternative splicing or gene fusion [55].

Additionally, metagenomics is an area of biological sciences concerned with acquisition of the whole genomic information of a biotope. Several microorganisms cannot be cultivated in a laboratory. To identify them without a cultivation process, metagenomic approaches can be performed. This can enhance knowledge of biodiversity or could lead to new biotechnological and pharmaceutical products [56]. Thus, next-generation sequencing can also raise metagenomics to a new level [56]. In this context, future analysis of mRNA levels under different conditions or in different cell types can be assessed by analysis of hybridization intensities and by application of methods using sequenced cDNA fragments, and both techniques will help to improve production output of cell cultures by optimizing cell growth and genetic activity.

5 Protein Microarray Technologies

Since the determination of the complete DNA sequences of a number of organisms, from bacteria to man, and the invention of new techniques like microarrays for monitoring biomolecular interactions, important milestones have been achieved in genomic and proteomic research. The results of such high-throughput screening approaches can change our fundamental understanding of life's cellular processes on a molecular level. However, gene expression analysis is not sufficient to predict the function of a protein. Monitoring protein interactions is an extremely complex issue because the proteome is the quantitative representation of the complete protein expression pattern of a cell, tissue, organ or organism under exactly defined conditions.

Ideally, the analysis of the proteome delivers the complete available set of all proteins currently present in an organism. This data cannot be obtained on the transcriptome level, since no straightforward correlation exists between the amount of mRNA and the actual amount of protein. Parameters like mRNA stability, protein degradation, posttranslational modifications and others prevent a statement of the actual amount of protein based on transcriptome analysis. However, this information is of utmost importance, making a high-throughput analysis of the proteome necessary. One attractive method is the use of protein microarrays, which consist of a solid support, e.g. glass or synthetic material, with a modified or coated surface. Using special printers (preferably non-contact printing heads), capture probes—which may be proteins, peptides, receptors, enzymes or antibodies—are transferred to this surface in the form of micro spots (<200 μm) in a regular manner. Every micro spot contains only one kind of capture probe (in most cases, antibodies). These immobilized capture probes are able to bind their corresponding target molecule from a complex solution.

Different formats of protein microarrays are available. In the forward phase format, antibodies immobilized on the microarray surface are used as capture probes for their target. Since it is possible to immobilize many different antibodies on one single microarray, the forward phase format enables the parallel detection of many different targets within a complex sample. One disadvantage of the forward phase format is the necessity to label the target proteins. This labelling procedure may alter the composition of the sample, different proteins are labelled with varying labelling efficiencies, and the labels introduced may mask the proteins' epitopes essential for binding to the immobilized antibody.

In contrast, the reversed phase microarray offers the possibility of detecting unlabelled proteins. In this format, the protein of interest is directly immobilized onto the microarray surface and probed with fluorescently labelled detection antibodies. While this method allows the detection of the protein of interest in hundreds of different samples in parallel, its major limitation is the binding capacity of the microarray surface, resulting in a low dynamic range of the assay.

In the so-called sandwich format, capture antibodies are immobilized on the microarray surface and the binding of the corresponding protein is detected via labelled detection antibodies (Fig. 1.6). Therefore, for each target protein, two antibodies binding to different epitopes of the target are required. Sandwich-based microarrays avoid the difficulties associated with labelling reactions and exhibit high sensitivity [57]. Moreover sandwich assays are known to be highly specific, since the target must be recognized by two different antibodies.

The direct extrapolation of DNA microarray techniques to protein microarrays is limited due to the sensitive nature of the printed antibodies, which have to keep their native conformation in order to maintain activity. The fabrication of protein arrays is therefore particularly challenging and protein arrays lag behind in development because of the instability of the immobilized protein [58]. One

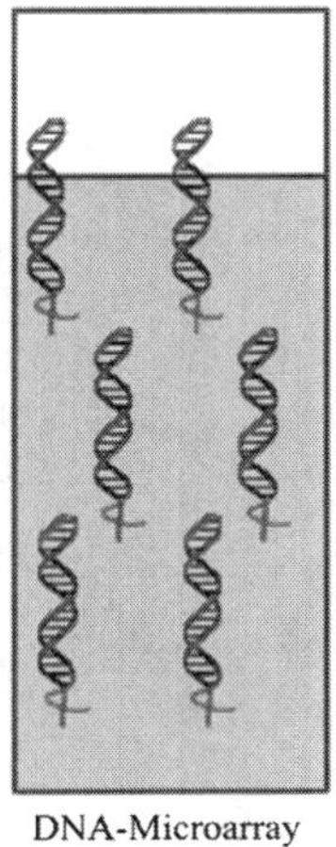

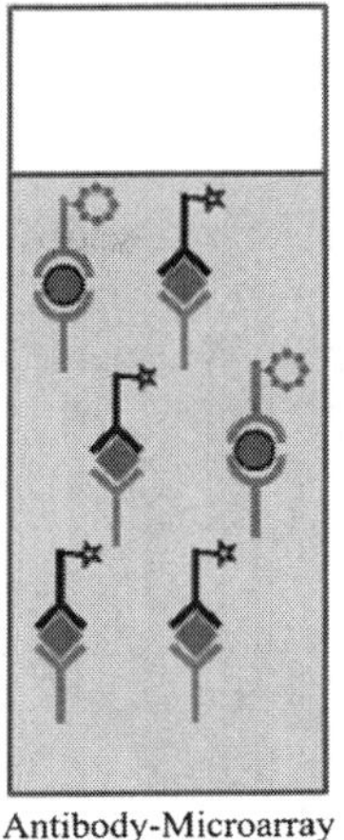

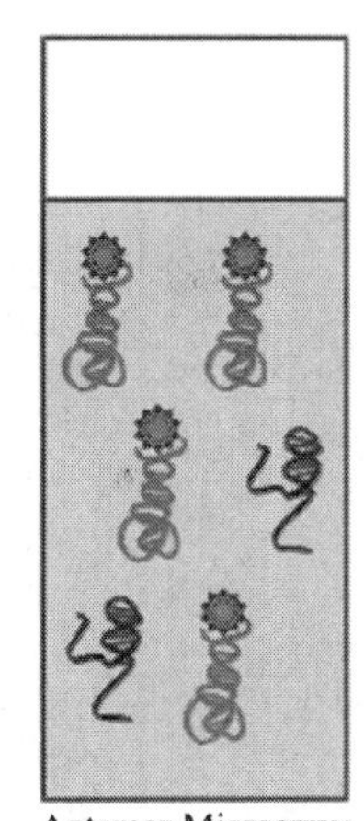

Fig. 1.6 Comparison of DNA, antibody and aptamer microarrays. An aptamer microarray consists of spotted DNA probes as capture molecules for protein detection

approach to overcoming this restriction is the utilization of a three-dimensional matrix for immobilization of the capture antibodies [59, 60]. In this structural environment, proteins are more likely to maintain their active configuration than on planar glass supports [58, 61]. Nitrocellulose membranes have shown their suitability for protein immobilization in Western blotting and the long-term stability of immobilized proteins on this support is known from immuno-diagnostic tests [62, 63]. Nitrocellulose membranes are therefore becoming the microarray substrate of choice in protein microarray applications [64, 65].

Another approach to overcoming the limitations caused by the low stability of immobilized antibodies is the utilization of more stable capture probes. In this context, aptamers have been investigated as an alternative to antibodies [66]. Aptamers are short single-stranded synthetic DNA or RNA oligonucleotides that can bind to a wide range of target molecules, including proteins.

As nucleic acids, aptamers can undergo denaturation, but the process is reversible. As a result of this stability and the possibility of automated selection of aptamers via systematic evolution of ligands by exponential amplification (SELEX), these oligonucleotides are highly promising capture molecules for protein microarrays.

6 Impact of Transcriptome Analysis on Strain Improvement

Mammalian cells in culture can be differentiated into two groups: primary and secondary cells. The latter ones are also known as immortal cells or cell lines. Primary cells are isolated directly from blood or tissue samples. These cells have a restricted life span due to the fact that they undergo only a limited number of cell divisions. Primary cells better represent the tissue from which they are taken and are normally heterogeneous. These cells could be used for R&D applications, particularly for in vitro tests of new drugs and toxicity tests. Continuous cell lines rarely occur spontaneously from primary tissue cells; mostly they are developed by transformation with carcinogenic substances or viral genes. Besides their infinite growth, cell lines have further advantages including faster growth and the ability to be cultured in suspension. This makes them suitable for the production of recombinant proteins in large-scale cultivation [67]. Primary cells as well as cell lines are available as test systems for gene expression analysis, but there are a few drawbacks to both cell types. Primary cells display inter-individual differences, e.g. caused by age and gender of the donor, and the widely used cell lines are limited in their metabolic function because some pathways are different from those in normal tissues [68].

One important aim of strain improvement is to understand and characterize the functional heterogeneity in a given population at the cellular and molecular level. Another aim is to identify and isolate high-producing cells from the entire population and use them in production processes. Combining flow cytometric analysis and sorting of live cells with transcriptome analysis aids in relating molecular regulation processes within cellular subpopulations to the dynamics of the whole cell population [69]. Transcriptome analysis can thus be used to improve cell growth and to increase the productivity of mammalian cell cultures, e.g. by gene optimization.

6.1 High-Producing Cells

Numerous methods exist for developing high-producing populations via fluorescence activated cell sorting (FACS) and gene expression profiling of sorted subpopulations [70]. Co-expression of green fluorescent protein (GFP) is common. Cells which show high GFP fluorescence can be separated to obtain desired populations due to the fact that GFP expression is correlated with high productivity [71]. Simple surface staining is also accomplished [72]. In addition to this method, cultivation at lower temperature and treatment with chemicals such as butyrate can be used to increase the productivity of a cell line. Up to several hundreds of genes are upregulated in high producers; these genes may be involved in the secretory pathway including the Golgi apparatus and cytoskeleton, and may also include genes responsible for product formation [73].

6.2 Cell Cycle Studies

Cytometry is also applicable for cell cycle studies, since it is possible to stain DNA whose content can be correlated with the cell cycle. DNA replication occurs exclusively during the S phase, such that G2-phase cells have twice the cellular DNA content of G1 cells [74]. For further analysis the cells can be synchronized into similar phases. This is achieved by various methods based on biological or physical effects. Due to the fact that the cell size changes during the mitosis, a separation by cell size is one of the practical physical methods. This separation can be performed by FACS or the centrifugal elutriation technique [75, 76]. Afterwards the cells grow in a synchronized cell cycle. This enables transcriptome analysis to search for genes involved in regulating the cell cycle or other cell-cycle-dependent gene activities like productivity rates [77]. The productivity of recombinant products depends on the cell cycle phase and the product. For recombinant proteins produced in CHO cells, it was ascertained that the productivity maximum occurs in the G_1 phase [78].

7 Concluding Remarks

The development of microarrays with DNA probes for gene expression analysis or antibody probes for proteomic applications based on hybridization processes (DNA probes) as well as on the immunological binding process (antibody probes) opens new horizons for biomolecular research. It can result in the production of new proteins, changes in membrane formation and various other alterations concerning cellular assembly [32].

Sequenced genomes are the basis for constructing DNA microarrays representing the common genes in a genome. Furthermore, they enable the synthesis of labelled cDNA from mRNA templates allowing high-throughput detection of transcript levels [33]. Various high-density oligonucletide microarrays are now available commercially.

DNA microarrays have therefore already changed the way scientists study gene expression, but the real challenge starts with determining the function of all the genes discovered within the organisms. DNA microarrays are applied in industrial analytics and biomedical diagnostics, as well as in criminology. They can considerably simplify and accelerate a number of expensive diagnostic methods. Although conventional biosensors work well, their function can be validated and perhaps improved through a functional genomics study in which the induction of several thousand genes is detected simultaneously. Specifically, the use of microarrays can accomplish the following goals:

- Compare the time course of a sensor signal with the actual genomic response.
- Identify genes that respond earlier or more specifically to toxins.
- Identify gene induction patterns that can identify one toxin versus another (which can in turn be incorporated into multichannel sensors).

In the last few years, functional transcriptomics has been advanced by both microarray technology and genome sequencing. Certainly microarray technology has achieved its technical limits and is more and more complemented by high-throughput next-generation sequencing technologies. Unlike microarrays, transcriptome sequencing (RNA-Seq) can evaluate absolute transcript levels, and detect novel transcripts and isoforms. Microarrays have the power to measure the expression of thousands of genes in parallel, but they are not able to reveal the coding sequences of the transcripts. The derived results are calculated from indirect hybridization data, which poses reproducibility and comparability problems. In fact, studies using both microarrays and RNA-Seq show a good correlation between the different data so that it is possible to compare results from one technology with the other [79], and both techniques help to improve production output of cell cultures by optimizing cell growth and genetic activity.

References

1. Khan J et al (1999) DNA microarray technology: the anticipated impact on the study of human disease. Biochimica Biophysica Acta Rev Cancer 1423(2):M17–M28
2. Duggan DJ et al (1999) Expression profiling using cDNA microarrays. Nat Genet 21:10–14
3. Brown PO, Botstein D (1999) Exploring the new world of the genome with DNA microarrays. Nat Genet 21:33–37
4. DeRisi JL, Iyer VR, Brown PO (1997) Exploring the metabolic and genetic control of gene expression on a genomic scale. Science 278(5338):680–686
5. Harrington CA, Rosenow C, Retief J (2000) Monitoring gene expression using DNA microarrays. Curr Opin Microbiol 3(3):285–291
6. Park JH et al (2004) Oligonucleotide microarray-based mutation detection of the K-rasgene in colorectal cancers with use of competitive DNA hybridization. Clin Chem 50(9):1688–1691
7. Hegde MR et al (2008) Microarray-based mutation detection in the dystrophin gene. Hum Mutat 29(9):1091–1099
8. Walter G et al (2000) Protein arrays for gene expression and molecular interaction screening. Curr Opin Microbiol 3(3):98–302
9. Wang DG et al (1998) Large-scale identification, mapping, and genotyping of single-nucleotide polymorphisms in the human genome. Science 280(5366):1077–1082
10. Strauss KA et al (2008) Clinical application of DNA microarrays: Molecular diagnosis and HLA matching of an Amish child with severe combined immune deficiency. Clin Immunol 128(1):31–38
11. Gunn SR, Robetorye RS, Mohammed MS (2007) Comparative genomic hybridization arrays in clinical pathology—progress and challenges. Mol Diagn Ther 11(2):73–77
12. Sebat JL, Colwell FS, Crawford RL (2003) Metagenomic profiling: microarray analysis of an environmental genomic library. Appl Environ Microbiol 69(8):4927–4934
13. Wang RL et al (2008) DNA microarray application in ecotoxicology: experimental design, microarray scanning, and factors affecting transcriptional profiles in a small fish species. Environ Toxicol Chem 27(3):652–663
14. Richmond CS et al (1999) Genome-wide expression profiling in Escherichia coli K-12. Nucleic Acids Res 27(19):3821–3835
15. Oh MK, Liao JC (2000) DNA microarray detection of metabolic responses to protein overproduction in *Escherichia coli*. Metab Eng 2(3):201–209
16. Wang M et al (2009) Microarray-based gene expression analysis as a process characterization tool to establish comparability of complex biological products: scale-up of a whole-cell immunotherapy product. Biotechnol Bioeng 104(4):796–808

17. Fodor SP et al (1991) Light-directed, spatially addressable parallel chemical synthesis. Science 251(4995):767–773
18. McGall GH et al (1997) The efficiency of light-directed synthesis of DNA arrays on glass substrates. J Am Chem Soc 119(22):5081–5090
19. Gao X, Gulari E, Zhou X (2004) In situ synthesis of oligonucleotide microarrays. Biopolymers 73(5):579–596
20. Yang YH, Speed T (2002) Design issues for cDNA microarray experiments. Nat Rev Genet 3(8):579–588
21. Kerr MK, Churchill GA (2001) Statistical design and the analysis of gene expression microarray data. Genet Res 77(2):123–128
22. Dombkowski AA et al (2004) Gene-specific dye bias in microarray reference designs. FEBS Lett 560(1–3):120–124
23. Landgrebe J, Bretz F, Brunner E (2004) Efficient two-sample designs for microarray experiments with biological replications. In Silico Biol 4(4):61–70
24. Kim SY, Lee JW, Sohn IS (2006) Comparison of various statistical methods for identifying differential gene expression in replicated microarray data. Stat Methods Med Res 15(1):3–20
25. Klebanov L et al (2007) Statistical methods and microarray data. Nat Biotechnol 25(1):25–26 author reply 26–27
26. Brazma A et al (2001) Minimum information about a microarray experiment (MIAME)-toward standards for microarray data. Nat Genet 29(4):365–371
27. Stekel D (2003) Microarray bioinformatics. Cambridge University Press, Cambridge
28. Adams R, Bischof L (1994) Seeded Region Growing. IEEE Trans Pattern Anal Machine Intell 16(6):641–647
29. Quackenbush J (2002) Microarray data normalization and transformation. Nat Genet 32(Suppl):496–501
30. Eisen MB et al (1998) Cluster analysis and display of genome-wide expression patterns. Proc Natl Acad Sci USA 95(25):14863–14868
31. Bloom JS et al (2009) Measuring differential gene expression by short read sequencing: quantitative comparison to 2-channel gene expression microarrays. BMC Genomics 10:221
32. van Vliet AH (2009) Next generation sequencing of microbial transcriptomes: challenges and opportunities. FEMS Microbiol Lett 302(1):1–7
33. Hinton JCD et al (2004) Benefits and pitfalls of using microarrays to monitor bacterial gene expression during infection. Curr Opin Microbiol 7(3):277–282
34. Jiang H, Wong WH (2009) Statistical inferences for isoform expression in RNA-seq. Bioinformatics 25(8):1026–1032
35. Oshlack A, Wakefield MJ (2009) Transcript length bias in RNA-seq data confounds systems biology. Biol Direct 4:14
36. Sanger F et al (1977) Nucleotide sequence of bacteriophage phi X174 DNA. Nature 265(5596):687–695
37. Marsh S (2007) Pyrosequencing applications. Methods Mol Biol 373:15–24
38. Sanger F, Nicklen S, Coulson AR (1977) DNA sequencing with chain-terminating inhibitors. Proc Natl Acad Sci USA 74(12):5463–5467
39. Ronaghi M, Uhlen M, Nyren P (1998) A sequencing method based on real-time pyrophosphate. Science 281(5375):5363
40. Blow N (2009) Transcriptomics: the digital generation. Nature 458(7235):239–242
41. Denoeud F et al (2008) Annotating genomes with massive-scale RNA sequencing. Genome Biol 9(12):R175
42. Morozova O, Marra MA (2008) Applications of next-generation sequencing technologies in functional genomics. Genomics 92(5):255–264
43. Margulies M et al (2005) Genome sequencing in microfabricated high-density picolitre reactors. Nature 437(7057):376–380
44. Wang L et al (2009) DEGseq: an R package for identifying differentially expressed genes from RNA-seq data. Bioinformatics 26(1):136–138

45. Sultan M et al (2008) A global view of gene activity and alternative splicing by deep sequencing of the human transcriptome. Science 321(5891):956–960
46. Cloonan N et al (2008) Stem cell transcriptome profiling via massive-scale mRNA sequencing. Nat Methods 5(7):613–619
47. Mortazavi A et al (2008) Mapping and quantifying mammalian transcriptomes by RNA-Seq. Nat Methods 5(7):621–628
48. de Magalhaes JP, Finch CE, Janssens G (2010) Next-generation sequencing in aging research: emerging applications, problems, pitfalls and possible solutions. Ageing Res Rev 9(3):315–323
49. Zhou XG et al (2010) The next-generation sequencing technology: a technology review and future perspective. Sci China Life Sci 53(1):44–57
50. Nowrousian M (2010) Next-generation sequencing techniques for eukaryotic microorganisms: sequencing-based solutions to biological problems. Eukaryot Cell 9(9):1300–1310
51. Simon SA et al (2009) Short-read sequencing technologies for transcriptional analyses. Annu Rev Plant Biol 60:305–333
52. Morozova O, Hirst M, Marra MA (2009) Applications of new sequencing technologies for transcriptome analysis. Annu Rev Genomics Hum Genet 10:135–151
53. Morin RD et al (2008) Profiling the HeLa S3 transcriptome using randomly primed cDNA and massively parallel short-read sequencing. Biotechniques 45(1):81
54. Graf A et al (2009) Yeast systems biotechnology for the production of heterologous proteins. FEMS Yeast Res 9(3):335–348
55. Wang B et al (2010) Survey of the transcriptome of*Aspergillus oryzae*via massively parallel mRNA sequencing. Nucleic Acids Res 38(15):5075–5087
56. Simon C, Daniel R (2009) Achievements and new knowledge unraveled by metagenomic approaches. Appl Microbiol Biotechnol 85(2):265–276
57. Rubina AY et al (2005) Quantitative immunoassay of biotoxins on hydrogel-based protein microchips. Anal Biochem 340(2):317–329
58. Zhu H, Snyder M (2003) Protein chip technology. Curr Opin Chem Biol 7(1):55–63
59. Zong Y et al (2007) Forward-phase and reverse-phase protein microarray. Methods Mol Biol 381:363–374
60. Wu P, Castner DG, Grainger DW (2008) Diagnostic devices as biomaterials: a review of nucleic acid and protein microarray surface performance issues. J Biomater Sci Polym Ed 19(6):725–753
61. Kukar T et al (2002) Protein microarrays to detect protein–protein interactions using red and green fluorescent proteins. Anal Biochem 306(1):50–54
62. Tonkinson JL, Stillman BA (2002) Nitrocellulose: a tried and true polymer finds utility as a post-genomic substrate. Front Biosci 7:c1–c12
63. Grainger DW et al (2007) Current microarray surface chemistries. Methods Mol Biol 381:37–57
64. Reck M et al (2007) Optimization of a microarray sandwich-ELISA against hINF-gamma on a modified nitrocellulose membrane. Biotechnol Prog 23(6):1498–1505
65. Walter J-G, Reck M, Praulich I (2010) Protein microarrays: reduced autofluorescence and improved LOD. Eng Life Sci 10(2):103–108
66. Walter JG et al (2008) Systematic investigation of optimal aptamer immobilization for protein-microarray applications. Anal Chem 80(19):7372–7378
67. Doyle A, Griffiths J, Newel D (1994) Cell & tissue culture: laboratory procedures. Wiley, New York, pp 3:01–3:03
68. Wilkening S, Stahl F, Bader A (2003) Comparison of primary human hepatocytes and hepatoma cell line Hepg2 with regard to their biotransformation properties. Drug Metab Dispos 31(8):1035–1042
69. Castro-Melchor ML, Le H, Hu W-S (2011) Transcriptome data analysis for cell culture process. Adv Biochem Eng Biotechnol

70. Achilles J et al (2007) Isolation of intact RNA from cytometrically sorted Saccharomyces cerevisiae for the analysis of intrapopulation diversity of gene expression. Nat Protoc 2(9):2203–2211
71. Browne SM, Al-Rubeai M (2007) Selection methods for high-producing mammalian cell lines. Trends Biotechnol 25(9):425–432
72. Brezinsky SC et al (2003) A simple method for enriching populations of transfected CHO cells for cells of higher specific productivity. J Immunol Methods 277(1–2):141–155
73. Kantardjieff A et al (2009) Transcriptome and proteome analysis of Chinese hamster ovary cells under low temperature and butyrate treatment. J Biotechnol 145(2):143–159
74. Jayat C, Ratinaud MH (1993) Cell cycle analysis by flow cytometry: principles and applications. Biol Cell 78(1–2):15–25
75. Majore I et al (2009) Identification of subpopulations in mesenchymal stem cell-like cultures from human umbilical cord. Cell Commun Signal 7:6
76. Moretti P et al (2010) Characterization and improvement of cell line performance via flow cytometry and cell sorting. Eng Life Sci 10(2):130–138
77. Spellman PT, Sherlock G (2004) Reply: whole-culture synchronization—effective tools for cell cycle studies. Trends Biotechnol 22(6):270–273
78. Dutton RL, Scharer J, Moo-Young M (2006) Cell cycle phase dependent productivity of a recombinant Chinese hamster ovary cell line. Cytotechnology 52(2):55–69
79. Fu X et al (2009) Estimating accuracy of RNA-seq and microarrays with proteomics. BMC Genomics 10:161

Adv Biochem Engin/Biotechnol (2012) 127: 27–70
DOI: 10.1007/10_2011_116

Published Online: 23 December 2011

Transcriptome Data Analysis for Cell Culture Processes

Marlene Castro-Melchor, Huong Le and Wei-Shou Hu

Abstract In the past decade, DNA microarrays have fundamentally changed the way we study complex biological systems. By measuring the expression levels of thousands of transcripts, the paradigm of studying organisms has shifted from focusing on the local phenomena of a few genes to surveying the whole genome. DNA microarrays are used in a variety of ways, from simple comparisons between two samples to more intricate time-series studies. With the large number of genes being studied, the dimensionality of the problem is inevitably high. The analysis of microarray data thus requires specific approaches. In the case of time-series microarray studies, data analysis is further complicated by the correlation between successive time points in a series.

In this review, we survey the methodologies used in the analysis of static and time-series microarray data, covering data pre-processing, identification of differentially expressed genes, profile pattern recognition, pathway analysis, and network reconstruction. When available, examples of their use in mammalian cell cultures are presented.

Keywords Alignment · Clustering · Differential analysis · DNA microarrays · Gene expression · Mammalian cells · Network reconstruction · Pathway analysis · Transcriptome · Time-series

Contents

(Authors Marlene Castro-Melchon, Huong Le and Wei-Shou Hu are contributed equally to this work).

M. Castro-Melchor · H. Le · W.-S. Hu (✉)
Department of Chemical Engineering and Materials Science,
University of Minnesota, 421 Washington Avenue SE,
Minneapolis, MN 55455-0132, USA
e-mail: wshu@umn.edu

1 Introduction

In the past decade, genome science has drastically changed our approaches to studying biosciences and broadened our ability to harness the potential of industrial organisms for technological applications. Importantly, genome-wide gene expression profiling using DNA microarrays has become widely employed in biotechnological research. Through DNA microarrays, we are able to look at the dynamics at the transcript level of the entire set of genes in order to explore the intricate relationships among the biochemical reactions, the signaling and regulation, the physiological events in the cells, and the global gene expression. In the next few years, we anticipate a greatly expanded reach of transcriptome analysis in cell culture research due to the dramatic advances in sequencing technology. Until recently, the application of transcriptome analysis in cell culture bioprocess has been rather limited because the genome sequence information available for the most commonly used cells, Chinese Hamster Ovary (CHO) cells, is not extensive. With the cost of DNA sequencing drastically reduced compared to even three years ago and the readily accessible sequencing services, one can expect that genome sequences for reference species will become available in the very near future. Furthermore, we can also expect that sequencing the genome of individual cell lines will become commonplace in a few years. Therefore, the affordability of high-throughput sequencing technology will push DNA microarrays to the forefront of cell culture bioprocess characterization, along with many routinely used quantitative tools such as HPLC and ELISA. However, unlike the conventional variables typically measured in a cell cultivation process, transcriptome data is unique in its high dimensionality: each time point of measurement yields up to tens of thousands of transcript level data. In some ways, the examination of the data is like looking for patterns in a starry sky; the comparison of different datasets is as if comparing the skies in different seasons or on different days.

In this review, we summarize commonly used microarray platforms and experimental designs, and review methods used in differential expression analysis, profile pattern recognition, pathway analysis, and network reconstruction. In each section, an overview of the basic methodology is provided, followed by a sequence of specific modifications and associated software. Finally, several examples are presented in which the methodology has been successfully applied. When available, examples using antibody-producing recombinant cell lines are emphasized.

2 Platform Overview

Several microarray platforms are currently available, each of them offering certain advantages. As new platforms are introduced, a reduction in cost and an increase in flexibility have been observed. Microarray platforms are generally classified into two-dye or single-dye, referring to the number of fluorescently labeled samples applied to each chip.

2.1 Two-Dye Microarrays

Two-dye microarrays were first used by Schena et al. [1] to measure the expression level of 45 *Arabidopsis* genes, and were soon followed by studies at the genome-wide level in yeast [2]. Two-dye cDNA arrays are prepared by immobilizing long (>500 nucleotides [nt]) cDNA probes prepared by PCR amplification onto a glass slide. cDNA microarrays allow the direct comparison of genes in two samples, each labeled with a different fluorescent dye. The native intensities of the two dyes are indicative of the transcript levels in each sample. The probes can be designed against the genome sequence of the organism to minimize the segments which may cause cross-hybridization with transcripts from other genes. However, for mammalian cell applications, the large number of probes renders this approach very costly, as specific primers have to be designed for the amplification of specific segments of a sequence. Thus universal primers that amplify the entire cDNA region of an expressed sequence tag clone are more frequently used. However, they are prone to non-specific hybridization, especially for alternatively spliced transcripts. cDNA microarrays also suffer from imprecise control of the amount of DNA immobilized on the surface, making it difficult to compare the levels of different genes in the same sample.

With the much reduced cost in oligonucleotide synthesis, cDNA microarrays are now used less frequently. In the past few years, many synthetic oligo-DNA-based microarrays have evolved to be suitable for use as either single-dye or two-dye arrays. One such platform that can be used as either single-dye or two-dye is that by Agilent [3]. Similar to cDNA microarrays, short ($\sim$60 nt) oligonucleotides synthesized in situ are printed onto a glass surface. As little as individual slides are available for unique custom designs. Multiplexing, that is, the availability of testing multiple samples in a single slide, is also available in Agilent's microarrays.

2.2 Single-Dye Microarrays

In contrast to two-dye arrays, single-dye arrays are designed to provide "absolute" measurement of the relative transcript level of each gene within a sample. With a "relative" measurement in two-dye arrays, a multiple sample comparison is cumbersome, requiring either a myriad of pairings of samples or the use of a common reference. With an absolute measurement and one sample for each array, even meta-analysis using hundreds of microarrays can be performed.

An example of a single-dye array is that by Affymetrix, Inc. [4], which uses a photolithographic process for printing probes. Gene expression is interrogated by probe sets, which consist of eight to eleven probe pairs. Each probe pair consists of two 25-mers, one being a perfect match, the other containing a mismatch at the 13th base pair. The photolithographic process, however, requires the creation of a set of masks for each array design (essentially four masks for each base position, thus each 25-mer will require 100 masks). The cost of generating a new set of masks limits the frequency of modifying or updating probes.

Probes on another single-dye platform, commercialized by Roche NimbleGen, Inc., are synthesized by photo-mediated chemistry using a proprietary Maskless Array Synthesizer [5]. The use of digital mirrors creates "virtual masks", allowing for flexible designs that can be easily modified. With their ability to control the area of probe to be very small, a very large number (in the millions) of probes can be placed on a single slide. This presents an advantage for large genomes, such as those of mammalian species. For smaller genomes or with a subset of genes, array multiplexing, i.e., using a single array for multiple samples, can be implemented. Furthermore, without the use of a mask, the cost of production is reduced. Making an array for only a small number of samples and frequent updating of probe design thus becomes affordable.

2.3 Other Platforms and Technologies

In addition to the glass-slide chip-based microarrays, other types of arrays have been developed. One such technology is Illumina's BeadArray, which uses three-micron silica beads that self-assemble in microwells with uniform spacing. In this capture technology, each bead is covered with thousands of copies of specific oligonucleotides.

With the rapid advances in DNA sequencing technologies and the decrease in sequencing cost, transcriptomes can now be analyzed by direct cDNA sequencing. In RNA-Seq, a population of RNA is converted to a cDNA library, which is then fragmented and sequenced using high-throughput technologies [6]. The abundance level of a particular sequence fragment is indicative of the abundance level of the transcript from which it is derived. Unlike DNA microarrays which can be used only to probe the expression of genes represented on the arrays, RNA-Seq detects all RNA species, including novel RNAs and alternative transcripts. It can also identify transcript boundaries, and has a much wider dynamic range, over several orders of magnitude (>8,000 fold), as there is no saturation of highly expressed transcripts.

3 Static Studies vs Time-Series Studies

Although microarrays can be used to probe transcript profiles of a large array of genes in a cell sample, most applications involve the comparison of different cell samples, either the same cell line under different conditions or different cell lines. In other words, most studies involve two or more cell samples. The use of DNA microarrays in the study of cell culture processes can be categorized into static or dynamic (time-series) types according to how samples are taken and compared.

Static studies compare two samples to identify differences in gene expression between them. The samples may be different cells or tissues, such as when comparing cell lines of different levels of antibody production [7]. In other cases, different process variables or culture conditions might be under study. The following studies using NS0 cells include examples of the use of microarrays to assess the effect of cell density [8], to study cell proliferation in protein-free media [9], and to analyze the effect of hypoxic stress [10].

Cell culture process is intrinsically a time-evolving event, entailing various stages of culture, from early exponential and exponential phases followed by a transition to stationary phase. In most cases, the environmental conditions change over time, either due to the culture's self-evolution or due to process-imposed culture condition alterations such as temperature or pH shift. The gene expression profiles thus inevitably change with culture time. Static studies offer rich information on the difference in gene expression between two conditions or two cell populations but only as a snap-shot frozen at a point of a long process.

Time-series studies sample over different time points along the duration of the culture and aim at capturing the trends in gene expression changes originated by regulatory events and fluctuations in environmental conditions. Furthermore, the temporal information held in time-series microarray data also enables one to infer causality in gene regulatory networks. An aim of time-series data analysis is thus to identify genes which have different dynamic behaviors over time in the same sample or to identify the same genes whose transcripts follow different time trends under different treatments [11]. Time-series studies are particularly relevant for cellular processes exhibiting periodic behaviors, such as cell cycle and circadian clock, as well as other intrinsically dynamic processes such as development and differentiation. Although this type of studies is abundant in yeast, *C. elegans*, and stem cells, fewer examples have been demonstrated in antibody-producing mammalian cells. In one example, gene expression time-profiles were compared between fed-batch processes yielding high and low titers using the same CHO cell line [12]. Dynamic regulation of transcription in a Human Embryonic Kidney (HEK) cell line in protein-free batch and fed-batch cultures was also unraveled [13]. In addition, time-series transcriptome data was explored to elucidate cellular mechanisms leading to an increase in productivity in CHO cells under sodium butyrate treatment and temperature shift [14].

In DNA microarray studies of mammalian cell cultures, the number of differentially expressed genes and the degree of their differential expression are often lower than typical changes observed in other systems such as in developmental processes or in microbial cultures [12]. Using a fold-change cut-off of 1.4–2.0, and a p-value cut-off of 0.05–0.1, it is common to identify much less than 10% of the genes as significant. This number often decreases sharply when the fold-change cutoff is raised above 2.0. For example, in studying the productivity of antibody-producing cell lines, a relative small number of genes are consistently different between high- and low-productivity clones [7]. These modest changes in gene expression thus require careful experimental design and subsequent data analysis. This situation contrasts with most cases found in bacteria undergoing changes in nutritional or other environmental conditions, and stem cells under directed differentiation in which often a large number of genes change their expression and many show large differences in gene expression levels.

4 Experimental Design

Microarray and RNA-Seq studies can provide a wealth of information. However, even with the decrease in cost in the past few years, they are still not bargain-price. The number of conditions to be tested and the number of samples for each condition have to be planned. When single-dye microarray platforms are used, there is no limitation to which comparisons among multiple samples are made. For two-dye arrays, however, the experimental design is crucial. With two-dye microarray platforms, one aims to measure the ratio of each gene's transcript level between two samples. When only two samples are involved, direct comparison is obtained using a single chip. With three samples, loop designs in which three arrays are used to obtain direct pair-wise comparison of the three sample pairs (1–2, 2–3, 3–1) can be applied [15]. An alternative, often referred to as reference design, is to hybridize each of the three samples to a common reference, and obtain indirect comparisons for each sample pair in the experiment. An often-used

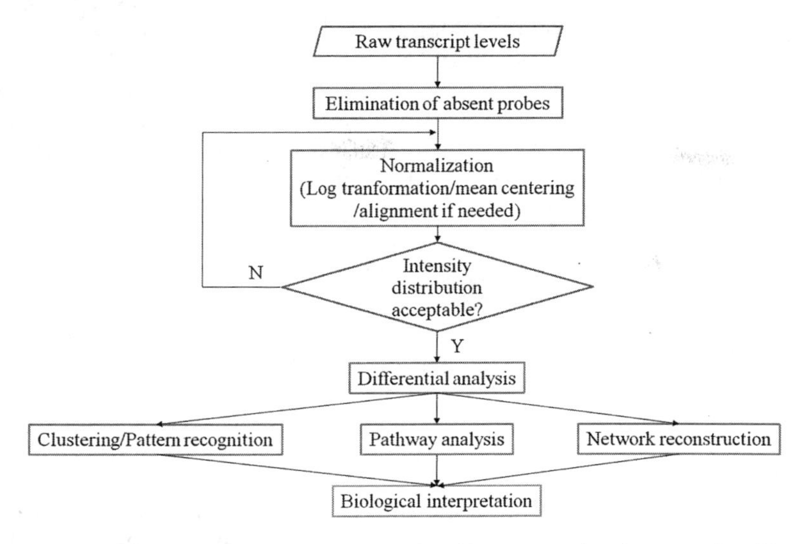

Fig. 1 General analysis process of microarray data. Raw expression data are often filtered to eliminate absent probes. The filtered data is subsequently normalized and processed using different methods if necessary. Genes exhibiting differential expression are identified using statistical tools. In addition to clustering, further analysis can be performed in a pathway/network context to finally interpret the biological meaning of differential expression

reference is a pool of RNA, either from all samples, to ensure that the transcripts of all genes on the array are present, or from sample(s) external to the experiment. An internal reference, for instance the first sample, can also be used to directly compare some of the pairs (1–2 and 1–3) and infer comparisons for the others (in this case 2–3). The amount of available reference sample might limit the number of arrays that can be done.

Time-series microarray studies present additional experimental design challenges. Frequently, the comparison is not only among data from different time points within the same treatment but also among series under different treatments. The number of samples to be collected and their distribution in time will define the ability of the experiment to capture the gene expression dynamics. The sample collection frequency should be high enough to capture the dynamics of genes with periodic behaviors or propensity for sudden changes in expression. This, however, might result in a very large number of samples, which is not always feasible due to cost or the amount of work involved [16]. If critical changes are suspected between the time points originally analyzed, additional microarrays can be performed. This is possible if samples were collected at intermediate time points. Another possibility is to fill these gaps using quantitative PCR measurements of transcripts of the target genes.

The general steps in analyzing gene expression data from microarrays are shown as a flowchart in Fig. 1. First, raw data is filtered to eliminate absent probes using intensity and/or detection p-value cutoffs. Filtered data is further normalized to generate a baseline for comparison across samples. Time alignment, log transformation, and scaling can be performed if necessary. Once the data has been properly processed, genes exhibiting differential expression can be identified using multiple statistical approaches. These significant genes are often further analyzed in a pathway/network context or by using clustering tools to infer the biological meanings of differential expression.

5 Data Pre-Processing

5.1 Normalization, Transformation, and Scaling

Gene expression levels measured using DNA microarrays are subject to a number of systematic biases, and hence should be globally adjusted (or normalized) to attain a common basis for all the microarrays to be compared. These variations in gene expression measures are often the result of differences in starting amounts of RNA, labeling, hybridization, and scanning efficiency [17]. Normalization is thus a necessary step regardless of the platform, or whether the experiment involves static or time-series samples. Different normalization methods (based on different sets of assumptions) often give different quantifications. Most normalization methods assume that the microarray contains a large and random set of genes. Furthermore, the number of differentially expressed genes is considered to be relatively small compared to the total number of genes present on the array. As a result, this differential expression does not affect the overall distribution of gene expression levels in each sample.

Linear and quantile normalization are most commonly used in microarray data processing. Linear normalization is often applied when gene expression measures in all arrays have similar distributions but different median values. Given the assumption that equal amounts of RNA are used in each sample, a normalization factor is calculated as the ratio of the median gene expression levels in two samples [17]. All gene expression measures are subsequently scaled using this factor such that these two samples have the same median gene expression level after normalization. A target median value can also be defined to linearly scale multiple samples. Linear normalization is thus conceptually simple, yet applicable to most cases in which the assumptions stated above are satisfied. However, possible lack of linearity between fluorescence intensity and the amount of DNA or RNA hybridized could introduce errors when linear normalization is applied.

Quantile normalization, on the other hand, assumes that all samples have the same gene expression level distribution [18]. Gene expression measures are adjusted such that each sample follows the same distribution, which is assumed to be the average distribution of all samples. This normalization method is frequently used to correct the gene expression level distribution in single-dye and two-dye arrays when genomic DNA is used in one channel. Sometimes, a drastic change in cell physiology may occur, causing a major shift in gene expression profiles. In such cases, the use of quantile normalization might not be appropriate. For example, as stem cells differentiate or cells enter different phases of growth, their transcriptional responses or cellular RNA composition may change drastically. Large variations in cellular RNA composition among samples violate the assumption that all samples have the same gene expression level distribution.

It is important to note that, in most experimental protocols, the amount of total RNA (in the case of prokaryotic samples) or poly(A)-tailed transcripts (in the case of eukaryotic samples) applied to each array is kept equal, and thus normalization methods only adjust the data to equal quantities of RNA. However, the RNA content per cell does not always remains constant under different conditions. Fast-growing cells have far more RNA than cells in the stationary phase, and thus total RNA content per cell varies. It is therefore important to know whether differential expression calls are based on per cell or per unit amount of RNA.

After normalization, the data is usually log-transformed. The variance, which is inherently large in microarray data, is reduced in log-transformed data. Normalized gene expression values can also be scaled to a mean or median value of zero. This is equivalent to centering the gene expression level distribution over zero (mean- or median-centering). Additionally, a standard deviation of one can be achieved using z-transformation. These data pre-processing steps can be performed using several software including Expressionist, GeneSpring, and R packages such as *affy*, *limma*, *beadarray* and *oligo*. Although data normalization, transformation, and scaling have become routine, these steps remain vital to all subsequent stages along the analysis pipeline of gene expression data.

5.2 Time Alignment

When comparing time-series experiments, it is important to control the starting cell population in different treatments to be identical, or at least as similar as possible. Under some conditions, variability is difficult to eliminate, resulting in somewhat different kinetic profiles even among biological replicate cultures. When applying microarrays to time-series studies, the aim is to identify the genes whose transcript dynamics change beyond the fluctuations in biological replicate cultures, and where the change can be attributed to experimental treatment. In assessing the similarity or difference between two cultures under different treatments, a direct comparison of time profiles is an obvious first approach. This is sound in the cases where the trends of growth and other growth-related variables (such as chemical profiles) are mostly identical. Often growth and other culture indicators reveal a difference, strongly hinting that the identical time points in two cultures may not correspond to identical "culture stage". In other words, the time frame of one culture has shifted from the reference time frame of the other culture. Direct comparison of time profiles of gene expression may give rise to many

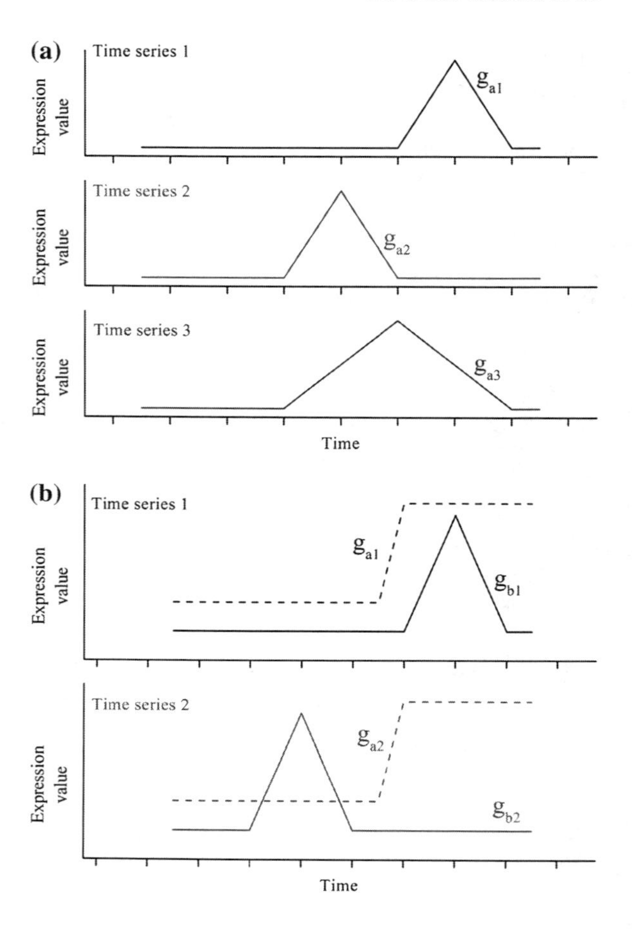

Fig. 2 Possible forms of time asynchronization in different series. **a** Expression profile of gene g_a in three different series. The expression profile in series 2, g_{a2}, shows a frame shift with respect to series 1, g_{a1}. The expression profile in series 3, g_{a3}, shows an expansion with respect to g_{a1}. These types of asynchronization can be adjusted globally. **b** Expression profiles of two genes, g_a and g_b, in two different series. Gene g_a displays the same expression profile in the two series. The peak observed in the expression profile of gene g_b in series 1 appears earlier in series 2. This time flip often requires local adjustment

falsely identified genes with different kinetic behaviors. Time alignment aims to identify potential time misalignments and correct them.

The change in time dynamics could be global, i.e., all the transcripts change their temporal profiles similarly. This change may also be segmented and local, i.e., only some sets of genes change coordinatedly apart from the rest of the genes or different sets of genes which change their dynamics differently. Such asynchronous behaviors need to be dealt with using some form of time alignment. Asynchronization between transcriptome time profiles appears in multiple forms, which can be largely divided into four types: frame shift, elastic compression or expansion, and time flip [19]. Frame shift occurs when one of the series experiences a lag phase with respect to the others. If the growth rate differs significantly between the series, their gene expression profiles may display elastic compression or expansion. Examples of frame shift and expansion are shown in Fig. 2a. These types of asynchronization are often adjusted globally. In addition, changes in a few subsets of genes can result in a flip in time order between different subsets of genes

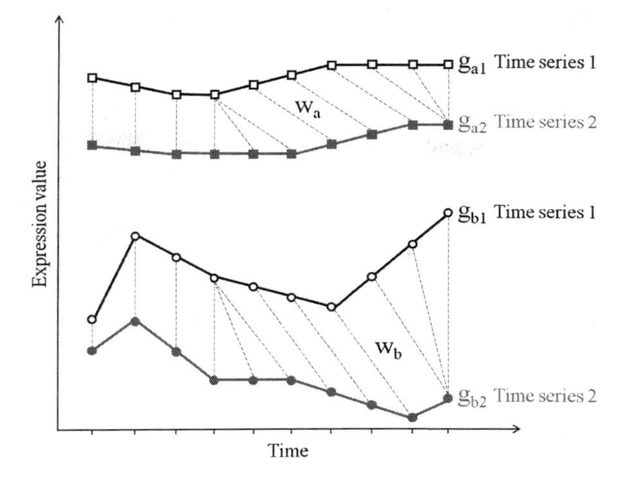

Fig. 3 Alignment of gene expression profiles using DTW. Two genes (g_a and g_b) are shown in two different series. For each gene, discrete time points in series 1 are mapped to those in series 2. The weighting factors (w_a and w_b) indicate the contribution of each gene to the final global adjustment. Gene g_a with a relatively flat profile is given a lower weight ($w_a < w_b$). The same alignment is imposed on both genes

(Fig. 2b). This time flip suggests the existence of multiple biological clocks controlling varied cellular processes in the experimental system and thus requires local alignment. As a result, when multiple treatments are being compared, gene expression data sets should be examined and, if necessary, properly aligned before subsequent analyses can be performed.

Conceptually, aligning time-series microarray data entails matching two patterns by locally compressing, expanding, or translating one with respect to the other such that their similar characteristics are aligned without altering the ordering of each sample. This can be performed on either the continuous representation of each series or the discrete values of gene expression. The alignment between time series can be achieved at a global level or at a local level to allow different subsets of genes to follow varied biological clocks.

An example of global alignment is the B-spline-based alignment method, which presents each gene expression profile as a spline curve of multiple low-degree polynomials [20]. To align different time series, one of the series is chosen as the reference, and the time points of the other series are mapped to the reference series by stretching and shifting the continuous representation of gene profiles. This method is particularly suited for long time series (e.g., $\geq$10 time points) [21]. The use of B-splines for alignment was demonstrated by aligning three yeast time series that begin in different phases and occur in different time scales [20].

A second example of global alignment, dynamic time warping (DTW), involves non-linear mapping between discrete time points of two series along the time dimension such that the distance between them is minimized [22]. In the case of transcriptome time series, the overall distance between the two series is computed as the weighted sum of distances contributed by all genes. The use of a weighting factor for each gene allows higher contribution to the overall distance measure to be given to genes with consistent expression profiles across two treatments, or to genes important to the biological activities being considered. In Fig. 3, the algorithm is exemplified with two genes (g_a and g_b) in two different series. The weighting factors (w_a and w_b) indicate the contribution of each gene to the final adjustment. g_a with a less dynamic profile thus has a lower weighting factor ($w_a < w_b$). In addition to alignment of transcriptome data, DTW has also been used to synchronize offline and online data of batch processes [23, 24].

Global alignment algorithms assume that all genes share the same alignment, that is, that all genes were affected in the same manner. The existence of multiple biological clocks within the same cell, however, can result in sets of genes being affected independently. In other words, genes in one set correspond to genes that follow a particular biological clock, sharing the same alignment, but they need to be warped separately from the rest of the genes. Recently, Smyth et al. [25] have proposed an algorithm capable of identifying sets of genes that present similar alignments when aligned independently. The resulting sets include genes that follow similar warpings, even though their expression profiles might be very different.

6 Identification of Differentially Expressed Genes

After transcriptome data has been pre-processed, a number of statistical approaches can be used to identify differentially expressed genes. Commonly used analytical methods for static transcriptome data include *t*-tests, ANalysis Of VAriance (ANOVA), Significance Analysis of Microarray data (SAM), and LInear Models for MicroArray data (*limma*). These methods are not all directly applicable for dynamic studies involving a chronological set of samples collected over time since a change in the time order will result in a different statistical inference [26]. Recently several methods based on regression, ANOVA, and Bayesian models have been adapted to handle time-series microarray data. In addition, a distance calculation approach has also been proposed for identification of kinetically differentially expressed genes.

6.1 Statistical Analysis of Gene Expression Data

The estimates of gene expression levels provided by microarray data are generally prone to two types of errors—systematic and random errors. Systematic error resulting from several factors such as RNA concentration measurement or dye-labeling efficiency can give rise to a systematic bias in the expression level estimates of all genes on the same array. This bias is often corrected using one of the normalization methods presented in the previous section on data pre-processing. Random error in the measurement of gene expression levels arises from random fluctuations in other steps, for instance array scanning. Inferential statistics is used to ensure that the observed change in gene expression did not occur by random chance.

Inferential statistics is applied to microarray data by invoking a null hypothesis. The null hypothesis holds true when all samples have the same average expression value for the gene of interest. Conversely, if the gene is expressed at a different level in at least one sample, the alternative hypothesis becomes valid. In order to assess the validity of either hypothesis, a test statistic is often estimated as the ratio between the change in a gene's expression values among samples and the variability in those measurements. Furthermore, a *p*-value computed using this test statistic is compared to an acceptable significance level α. The smaller the *p*-value is compared to α, the stronger the evidence is against the null hypothesis, and in support of the gene being differentially expressed in at least one sample.

In a typical microarray experiment, tens of thousands of genes are tested simultaneously, and a large number of them are likely to be identified as differentially expressed. Even with a small *p*-value, such as 0.01 that is normally considered to be rather stringent, a significant number of those genes identified as differentially expressed might be by random chance. For example, if 1,000 genes out of 10,000 in total are identified as differentially expressed, each with a *p*-value <0.05, then 500 of these 1,000 genes might have been identified by chance. One way to control the potentially high error rate is to set each gene's *p*-value to an *n*-fold lower significance level, α/n, where n is the total number of genes. This is often referred to as the Bonferroni correction for the family-wise error rate—(FWER) [27]. However, this correction imposes an extremely stringent criterion. In the previous example, the *p*-value will have to be set at less than 0.000005. This would likely result in failure to identify the majority of genes that are indeed differentially expressed. An alternative is to control the number of false positives among the number of genes declared as differentially expressed rather than the total number of genes. This statistic, referred to as false discovery rate (FDR), is less stringent than the FWER and thus offers more power than the FWER to detect differential expression [28]. Therefore, in multiple hypothesis testing, FDR is often used in place of *p*-value.

6.1.1 Statistical Analysis of Static Gene Expression Data

A variety of methods are available for hypothesis testing. A *t*-test is often used when only two samples are compared for differential gene expression. When three or more samples are involved, ANOVA is recommended to avoid performing multiple *t*-tests, which will most likely result in an increased false-positive rate. Both methods assume the expression levels of a gene in different samples follow a normal distribution. When this assumption does not hold true, non-parametric tests including the Wilcoxon rank-sum test and permutation-based test are often the methods of choice.

t-Test

t-Tests are considered the simplest statistical methods to identify differentially expressed genes. A *t*-statistic is calculated as the ratio between the difference in gene expression levels of two samples and the pooled variance. Furthermore, a degree of freedom is calculated from the sample sizes—with more penalties if the two samples have unequal variances (Welch's *t*-test), and no penalties if the assumption of equal variances holds true (Student's *t*-test). A *p*-value, which can be obtained using the *t*-statistic and the degree of freedom, is compared to a pre-defined significance level α to detect differential expression. *t*-Tests can be easily performed in Microsoft Excel, several R packages, and a variety of software including Spotfire and Expressionist.

Gene expression responses during metabolic shift in a hybridoma cell culture have been investigated using the Student's *t*-test on cDNA microarray data [29]. 123 probes were identified as changing their expression levels (fold-change $\geq$1.4 and *p*-value $\leq$0.1) when the cells shifted to a lactate consumption state. Another example involves the survey of global gene expression changes in a recombinant antibody-producing CHO cell line and a mouse hybridoma cell line under sodium butyrate treatment [30]. Using a fold-change cutoff of 1.4 and a *p*-value cutoff of 0.05, most transcripts were found to be expressed at similar levels in both cell lines, indicating that the transcriptional responses under exposure to sodium butyrate are rather conserved.

Analysis of Variance

When more than two samples are involved, single-factor ANOVA is often used. The overall variance in gene expression among different samples is partitioned into separate sources of variations. The total variation, as evaluated by sum of squares (SS_{Total}), arises from two sources—the actual differential expression among these samples ($SS_{Treatment}$) and the random error (SS_{Error}). The means sum of squares (MS) for treatment and error can be estimated by dividing each SS by the corresponding degree of freedom. The quotient of these two MSs is taken as the *F*-statistic, which further provides a *p*-value for inference of differential expression.

When the experiment involves several factors (or variables; in ANOVA they are referred to as "factors"), and one wishes to segregate the effects of those factors, multiple-factor ANOVA is used. Based on the same working principles described above, multiple-factor ANOVA also partitions the total variation into different sources—the actual effect of each experimental factor, their interactions, and the random error. A *p*-value for each term can be derived similarly, and whether these factors significantly affect the change in gene expression can thus be concluded. Both single-factor and multiple-factor ANOVA can be performed easily using Microsoft Excel, as well as several R packages.

Variation in gene expression within and between two populations of the genus *Fundulus* was uncovered using ANOVA on $\log_2$-normalized microarray data on 907 genes [31]. 161 genes were differentially expressed among individuals within a population, whereas only 15 genes differed between populations, suggesting that substantial natural variation exists in gene expression. A linear ANOVA model was also fitted to the expression levels of more than 3,000 genes expressed during embryonic development of six *Drosophila* species [32]. More than 80% of genes best fit to models incorporating stabilizing selection, and maximal similarity is observed during mid-embryogenesis rather than early or late stages of development. This result thus supports the developmental hourglass model, and the theory that natural selection acts to conserve gene expression patterns during the phylotypic period.

Significance Analysis of Microarray (SAM)

Similar to *t*-tests, SAM also calculates a "relative difference" (d), which resembles the ratio between difference in average gene expression values and the pooled variance in two treatments for each gene [33]. The expression levels in all replicated samples of these two treatments are then permuted, and an average "relative difference" over these permutations (d_E) is estimated. For the majority of genes, which are assumed not to be differentially expressed, the average difference obtained from permutation (d_E) is largely the same as the observed one (d). If the discrepancy between d_E and d exceeds a threshold, the gene is considered differentially expressed. In order to calculate the FDR for each gene, two horizontal cutoffs are defined—one as the smallest observed difference of up-regulated genes, and the other as the least negative of down-regulated genes. The average number of genes with d_E exceeding these cutoffs in all permutations can be considered as the number of false positives, and is used to assess FDR. A convenient Microsoft Excel add-in for SAM is available, and the packages *siggenes* and *samr* in R are also publicly accessible.

The advantage of SAM over other statistical methods was demonstrated when examining the transcriptional responses of human lymphoblastoid cells under irradiation [33]. 34 genes were identified as significant at an FDR of 12% using SAM compared to more than 60% using other methods. In another example, SAM was used to identify about 400 genes contributing to the impaired differentiation capacity of murine neural stem cells (NSCs) defective in p53 and PTEN genes [34]. The majority of genes involved in cell cycle regulation were also found to be significantly down-regulated when HeLa cells were transfected with siRNA against PHF8, an H4K20me1 demethylase [35].

Linear Models of MicroArray data

In this approach, a linear hierarchical model with arbitrary coefficients and contrasts across multiple samples for each gene is developed [36, 37]. Furthermore, marginal distributions of the observed statistics are used to estimate the hyperparameters under consistent and closed forms. In addition, the ordinary *t*-statistic can be replaced by a moderated one, which implicitly results in shrinkage of all gene-wise variances into a common value. This moderate *t*-statistic follows a *t*-distribution with augmented degrees of freedom, and thus can be extended for multiple-sample comparisons by using the corresponding *F*-statistics. The R package *limma* is publicly available.

Transcriptional responses upon restoration of p53 in adenocarcinomas were revealed using *limma* [38]. p53-restored samples were shown to cluster with adenomas rather than carcinomas, suggesting that adenocarcinoma cells can be specifically removed from the tumors. *limma* was also used to compare gene expression signatures between cultured thymic epithelial cells (TECs) and multipotent hair follicle (HF) stem cells [39]. 119 genes were identified as being differentially expressed between these two samples with a fold-change cut-off greater than four and a p-value less than 0.001.

6.1.2 Statistical Analysis of Dynamic Gene Expression Data

Time-series transcriptome data offer a great advantage when exploring transcription as a dynamic process, yet their analysis is more complicated than analyzing multiple samples unrelated in time. Transcriptional responses at a certain time point often carry information about cellular behaviors in previous stages. Thus samples within a series are mutually dependent, and should not be analyzed using traditional statistical approaches. Rather, methods taking this interdependency into consideration such as Extraction of Differential Gene Expression (EDGE), Microarray Significant Profiles (maSigPro), ANalysis Of Variance–Simultaneous Component Analysis (ANOVA-SCA), and multivariate Bayesian models are more suitable. The number of time points in each series, the number of series, and the availability of replicates will guide the selection of algorithm to use in data analysis. This analysis can become even more challenging if the sampling frequency is not uniform across multiple series.

Extraction of Differential Gene Expression (EDGE)

In EDGE, differential analysis is also approached as a hypothesis-testing problem. The null hypothesis is that a gene's expression does not change both over time within a single treatment and across multiple treatments [26, 40]. The expression profile of each gene is modeled using a p-dimensional basis, usually a pth-order polynomial, or a natural cubic spline function. The parameters of these functions are then estimated by minimizing the sum of squared errors (SSE) between the model-fitted expression values and the corresponding actual ones. The parameterization of gene expression profiles allows the hypothesis testing to be performed by comparison of the fitted parameters. As such, an F-statistic is calculated for each gene to reflect the relative difference in SSE of the model-fitted gene expression profiles under the null and the alternative hypotheses, respectively. This statistic is used alongside a null distribution generated using a resampling method to estimate a q-value, which accounts for the FDR incurred in multiple hypothesis testing [41].

The open-source software EDGE [42] has facilitated the use of this methodology in analyzing time-course gene expression data. Differential expression can be surveyed along the time axis within each treatment or across multiple treatments. EDGE was used to define the transcriptomic signatures of aging in several tissues in *Drosophila melanogaster* [43]. In a mouse model, a complex transcriptional hierarchy comprising more than one thousand genes regulated during endocrine differentiation was also identified using EDGE [44].

Microarray Significant Profiles (maSigPro)

Microarray Significant Profiles, maSigPro [45], uses a two-step regression approach to identify differentially expressed genes in time-series microarray data. Single or multiple time series can be analyzed, with multiple time series being analyzed directly instead of performing multiple pair-wise analyses. This methodology not only detects kinetically differentially expressed genes, but also uncovers changes in gene expression trends. In the first step of gene selection, expression data is fitted using a global regression model which considers all experimental variables and their interactions. If there are n groups, $(n - 1)$ dummy variables are defined. Each dummy variable allows the distinction between each group and the reference group. Furthermore, an ANOVA table is generated for each gene. If the gene shows differences between any group and the reference group, the regression coefficients will be statistically significant as determined by an F-statistic and its associated p-value. In the second step of variable selection, the best model for each gene is obtained using a stepwise regression approach. The variables that best fit the data represent the time effects and their interactions with the dummy variables. For finding those genes with significant differences in group x with respect to the reference series, the genes with significant coefficient for the dummy variable $(x - 1)$ are selected.

The package maSigPro is available in R and includes several tools for result visualization. In addition, it is part of the oneChannelGUI package [46], which provides a graphical interface for the analysis of Affymetrix microarrays, and was included in the popular software Gene Expression Pattern Analysis Suite (GEPAS) [47]. An extension of maSigPro, maSigFun [48], is used to fit regression models for genes with the same functional class and for the functional assessment of time-course microarray data. maSigPro has also been implemented in Corra [49], an R package devoted to the analysis of LC–MS-based proteomics. maSigPro has been used to analyze data from intrinsically dynamic processes such as spatial differentiation in fungi [50], and plant development [51–53], as well as periodic responses such as the rhythmically expressed genes in mouse distal colon [54].

ANOVA-SCA

ANOVA-SCA (or ASCA for short) is considered a combination of a statistical method (ANalysis Of VAriance, ANOVA) and a dimensionality reduction approach (Simultaneous Component Analysis, SCA) [55–57]. ANOVA-SCA is particularly useful when two or more quantitative variables are involved, such as time and dose. In the first step, an ANOVA model is applied for each gene expression measure to separate the variability caused by these two different variables. The model parameters obtained for all genes under each experimental condition are subsequently organized into a matrix form. The second step involves applying principal component analysis simultaneously on all matrices obtained under all experimental conditions. A number of constraints can be further imposed such that the resulting matrices are mutually independent. Such constraints on orthogonality enable the ASCA model parameters to be estimated independently by solving a simple least-squares optimization problem. Statistical significance of these experimental variables and their interactions can be further inferred using a permutation approach [58]. In particular, all experimental conditions are permutated to obtain a no-effect distribution, thus providing a baseline to conclude whether the observed effect is indeed significant.

One of the earliest applications of ASCA was for analyzing a metabolomics experiment in which the effects of time and vitamin C dose on the NMR spectra of guinea pig urine samples were delineated [55]. Individual variations caused by time and doxorubicin dose on metabolite mass spectrometry profiles were also uncovered using ASCA in a toxicology study on rats [59]. Given the intrinsic generalizability of ASCA, it is not surprising to find this approach extended into discovery of kinetically differentially expressed genes [60]. Two statistics—SPE (Squared Prediction Error) and leverage—were proposed to evaluate the goodness of fit of the ASCA model, and the degree of agreement with which a gene profile follows the main expression patterns, respectively. This adapted version of the original algorithm, *ASCA-genes*, has been implemented in the R language. Furthermore, *ASCA-fun* was devised to perform functional analysis on time-series microarray data [48]. In this method, genes ranked according to their correlation to the principal time components identified by ASCA were used to assess functional enrichment in the dataset following Gene Set Analysis (GSA) procedures.

Bayesian Approaches

A multivariate empirical Bayes model was applied to time-series microarray data by Tai and Speed [11]. The algorithm, implemented in the R package *timecourse*, however, requires replicates of the full time-series. This algorithm calculates multivariate versions of the log-odds, or B-statistic (MB-statistic), and the Hotelling statistic ($\tilde{T}^2$). When the numbers of replicates are the same for all genes, the MB-statistic is equivalent to the $\tilde{T}^2$-statistic. The algorithm can be used in one-treatment problems and multi-treatment problems. Although this method ranks the genes, it does not provide a significance cutoff.

A fully Bayesian approach for microarray analysis was implemented in clustering [61] and later for the analysis of time series [62]. This fully Bayesian approach can handle short series, non-uniform sampling and missing data and does take into consideration the temporal structure of the time series. Gene expression profiles are modeled with Legendre or Fourier polynomials, and the coefficients and the degrees of these polynomials are estimated using a Bayesian approach. The differentially expressed genes identified in this Bayesian multiple-testing procedure are ranked, and their expression profiles are estimated. This estimation allows the visualization of each gene expression profile as a single smooth curve.

The fully Bayesian approach was demonstrated when analyzing the time series obtained by stimulating human breast cancer cells with estradiol after different time periods. The algorithm is implemented in the Bayesian user-friendly software for Analyzing Time Series (BATS) [63], a graphic user interface written in Matlab. The BATS package requires 5–6 time points and replicates are recommended but not required. At the moment, however, BATS can only handle one treatment time series. Its extension to multiple time series is under development.

6.2 Calculation of Distances Between Gene Expression Profiles

Just as in the calculation of the geometrical distance between any two vectors, a distance value can be computed to quantitatively describe the difference between two expression profiles of the same gene. By condensing all distance measures between the corresponding time points, the comparison of these two profiles is reduced into a single number. Two frequently used metrics are Euclidean distance and Pearson's correlation (Fig. 4). Euclidean distance, also known as $L - 2\ norm$, assesses the absolute difference between two time profiles. As a result, genes with the highest Euclidean distance between two treatments are often the ones with high expression levels, and are most likely to be identified as differentially expressed despite having similar expression trends in these treatments. Gene expression data can then be mean-centered or z-transformed to alleviate the dominance of these high-abundance transcripts. On the other hand, Pearson's correlation quantifies the overall similarity between the two trends regardless of the absolute values of gene expression. Small fluctuations in gene expression between low-abundance transcripts can thus be manifested as being markedly different since only expression trend is considered.

The choice of distance metric therefore depends on the question being asked. If the absolute values of expression measures are critical, the Euclidean metric is often preferred. Alternatively, the Pearson's correlation coefficient is a more suitable similarity measure if the overall trend of expression is pertinent to the analysis. A combination of both metrics is therefore recommended to integrate the differences in absolute expression magnitude and expression trend.

Following selection of a proper metric and distance calculation, a distribution of this representative difference can be plotted, and a threshold is often set to declare differential expression. Genes having distance measures between their expression profiles in two treatments above a certain threshold are considered to be differentially expressed. Manual inspection of gene expression profiles is often recommended to confirm the differential expression. In addition, if both treatments

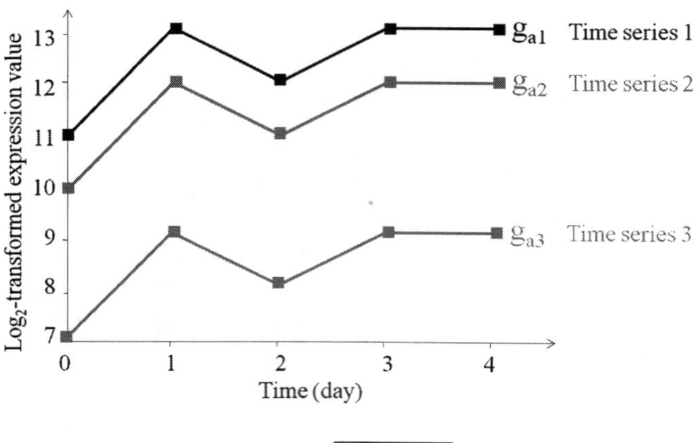

Euclidean distance
$$Eucl(g_{a1},g_{a2})=\sqrt{\sum_{i=0}^{4}(g_{a1,i}-g_{a2,i})^2}=\sqrt{(11-10)^2+(13-12)^2+(12-11)^2+(13-12)^2+(13-12)^2}=2.2$$
$$Eucl(g_{a1},g_{a3})=\sqrt{\sum_{i=0}^{4}(g_{a1,i}-g_{a3,i})^2}=\sqrt{(11-7)^2+(13-9)^2+(12-8)^2+(13-9)^2+(13-9)^2}=8.9$$

Pearson correlation
$$Corr(g_{a1},g_{a2})=\frac{\sum_{i=0}^{4}(g_{a1,i}-\overline{g_{a1}})(g_{a2,i}-\overline{g_{a2}})}{\sqrt{\sum_{i=0}^{4}(g_{a1,i}-\overline{g_{a1}})^2(g_{a2,i}-\overline{g_{a2}})^2}}=1 \rightarrow \text{Pearson distance}\ (g_{a1}, g_{a2})=0$$
$$Corr(g_{a1},g_{a3})=\frac{\sum_{i=0}^{4}(g_{a1,i}-\overline{g_{a1}})(g_{a3,i}-\overline{g_{a3}})}{\sqrt{\sum_{i=0}^{4}(g_{a1,i}-\overline{g_{a1}})^2(g_{a3,i}-\overline{g_{a3}})^2}}=1 \rightarrow \text{Pearson distance}\ (g_{a1}, g_{a3})=0$$

Fig. 4 Calculation of distance between the expression profiles of a gene in two series. The distance between the expression profiles of gene g_a in three series can be measured using different metrics. The Euclidean metric quantifies the absolute geometric distance between the profiles, whereas the Pearson metric evaluates the correlation of trends in expression. Thus even though the Euclidean distance of g_a between series 1 and series 3 (Eucl (g_{a1}, g_{a3})) is much higher than that between series 1 and series 2 (Eucl (g_{a1}, g_{a2})), their Pearson correlations (Corr(g_{a1}, g_{a2}) and Corr(g_{a1}, g_{a3})) are indeed the same

are replicated, a statistic can be derived by permuting replicated samples between the two treatments. An average distance over all permutations is calculated, and compared to the actual distance to infer a statistical significance level. However, optimizing the difference threshold between the average and the actual distance can be indeed challenging.

This approach was used in a number of studies conducted in *Streptomyces coelicolor*. Genes involved in regulatory circuits related to antibiotic production were identified using Euclidean distance as criterion for differential expression [19]. Euclidean distance was also used in conjunction with principal component analysis (PCA) to reveal genes kinetically perturbed when the *Streptomyces coelicolor* sigma-like protein AfsS was disrupted [64]. In a recent study, more than 900 genes were identified as differentially expressed in an antibody-producing CHO cell line between the butyrate-treated 33°C culture and the non-treated culture [14].

7 Profile Pattern Recognition

Microarray data, with their large size and high dimensionality, are inherently complex. Compared to the number of genes (i.e., dimensionality), the number of samples is almost always small, making it difficult to find an answer to the question being asked. Often, an objective in a microarray experiment is to identify genes with a certain profile or pattern. Sometimes, however, which patterns are present in the data are not even known. In order to identify patterns that exist in the data, two types of techniques can be used: unsupervised and supervised algorithms.

7.1 Unsupervised Classification Methods

Unsupervised pattern recognition consists of organizing data based on the properties of the data themselves without reference to additional information [65]. Mathematical algorithms determine the search for natural patterns existing in the data [66]. The goal of unsupervised pattern recognition is to identify small subsets of genes that display similar expression patterns [67]. Instead of clustering genes, clustering samples based on their expression profiles can also be a goal in clustering analysis. In this case samples with similar expression profiles might help identifying groups, or labels, that can be given to those samples.

Although the term unsupervised pattern recognition is commonly used as a synonym for clustering, it actually encompasses other techniques, such as non-negative matrix factorization (NMF) and principal component analysis (PCA).

7.1.1 Dimensionality Reduction Techniques

Because microarray data is often obtained from only a small number of samples and entails thousands of genes, dimensionality reduction can be helpful for visualization, clustering, and classification. When transcriptome data is represented as an n by m matrix, in which n is the number of genes and m the number of samples ($n \gg m$), dimensionality reduction techniques can be used to identify a smaller number of principal gene expression patterns k (Fig. 5). This can be done by factorizing the original gene expression matrix (A) into two sub-matrices: one containing eigenarrays (W) and the other containing k eigengenes (H). The expression level of each gene in these m samples can be represented as a linear combination of the k eigengenes. Similarly, the overall expression pattern in each sample can be represented as a linear combination of the k eigenarrays.

In PCA, the data is transformed into a new set of variables called principal components (PCs). The principal components are uncorrelated, and, furthermore, they are ranked so that the first PCs contain most of the variation present in all of

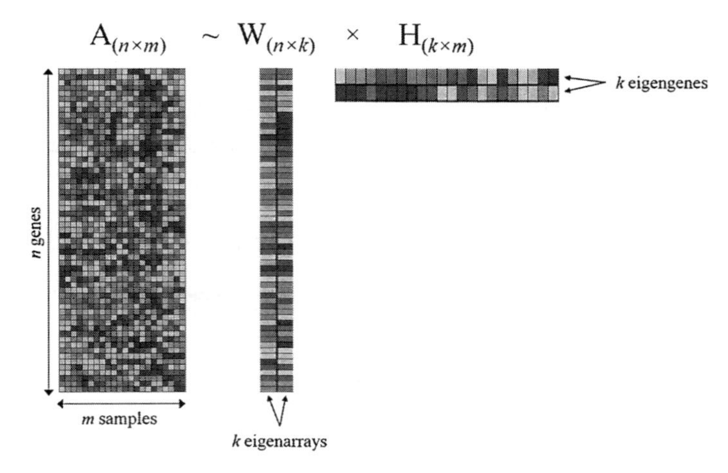

Fig. 5 Matrix factorization in dimensionality reduction techniques: NMF and PCA. Microarray data is organized into a matrix (A) with each row representing the expression levels of a gene in m samples. This original matrix can be decomposed into two sub-matrices: one containing k eigenarrays (W), and the other containing k eigengenes (H). The expression levels of each gene in these m samples can be represented as a weighted combination of the k eigengenes. Similarly, the overall gene expression pattern in each sample can be represented as a weighted combination of k eigenarrays

the original variables [68]. Since the first few PCs capture most of the variation in the original data, it is customary to use only the first few PCs [69]. When the data are projected along the first few PCs (most commonly the first two or three), in many cases it is possible to identify groups.

In PCA, the gene expression values can be reconstructed by a weighted sum of the eigengenes; however, there is no restriction on the sign of the weights. This can cause some variability due to cancellations, if eigengenes with both negative and positive weights are added. In a similar technique, NMF, the coefficients are forced to be non-negative, which ensures that the contributions from principal gene expression patterns are positive and thus additive [70, 71].

Both techniques, PCA and NMF, have been used in the identification of biomarkers; for an example see [72]. PCA has been used to characterize the gene expression of stem cells in different phases [73] and different types of stem cells [74]. As NMF has been found superior to PCA in reducing microarray data [75], it has been used more extensively in the identification of cancer molecular patterns for gene expression data [70, 76, 77].

7.1.2 Clustering

Clustering is one of the most widespread tools for grouping transcripts in microarray data. The concept of clustering is based on the simple idea of grouping similar objects. The goal is to maximize the similarity between objects in the same cluster, and minimize the similarity of objects in different clusters. How similarity is measured is thus a key part of clustering algorithms. In the case of microarray data, the expression profile of a gene, made up by the different samples, is seen as a series of coordinates that define a vector [78]. Distance metrics can thus compare the similarity of the direction and/or magnitude of two or more vectors.

Traditional clustering algorithms have existed since the 1950s and have been applied to a number of problems, including image analysis, marketing, for document classification (such as books), and for population studies. These traditional algorithms have also been used to cluster transcriptome data. In addition, specialized clustering algorithms have been developed for time-series data.

Clustering for Static Sampling

In the case of static sampling, transcriptome data can be represented as a matrix, with each row representing a gene, and each column representing a single condition. The data can thus be represented as vectors and the distance between these vectors can be determined. Note that there are two ways to organize the data. One is to take the expression value of each gene across different samples as a vector. The other one is to take the expression of all genes in a sample as a vector. Clustering can thus be used to find genes behaving similarly in different samples or samples which are "similar" in overall gene expression. In the following section, all examples are illustrated as clustering genes with similar transcriptional behaviors in different samples. The alternative of classifying samples based on their overall gene expression data is demonstrated in the supervised classification topic.

A distance measure (such as Euclidean, Manhattan, Chebyshev, Mahalanobis, Pearson, cosine, Spearman, or Kendall) is used to assess similarity and the data is then organized into clusters according to clustering rules. These clusters can be of fixed size, the number of clusters determined a priori) or natural clusters can be discovered in the data. The most commonly used clustering algorithms broadly correspond to two categories: hierarchical clustering and partitional clustering.

Hierarchical clustering can be bottom-up, starting with single-gene clusters and joining the most similar clusters until a single cluster with all genes is obtained; or top-down, starting with all genes in a single cluster and dividing them into smaller clusters [79]. In both cases, the result is represented as a hierarchical tree, or dendrogram. Most commonly, the bottom-up approach is used (Fig. 6). Initially, two closest genes (1 and 2; then 3 and 4) are joined using one of the distance metrics. In the next iteration, a linkage or amalgamation rule is needed to join these multiple-gene clusters [80]. This rule can be single linkage, complete linkage, or average linkage. In single linkage (also known as nearest neighbor), the similarity of these two clusters is the shortest distance of all pair-wise comparisons of the genes in one cluster to the other; in this example, the distance between gene 1 and gene 3. In complete linkage (also known as furthest neighbor), the similarity of these two clusters is defined as the largest distance of these pair-wise

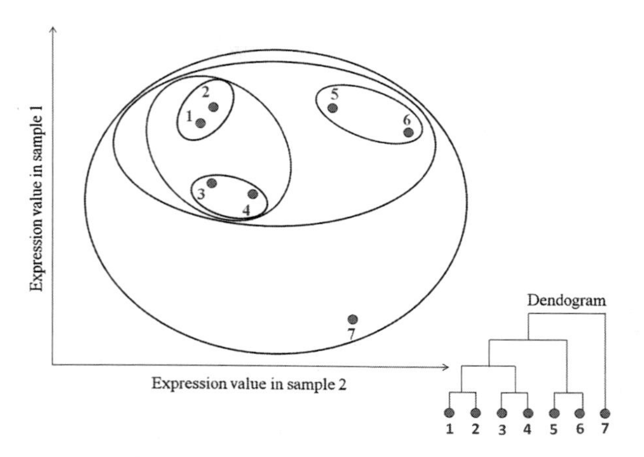

Fig. 6 Hierarchical clustering. The algorithm starts with each gene belonging to its own cluster, followed by joining the two closest genes: 1 and 2. Subsequently, individual genes or multi-gene clusters are joined using single linkage, complete linkage, or average linkage. In this case, single linkage is used, i.e., the distance between two clusters is taken as the shortest distance between any two members of the clusters. Thus the distance between cluster 1–2 and cluster 3–4 is the distance between genes 1 and 3. The two closest clusters are joined accordingly, in this case cluster 1–2 and cluster 3–4. This grouping is continued until all genes are joined into one cluster, and the whole process can be visualized as a dendrogram

comparisons; in this case, the distance between gene 2 and gene 4. In average linkage, the distance between these two clusters is that between their centroids [65]. In this instance, the centroid of the first cluster is a hypothetical gene "in the middle of" gene 1 and gene 2, and thus its expression level is taken as the average expression level of these two genes.

Hierarchical clustering has been used extensively to compare cell types and tissues, including diseased vs. healthy cells, and drug effects, for example [81–85]. Hierarchical clustering has also been used to classify proteomic profiles of serum, plasma, and modified media supplements used in cell culture [86], and metabolomic profiles of extracellular metabolites in recombinant CHO fed-batch cultures [87].

In partitional clustering, data points are separated into a pre-defined number of clusters. In the first step of these iterative algorithms, data points are randomly assigned to clusters. The distance between individual data points and the cluster is then calculated and used to reassign the data points to the cluster to which they are closest. This process continues until all data points are assigned to the closest cluster [88]. *K*-means clustering, Self-Organizing Maps (SOM), and Fuzzy C-means (FCM) clustering are among the best known clustering algorithms in this category. One limitation of these algorithms is that the number of clusters has to be fixed from the beginning, and thus the results are dependent on it [89].

In *k*-means clustering [90], *k* is the number of clusters, and is a required input. *k* random points are used as cluster centers (or means) at initialization. All data points are assigned to these initial clusters by finding the one with the closest distance. In iterative steps, the mean of each cluster is recalculated and the data points reassigned to new clusters [91]. This process continues until the assignment does not change markedly. As the value of k greatly influences the final outcome, several algorithms include a procedure to determine the best k. k-means clustering has been used to analyze transcriptome data of cancer cells [92] and stem cells [74, 93] among others.

Similar to k-means clustering, in the case of SOMs [94], the number of clusters is also a required input. In addition, their geometry must be specified (grid size). Thus not only the number of clusters but also their geometry has an effect on the final clustering result. A seed vector is first assigned to each cluster, and data assigned to these clusters in an iterative process. In each iteration, randomly selected gene expression data is compared to the seed vectors. The gene is assigned to the cluster that has the more similar seed vector. The value of the seed vector is updated, so that it is more similar to the expression of the gene used in the comparison. Because the cluster centers are part of a grid, the values of the other seed vectors are also modified, although to a lower extent. SOMs have been used to analyze monolayers of cultured rat hepatocytes [95], to study hematopoietic differentiation [96], to investigate saline osmotic tolerance in yeast [97], and to investigate hepatic differentiation [98], among others.

Whereas k-means and SOM assign each gene to a single cluster (hard clustering), FCM [99] links each gene to all clusters using a series of values. Values close to one indicate strong association to a cluster, and values close to 0 indicate absence of association. These indexes define the membership of each gene with respect to all clusters [100]. In addition to the number of clusters, the fuzziness parameter is also a required input. Kim et al. [101] have reported that the fuzziness parameter is sensitive to the normalization method used, and thus the clustering results vary with the normalization method. Recently, a method for the determination of the optimal parameters for FCM has been proposed [102]. FCM has been used to analyze gene expression profiles in high-grade gliomas [103] and in tumor sample classification [104].

Clustering for Dynamic Sampling

Clustering algorithms such as hierarchical clustering, k-means, and SOM are also commonly used to analyze time-series data. However, these algorithms do not take into account the sequential aspect of time-series data [105]. Thus clustering of time series requires specialized algorithms. Some of the specialized algorithms require long series (>10 time points), whereas others have been developed specifically for short time series.

B-splines [20, 106, 107], linear splines [108], ordered restricted inference [109], hidden Markov models [110], and gene expression dynamics using regression [111] are examples of clustering algorithms that can be used for long time-series data. Fuzzy C-Varieties with Transitional State Discrimination preclustering

(FCV-TSD) [112], ASTRO and MiMeSR [105], and Short Time-series Expression Miner (STEM) [113] are examples of clustering algorithms developed specifically for short time-series data.

STEM selects a set of potential expression profiles, each representing a unique pattern. Each gene is then assigned to the profile that best represents it. The significance of each profile is determined using hypothesis testing. The number of genes assigned to each profile under the true ordering is compared to the average number of genes assigned to each profile when permutated data is used. The significant profiles can then be analyzed independently or grouped into clusters. STEM has been used to cluster time-course microarray data collected in the study of egg development in *Drosophila melanogaster* [114], salt stress in *Medicago truncatula* [115], and muscle differentiation [116].

Biclustering takes clustering algorithms a step further. It consists of simultaneous clustering of both genes (rows) and conditions (columns) [117]. The goal in biclustering is to find submatrices [118], that is, to identify subgroups of genes and/or subgroups of conditions with highly correlated behaviors. Thus biclustering can find correlations in certain datasets where other algorithms cannot. Biclusters can be of constant row, constant column, or both constant row and column.

Among the software that can perform biclustering are Gene Expression Mining Server (GEMS) [119], Expression Analyzer and DisplayER (EXPANDER) [120], Phase-shifted Analysis of Gene Expression (PAGE) [121], Biclustering Gene Expression Time Series (BIGGEsTS) [122], Biclustering algorithm and Visualization (BiVisu) [123], and Biclustering Analysis Toolbox (BicAT) [124], which integrates several biclustering algorithms.

7.2 Supervised Classification Methods

Unsupervised classification methods are used for the identification of naturally existing clusters within the data. Supervised approaches, on the other hand, are designed to address the following question: given a set of samples categorized into pre-defined groups (training set), can we use the gene expression data of these samples to construct a rule, or a function, to differentiate these groups? This also implies the ability to use this rule for classification of new, uncategorized samples (test set) based on their expression data.

Since the classification rule is built upon the training set, it may fit this dataset "too well" and thus have poor performance on unclassified samples in the test set (Fig. 7a). In this example, the high producer clones (blue circles) and the low producer clones (black squares) can be simply separated by a linear model (solid line), allowing several samples to be misclassified (outliers). Yet the model can become over-complicated (dashed line) when trying to classify correctly all outliers and thus often results in a higher error rate in classifying regular samples. This is known as "overfitting", and ideally should be assessed using an independent test set. However, in situations where acquiring additional data is

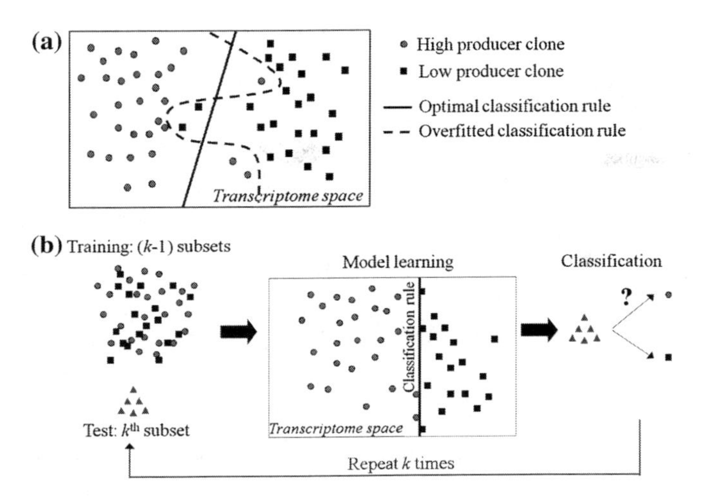

Fig. 7 Overfitting of training data and k-fold cross-validation scheme. **a** High producer clones (*blue circles*) and low producer clones (*black squares*) can be separated using a linear model (*solid line*) with a few outliers. Yet the model can become complex (*dashed line*) when all outliers are taken into consideration. This overfitted model will have a high error rate when classifying new samples. **b** The data is split into k subsets: $(k - 1)$ subsets are used for training the model, and testing is performed on the kth subset. This process is repeated k times until all data have been used for testing

expensive or not feasible, various cross-validation schemes can be used. The leave-one-out scheme allocates one sample for testing whereas the rest are used to train the classification model. In the hold-out scheme, the data is split into two equal sets—one is used for training, and the other for testing. Another frequently used method is the k-fold cross-validation, in which the data is divided into k sets—the first $(k - 1)$ sets are used for training, and the last one for testing (Fig. 7b). This process is repeated until all data have been used for testing. Commonly used supervised classifiers for gene expression data include K-Nearest Neighbors (KNNs), decision trees, Artificial Neural Networks (ANNs), and Support Vector Machines (SVMs). These algorithms have been implemented in several code libraries and various downloadable packages in Matlab and R.

7.2.1 K-Nearest Neighbors

KNN is among the simplest and most fundamental classification methods, and is often the first choice when prior knowledge about the dataset is minimal. Given that a set of samples have been classified into different groups, a new sample will be assigned into the group whose members constitute the majority in the neighborhood of the sample [125, 126]. The choice of distance metric thus becomes vital in this case —a new sample can be assigned to a different group when a different distance metric is applied. In addition, if a certain group is dominant in size compared to the others,

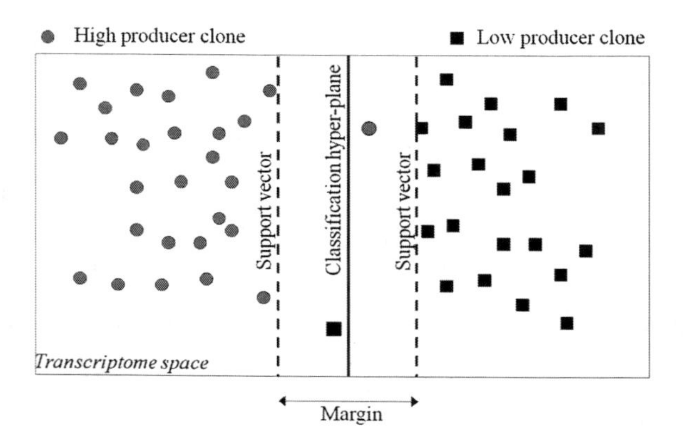

Fig. 8 Support Vector Machines with soft margin. Binary SVM algorithms search for a separation hyperplane that maximizes the margin (or distance) between two groups: in this case, high producer clones and low producer clones. Samples on the margins are referred to as "support vectors". A few samples can be misclassified in order to obtain a maximal margin ("soft margin")

a bias in assigning new samples into that group is likely to occur. One way to circumvent this problem involves giving each "neighbor" a weight inversely proportional to its distance to the new sample. Furthermore, the distance threshold and the number of "neighboring" samples k also have an effect on the final classification, and thus should be optimized using cross-validation.

In the FDA MicroArray Quality Control (MAQC) project, a KNN data analysis protocol was developed to predict the clinical outcome of about 500 new neuroblastoma patients [127]. These KNN models were built using a large gene expression dataset obtained from approximately 700 breast cancer, neuroblastoma, and multiple myeloma samples. In another example, gene expression signatures from 4413 probes in 37 colorectal cancer samples were also used to train a KNN model which was further validated using a leave-one-out scheme [128]. This model successfully classified these samples into serrated and conventional colorectal cancer samples using the expression data of 10 genes.

7.2.2 Decision Trees

Decision trees are built using an iterative scheme in which a question about the gene expression signatures of the training samples is posed at each node [126, 129–131]. The entire tree is obtained by repeated splitting of those samples into two or multiple descendant subsets. The training samples will guide the choice of splitting rules such that each terminal node of the tree, i.e., leaf, is assigned a group label. Thus decision trees are often more interpretable than other classifiers, and naturally support multiple-group assignment. Furthermore, multiple decision trees can be combined into an ensemble, e.g., random forest, to increase the classification accuracy [132, 133]. When applying decision trees, it is critical to control the complexity of the tree, i.e., avoid overfitting the training data. In addition to using cross-validation, one can also prune the tree by collapsing several internal nodes into one leaf, or stop branching the tree when there is no substantial improvement in the homogeneity of the final group assignment.

Several decision-tree algorithms were applied to 869 genes differentially expressed in earthworms in response to explosive compounds TNT or RDX [134]. 354 genes were subsequently selected by these algorithms as classifiers, and ranked according to their significance in the assembled tree. In another application, hierarchical clustering results of gene expression data from three different cohorts of 481 breast cancer samples were further analyzed using decision trees [135]. Four groups with different expression levels of osteopontin (OPN), activated leukocyte cell adhesion molecule (ALCAM), human epidermal growth factor 2 (HER2), and estrogen receptor (ER) were found. Patients with high OPN and low ER, HER2 and ALCAM were placed in a particularly high-risk group.

7.2.3 Artificial Neural Networks

ANNs were developed based on the computation principles occurring in the network of neurons within the human brain [126, 136, 137]. An ANN model can be considered as an assembly of interconnected nodes in which all input sources, in this case the expression values of all genes on the array, are weighted and combined. This weighted average is compared to a threshold, yielding an output value based on a step function. If the average exceeds the threshold, the output value will be one, corresponding to one group; zero, which corresponds to the other group. During the training process, the weighting factors and the threshold can be estimated iteratively, and a linear decision boundary (i.e., separating hyper-plane) can be obtained. Yet when the data are not linearly separable, hidden layers of intermediate nodes can be added to the network. A partial classification is performed at each layer, and assembled to achieve the final classification at the output node. Furthermore, alternative functions such as sigmoid or linear model can be utilized in place of the simple step function in these feed-forward neural networks.

Using gene expression data obtained from 63 training samples of small, round blue cell tumors (SRBCTs), 3750 ANNs have been constructed and cross-validated [138]. Without overfitting, these models successfully classified the samples into four diagnostic categories of tumors. ANNs have also proven efficient in tracking transcriptional changes responsible for progression from the chronic stage to a highly aggressive acute stage of adult T-cell leukemia (ATL) [139]. Using gene expression data from more than 44,000 probe sets and 10-fold cross-validation on 37 samples, 44 “predictor” genes could be identified, offering the possibility of diagnosing different ATL stages.

7.2.4 Support Vector Machines

In binary SVM, two groups (for example, high producer clones and low producer clones) are separated in such a way that the distance between the training samples and the decision boundary is maximized [126, 140, 141] (Fig. 8). This optimization process results in the construction of a separating hyper-plane, i.e., a linear line in 2-dimensional space, which maximizes the margin between the two groups. In several cases where the samples are not linearly separable in the original space, a kernel function can be chosen to transform the data to a higher-dimensional space in which a "linear" hyper-plane can be found. Furthermore, a few anomalous samples are often allowed to be misclassified to achieve a larger margin. Thus a cost function has to be selected and optimized such that the size of this "soft" margin is balanced with the allowable degree of hyper-plane violation.

Gene expression data from 97,802 clones was used to construct several SVM models using the simple dot-product kernel and validated through the leave-one-out scheme [142]. 31 human tissue samples were successfully classified by these models into cancerous ovarian and normal tissues. Interestingly, an SVM model was also built using gene expression profiles from seven high and four low recombinant IgG-producing NS0 cell lines. Through the leave-one-out cross-validation process, the transcriptomic differences between these high and low producers were indeed highlighted, supporting the molecular basis of productivity trait [143].

8 Pathway Analysis

Microarray analysis results in a list of differentially expressed genes or genes with a dynamic trend over time. It is possible that the transcriptional changes seen on those genes might not be independent, but rather have occurred in a coordinated manner. Thus understanding the physiological relevance of these changes requires analysis in a biological context, beyond what differential expression analysis can determine. Furthermore, examining genes in each pathway as a whole allows one to detect subtle, yet consistent, transcriptional changes that would otherwise be neglected by differential gene expression analysis.

Pathway analysis involves mapping the list of differentially expressed genes onto known pathways in order to elucidate a whole chain of events which might have occurred during the experiment. Depending on the microarray platform, probe identifiers can be linked to different sources of annotation, for instance, Gene Ontology (GO) [144], Kyoto Encyclopedia of Genes and Genomes (KEGG) [145], and Gene Map Annotator and Pathway Profiler (GenMAPP) [146]. This retrieval of pathway information allows all differentially expressed genes in a certain pathway to be highlighted. However, statistical tests need to be performed to confirm whether the entire pathway is indeed enriched or under-represented rather than occurring by random chance. A number of methods and software have been developed to assess the statistical significance of this functional enrichment/under-representation, including Ingenuity's IPA [147], GeneGo's MetaCore [148], GenMAPP's MAPPFinder [146], Gene Set Enrichment Analysis (GSEA) [149], and Gene Set Analysis [150]. Those methods differ in the calculation of the enrichment score and the corresponding significance level, usually p-value or FDR. For illustrative purposes, two representative methods, MAPPFinder and GSEA, are described in the following section.

8.1 MAPPFinder

In order to assess the degree of enrichment for each pathway (or gene set), MAPPFinder calculates a *z*-score using the number of differentially expressed genes in the set, the number of genes in the set, the number of differentially expressed genes in total, and the total number of genes on the array [151–153]. A high positive *z*-score indicates that the pathway of interest is significantly enriched, and an extreme negative *z*-score suggests that it is under-represented. Furthermore, if a *p*-value is desired, a *z*-score of 1.96 or −1.96 can be converted to a *p*-value of 0.05 given that the data strictly follows a hyper-geometric distribution. It is important to note that, similar to several other pathway analysis tools, MAPPFinder also requires a pre-defined list of differentially expressed genes. This is sometimes challenging since the list can vary considerably depending on the selected fold-change and the *p*-value cutoff.

Prickett et al. have demonstrated the use of MAPPFinder in uncovering several immune-system pathways affected in chicken infected with a protozoan parasite [154]. 1,175 genes, accounting for about 10% of the total unique Ensembl genes present on the array, were mapped to 85 inferred chicken pathways in GenMAPP, 18 of which were either up- or down-regulated at a *p*-value cut-off of 0.05. In another study, functional enrichment information obtained from MAPPFinder was linked automatically to the original gene expression data to calculate the average intensity or ratio of all differentially expressed genes in each pathway [155]. This quantitative evaluation of dose- and time-dependent microarray data in rats exposed to toxicants thus allows one to calculate an effective dose (ED_{50}) for each pathway, which plays an important role in risk assessment.

8.2 Gene Set Enrichment Analysis

GSEA is a powerful tool for pathway analysis which calculates gene set enrichment using all genes present on the array instead of a pre-defined set of differentially expressed genes [149, 156, 157]. An ordered list is first generated by ranking all genes in the dataset based on their signal-to-noise ratio (Fig. 9). This ratio is often the quotient between the difference in average expression levels and the overall variability of measurement. In the second step, a running-sum statistic is measured for each pathway (or gene set S) by travelling down the ordered list. If the gene encountered is a part of the gene set of interest, the statistic is increased; otherwise it is decreased. The magnitude of this change is set to be proportional to the signal-to-noise of that gene and the size of the gene set it belongs to. The maximum deviation from zero of the running-sum statistic is chosen as the enrichment score (ES), and an associated statistical significance (*p*-value) can be calculated using a permutation scheme. Concurrently, a leading-edge subset of genes which are key contributors to enrichment of the function represented by the gene set can also be exported.

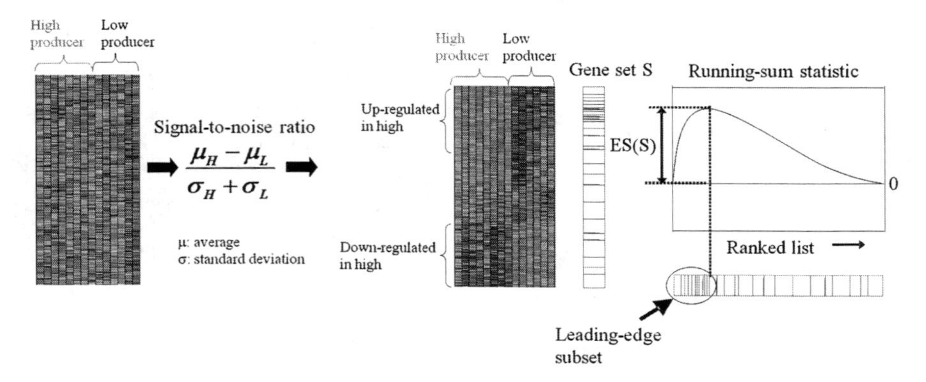

Fig. 9 Gene Set Enrichment Analysis (GSEA). Genes are ranked based on their signal-to-noise ratios to create an ordered list. A running sum statistic is calculated by walking down this list. If the gene encountered is part of the gene set of interest, the running sum statistic is increased; otherwise, it is decreased. The enrichment score (ES) of each gene set (S) is chosen as the maximum deviation of this statistic from zero. Genes with key contributions to the enrichment of the gene set are listed in the leading-edge subset

Deregulated functional categories in Ewing's sarcoma family tumors (ESFT) cell lines under hypoxia were identified by applying GSEA with three different gene sets [158]. Hypoxia-related functions such as angiogenesis, vasculature development, and glucose metabolism were shown to be up-regulated under hypoxic conditions. GSEA was also used alongside other pathway analysis tools to investigate the biological relevance of transcriptional differences between neurofibromatosis type 1 (NF1)-haploinsufficient lymphoblastoid cell lines (LCLs) and mouse B lymphocytes [159]. Despite the modest changes in gene expression detected using the *t*-test, several pathways were shown to experience perturbations including cell cycle, DNA replication and repair, transcription and translation, and immune response.

9 Network Reconstruction

Gene network inference attempts to reconstruct gene networks reflecting their interactions from high-throughput data, especially microarray data. Network reconstruction is a challenging task as gene interactions are dynamic and membership of particular elements in a network is not always permanent. In this regard, the use of microarray data compiled under a wide range of conditions, or from a variety of mutants, can help unveil interactions. Also, time-series microarray data is of particular relevance in reverse engineering regulatory networks. In addition to algorithms for constructing regulatory networks using static gene expression data, special algorithms have also been developed for data obtained from time-series microarrays.

9.1 Network Reconstruction From Static Gene Expression Data

9.1.1 Information Theoretic Methods

Several methods based on information theory have been used for reverse engineer cellular networks from microarray expression profiles. These methods calculate mutual information (MI) between pairs of gene expression profiles. An advantage of MI over other measures of relatedness is that it can detect non-linear interactions. Although these algorithms can be used on time-series data, the sequential aspect is lost, as each sample time point would be considered a different condition.

The original algorithm, relevance networks (RELNET) [160], infers an interaction if MI for a pair is larger than a threshold. RELNET has been applied to reconstruct networks in yeast [160], in cancer cell lines [161], in human hepatoma cells [162], and to identify hub cancer genes [163]. This approach, however, can result in many false positives, and thus extensions which discriminate between direct and indirect interactions have been developed.

Extensions to RELNET proceed in two steps. The first is common to all methods, and consists of calculating MI between pairs of gene expression profiles. In the second step the MI values are assessed and compared, and interactions inferred. The second step is unique to each method.

Context Likelihood of Relatedness (CLR) [164] is an algorithm that removes false correlations by comparing MI for each pair with a background distribution of MI scores. CLR was used to reconstruct parts of the transcriptional regulatory network of the pathogen *Salmonella typhimurium* [165].

A second algorithm based on relevance networks, the Algorithm for the Reconstruction of Accurate Cellular Networks (ARACNE) [166–168], eliminates indirect relationships by using data process inequality (DPI), a characteristic of mutual information. ARACNE has been used in reverse engineering the regulatory networks of human B cells [166], in the identification of the targets of the transcriptional repressor BCL6 [169], in the reconstruction of red blood cell metabolism from metabolic data [170], and in the genome-wide reconstruction of the regulatory networks of *Streptomyces coelicolor* [171], an antibiotic producer. CLR and ARACNE were both used to identify genes regulated by Nrf2 in response to oxidative stress [172], and to infer the connectivity of phosphorylation sites in receptor tyrosine kinases [173].

A third algorithm, Minimum Redundancy Networks (MRNET) [174], performs a series of maximum relevance/minimum redundancy (MRMR) selection procedures for each gene and selects the genes having the highest MI with the target.

RELNET, CLR, ARACNE, and MRMR are included in the R package minet (Mutual Information NETwork inference) [175]. The networks resulting from these algorithms can be visualized using the R package Rgraphviz [176]. In addition, the Java implementation of ARACNE includes Cytoscape [177] for network visualization.

9.1.2 Bayesian Networks

Bayesian networks have recently emerged as promising approaches for inferring gene regulatory networks using microarray data. These methods are particularly suitable for the reconstruction of cellular networks due to their ability to capture the stochastic nature of gene regulation and allow causality inference [178, 179]. Furthermore, prior knowledge can be incorporated to improve the accuracy of the final network structure.

A Bayesian network can be represented as a directed acyclic graph, in which each node is a gene, and the edge between two nodes denotes the dependency between two corresponding genes [180, 181]. A joint probability for the network is thus calculated as a product of multiple conditional probabilities for each gene, given that it is regulated by a defined set of parent genes. These probability functions can be either discrete, e.g., binomial distributions, or continuous, e.g., normal density function. Among the several possible networks being reconstructed, an optimal network can be chosen by maximizing the corresponding posterior probability.

Bayesian predictive networks have been constructed using gene expression in combination with genotypic, transcription factor binding site, and protein–protein interaction data in yeast [182]. These networks were shown to successfully predict regulators causing hot spots of gene expression activity in a dividing yeast population. Molecular mechanisms underlying transcriptome reprogramming in cyanobacteria under altered environments were also revealed using Bayesian networks [183]. A large number of genes in the core transcriptional response (CTR) are associated with oxidative stress under most perturbations, indicating the important role of reactive oxygen species in the regulation of these genes.

9.2 Network Reconstruction from Dynamic Gene Expression Data

9.2.1 Information Theoretic Method: TimeDelay-Aracne

Some of the algorithms originally used for network reconstruction using static sampling data have been extended to take advantage of the dependency information contained in time-series data. One such example is an extension of ARACNE. This extension, implemented in the TimeDelay-ARACNE algorithm [184], uses time-course data to retrieve time statistical dependencies between gene expression profiles. This algorithm considers the possibility that the expression of a gene at a certain time could depend on the expression level of another gene at an earlier time point; that is, it detects time-delayed dependencies. The algorithm performs three steps: firstly, it detects the time point of the initial changes in the expression for all genes; secondly, it constructs networks by calculating time-dependent MIs; and thirdly, it performs network pruning using DPI. TimeDelay-ARACNE, which has been implemented in R, also attempts to infer edge directionality.

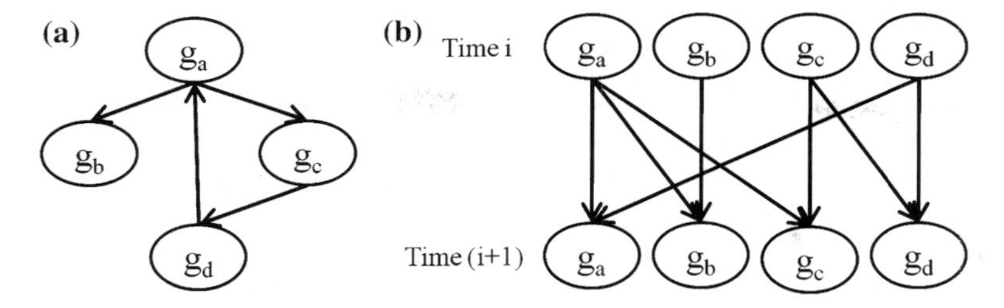

Fig. 10 Gene regulatory network with feedback loop deciphered using DBN. A regulatory network containing four genes (g_a, g_b, g_c, and g_d), three of which form a feedback loop ($g_a \rightarrow g_c \rightarrow g_d \rightarrow g_a$). **a** The feedback loop among g_a, g_c, and g_d is deciphered by allowing cross-interactions along the time axis. **b** The expression level of g_c at time point (i + 1) is dependent on that of g_a at time point i. Similarly, the expression level of g_d at time point (i + 1) is dependent on that of g_c at time point i. The loop is closed by allowing g_d's expression level at time point i to have an effect on that of g_a at time point (i + 1). Note that each gene's expression level at a certain time point is always dependent on its own expression level at the previous time point

9.2.2 Dynamic Bayesian Networks

Built upon Bayesian networks, Dynamic Bayesian Networks (DBNs) also calculate a joint probability using the conditional probability of each gene, and select the optimal network based on the posterior probability. DBNs further allow time delay and modeling of feedback loops by incorporating temporal information associated with time-series data. For instance, the cyclic regulation among genes g_a, g_c, and g_d shown in Fig. 10a can be represented by allowing these genes to cross-interact from time point i to time point (i + 1) (Fig. 10b). To further enhance the prediction accuracy and reduce the computational complexity of DBNs, a number of modifications have also been proposed [185]. For example, potential regulators are limited to those genes with either preceding or simultaneous expression changes. Transcriptional time lags between regulators and target genes can also be estimated, and statistical analysis is thereby restricted within that time frame to improve the accuracy of the prediction.

DBNs have been used successfully to construct gene regulatory networks in yeast using cell cycle time-series microarray data in two independent studies [179, 185]. Main regulatory nodes in the S.O.S DNA repair network in *E. coli* were also extracted using DBNs [186]. Compared to other methods for inferring gene regulatory networks such as Granger causality and probabilistic Boolean network, DBNs consistently displayed enhanced performance. This was especially the case for short time series, as exemplified with data obtained from muscle development in fruit fly [187], normal and infected *Arabidopsis* leaves [188], and food intake effect on human blood [189]. Furthermore, the causality inference power of DBNs was substantially improved when time-series gene perturbation data was also incorporated [190].

10 Concluding Remarks

In this review, methods for the analysis of microarray data are summarized, with a focus on their use in mammalian cell culture. Whereas specific algorithms used for each step depend on the type of data and the question being asked, the general steps for microarray data analysis remain valid. These steps include data pre-processing followed by identification of differentially expressed genes at a minimum, but greater biological insight can be gained by using other types of analysis such as profile pattern recognition, pathway analysis, and network reconstruction.

Even though transcriptome studies of antibody-producing cell lines have been few compared to other cell types, the next few years will see an increase in the resources available for studying genomes and transcriptomes, and this will greatly benefit the understanding of these relevant cell lines.

References

1. Schena M, Shalon D, Davis RW, Brown PO (1995) Quantitative monitoring of gene expression patterns with a complementary DNA microarray. Science 270:467
2. Lashkari DA, DeRisi JL, McCusker JH, Namath AF, Gentile C, Hwang SY, Brown PO, Davis RW (1997) Yeast microarrays for genome wide parallel genetic and gene expression analysis. Proc Natl Acad Sci USA 94:13057
3. Agilent http://www.genomics.agilent.com/GenericB.aspx?PageType=Custom&SubPageType=Custom&PageID=2011
4. Affymetrix http://www.affymetrix.com/browse/brand/affymetrixMicroarraySolutions/brandAffymetrixMicroarraySolutions-overview.jsp?category=35722&categoryIdClicked=35722&rootCategoryId=35677&navMode=35722&parent=35722&aId=affymetrixmicroarraybrandsNav
5. Nimblegen http://www.nimblegen.com/products/expression/index.html
6. Wang Z, Gerstein M, Snyder M (2009) RNA-Seq: a revolutionary tool for transcriptomics. Nat Rev Genet 10:57
7. Seth G, Charaniya S, Wlaschin KF, Hu W-S (2007) In pursuit of a super producer–alternative paths to high producing recombinant mammalian cells. Curr Opin Biotechnol 18:557
8. Krampe B, Swiderek H, Al-Rubeai M (2008) Transcriptome and proteome analysis of antibody-producing mouse myeloma NS0 cells cultivated at different cell densities in perfusion culture. Biotechnol Appl Biochem 50:133
9. Spens E, Häggström L (2009) Proliferation of NS0 cells in protein-free medium: the role of cell-derived proteins, known growth factors and cellular receptors. J Biotechnol 141:123
10. Swiderek H, Logan A, Al-Rubeai M (2008) Cellular and transcriptomic analysis of NS0 cell response during exposure to hypoxia. J Biotechnol 134:103
11. Tai YC, Speed TP (2006) A multivariate empirical Bayes statistic for replicated microarray time couse data. Ann Stat 34:6
12. Schaub J, Clemens C, Schorn P, Hildebrandt T, Rust W, Mennerich D, Kaufmann H, Schulz TW (2010) CHO gene expression profiling in biopharmaceutical process analysis and design. Biotechnol Bioeng 105:431
13. Lee YY, Wong KTK, Nissom PM, Wong DCF, Yap MGS (2007) Transcriptional profiling of batch and fed-batch protein-free 293-HEK cultures. Metab Eng 9:52

14. Kantardjieff A, Jacob NM, Yee JC, Epstein E, Kok Y-J, Philp R, Betenbaugh M, Hu W-S (2010) Transcriptome and proteome analysis of Chinese hamster ovary cells under low temperature and butyrate treatment. J Biotechnol 145:143
15. Kerr MK, Churchill GA (2001) Statistical design and the analysis of gene expression microarray data. Genet Res 77:123
16. Wang X, Wu M, Li Z, Chan C (2008) Short time-series microarray analysis: methods and challenges. BMC Syst Biol 2:58
17. Quackenbush J (2002) Microarray data normalization and transformation. Nat Genet 32:496–501
18. Bolstad BM, Irizarry RA, Astrand M, Speed TP (2003) A comparison of normalization methods for high density oligonucleotide array data based on variance and bias. Bioinformatics 19:185
19. Mehra S, Lian W, Jayapal K, Charaniya S, Sherman D, Hu W-S (2006) A framework to analyze multiple time series data: a case study with Streptomyces coelicolor. J Ind Microbiol Biotechnol 33:159
20. Bar-Joseph Z, Gerber GK, Gifford DK, Jaakkola TS, Simon I (2003) Continuous representations of time-series gene expression data. J Comput Biol 10:341
21. Bar-Joseph Z (2004) Analyzing time series gene expression data. Bioinformatics 20:2493
22. Sakoe H, Chiba S (1978) Dynamic programming algorithm optimization for spoken word recognition. IEEE Trans Acoust Speech Signal Process 26:43
23. Kassidas A, MacGregor JF, Taylor PA (1998) Synchronization of batch trajectories using dynamic time warping. AIChE J 44:864
24. Ramaker H-J, van Sprang ENM, Westerhuis JA, Smilde AK (2003) Dynamic time warping of spectroscopic BATCH data. Anal Chim Acta 498:133
25. Smith AA, Vollrath A, Bradfield CA, Craven M (2009) Clustered alignments of gene-expression time series data. Bioinformatics 25:i119
26. Storey JD, Xiao W, Leek JT, Tompkins RG, Davis RW (2005) Significance analysis of time course microarray experiments. Proc Natl Acad Sci USA 102:12837
27. Bonferroni CE (1936) Teoria statistica delle classi e calcolo delle probabilità. Pubblicazioni del R Istituto Superiore di Scienze Economiche e Commerciali di Firenze 8:3
28. Benjamini Y, Hochberg Y (1995) Controlling the false discovery rate: a practical and powerful approach to multiple testing. J R Stat Soc Ser B (Methodological) 57:289
29. Korke R, Gatti MDL, Lau ALY, Lim JWE, Seow TK, Chung MCM, Hu W-S (2004) Large scale gene expression profiling of metabolic shift of mammalian cells in culture. J Biotechnol 107:1
30. De Leon Gatti M, Wlaschin KF, Nissom PM, Yap M, Hu W-S (2007) Comparative transcriptional analysis of mouse hybridoma and recombinant Chinese hamster ovary cells undergoing butyrate treatment. J Biosci Bioeng 103:82
31. Oleksiak MF, Churchill GA, Crawford DL (2002) Variation in gene expression within and among natural populations. Nat Genet 32:261
32. Kalinka AT, Varga KM, Gerrard DT, Preibisch S, Corcoran DL, Jarrells J, Ohler U, Bergman CM, Tomancak P (2010) Gene expression divergence recapitulates the developmental hourglass model. Nature 468:811
33. Tusher VG, Tibshirani R, Chu G (2001) Significance analysis of microarrays applied to the ionizing radiation response. Proc Natl Acad Sci USA 98:5116
34. Zheng H, Ying H, Yan H, Kimmelman AC, Hiller DJ, Chen A-J, Perry SR, Tonon G, Chu GC, Ding Z, Stommel JM, Dunn KL, Wiedemeyer R, You MJ, Brennan C, Wang YA, Ligon KL, Wong WH, Chin L, DePinho RA (2008) p53 and Pten control neural and glioma stem/progenitor cell renewal and differentiation. Nature 455:1129
35. Liu W, Tanasa B, Tyurina OV, Zhou TY, Gassmann R, Liu WT, Ohgi KA, Benner C, Garcia-Bassets I, Aggarwal AK, Desai A, Dorrestein PC, Glass CK, Rosenfeld MG (2010) PHF8 mediates histone H4 lysine 20 demethylation events involved in cell cycle progression. Nature 466:508
36. Lonnstedt I, Britton T (2005) Hierarchical Bayes models for cDNA microarray gene expression. Biostatistics 6:279

37. Smyth GK (2004) Linear models and empirical bayes methods for assessing differential expression in microarray experiments. Stat Appl Genet Mol Biol 3:article3
38. Feldser DM, Kostova KK, Winslow MM, Taylor SE, Cashman C, Whittaker CA, Sanchez-Rivera FJ, Resnick R, Bronson R, Hemann MT, Jacks T (2010) Stage-specific sensitivity to p53 restoration during lung cancer progression. Nature 468:572
39. Bonfanti P, Claudinot S, Amici AW, Farley A, Blackburn CC, Barrandon Y (2010) Microenvironmental reprogramming of thymic epithelial cells to skin multipotent stem cells. Nature 466:978
40. Storey JD, Dai JY, Leek JT (2007) The optimal discovery procedure for large-scale significance testing, with applications to comparative microarray experiments. Biostatistics 8:414–432
41. Storey JD, Tibshirani R (2003) Statistical significance for genome wide studies. Proc Natl Acad Sci USA 100:9440
42. Leek JT, Monsen E, Dabney AR, Storey JD (2006) EDGE: extraction and analysis of differential gene expression. Bioinformatics 22:507
43. Zhan M, Yamaza H, Sun Y, Sinclair J, Li H, Zou S (2007) Temporal and spatial transcriptional profiles of aging in Drosophila melanogaster. Genome Res 17:1236
44. White P, Lee May C, Lamounier RN, Brestelli JE, Kaestner KH (2008) Defining pancreatic endocrine precursors and their descendants. Diabetes 57:654
45. Conesa A, Nueda MJ, Ferrer A, Talon M (2006) maSigPro: a method to identify significantly differential expression profiles in time-course microarray experiments. Bioinformatics 22:1096
46. Sanges R, Cordero F, Calogero RA (2007) oneChannelGUI: a graphical interface to Bioconductor tools, designed for life scientists who are not familiar with R language. Bioinformatics 23:3406
47. Tarraga J, Medina I, Carbonell J, Huerta-Cepas J, Minguez P, Alloza E, Al-Shahrour F, Vegas-Azcarate S, Goetz S, Escobar P, Garcia–Garcia F, Conesa A, Montaner D, Dopazo J (2008) GEPAS, a web-based tool for microarray data analysis and interpretation. Nucleic Acids Res 36:W308
48. Nueda MJ, Sebastian P, Tarazona S, Garcia-Garcia F, Dopazo J, Ferrer A, Conesa A (2009) Functional assessment of time course microarray data. BMC Bioinformatics 10(6):S9
49. Brusniak MY, Bodenmiller B, Campbell D, Cooke K, Eddes J, Garbutt A, Lau H, Letarte S, Mueller LN, Sharma V, Vitek O, Zhang N, Aebersold R, Watts JD (2008) Corra: computational framework and tools for LC-MS discovery and targeted mass spectrometry-based proteomics. BMC Bioinformatics 9:542
50. Levin AM, de Vries RP, Conesa A, de Bekker C, Talon M, Menke HH, van Peij NN, Wosten HA (2007) Spatial differentiation in the vegetative mycelium of Aspergillus niger. Eukaryot Cell 6:2311
51. Wong CE, Singh MB, Bhalla PL (2009) Molecular processes underlying the floral transition in the soybean shoot apical meristem. Plant J 57:832
52. Wong CE, Singh MB, Bhalla PL (2009) Floral initiation process at the soybean shoot apical meristem may involve multiple hormonal pathways. Plant Signal Behav 4:648
53. Pascual L, Blanca JM, Canizares J, Nuez F (2009) Transcriptomic analysis of tomato carpel development reveals alterations in ethylene and gibberellin synthesis during pat3/pat4 parthenocarpic fruit set. BMC Plant Biol 9:67
54. Hoogerwerf WA, Sinha M, Conesa A, Luxon BA, Shahinian VB, Cornelissen G, Halberg F, Bostwick J, Timm J, Cassone VM (2008) Transcriptional profiling of mRNA expression in the mouse distal colon. Gastroenterology 135:2019
55. Smilde AK, Jansen JJ, Hoefsloot HCJ, Lamers ANR-J, van der Greef J, Timmerman ME (2005) ANOVA-simultaneous component analysis (ASCA): a new tool for analyzing designed metabolomics data. Bioinformatics 21:3043
56. Jansen JJ, Hoefsloot HCJ, Greef JVD, Timmerman ME, Westerhuis JA, Smilde AK (2005) ASCA: analysis of multivariate data obtained from an experimental design. J Chemometr 19:469

57. Smilde AK, Hoefsloot HCJ, Westerhuis JA (2008) The geometry of ASCA. J Chemometr 22:464
58. Vis D, Westerhuis J, Smilde A, van der Greef J (2007) Statistical validation of megavariate effects in ASCA. BMC Bioinformatics 8:322
59. Wang J, Reijmers T, Chen L, Van Der Heijden R, Wang M, Peng S, Hankemeier T, Xu G, Van Der Greef J (2009) Systems toxicology study of doxorubicin on rats using ultra performance liquid chromatography coupled with mass spectrometry based metabolomics. Metabolomics 5:407
60. Nueda MJ, Conesa A, Westerhuis JA, Hoefsloot HCJ, Smilde AK, Talon M, Ferrer A (2007) Discovering gene expression patterns in time course microarray experiments by ANOVA SCA. Bioinformatics 23:1792
61. Heard NA, Holmes CC, Stephens DA (2006) A quantitative study of gene regulation involved in the immune response of anopheline mosquitoes. J Am Stat Assoc 101:18
62. Angelini C, De Canditiis D, Mutarelli M, Pensky M (2007) A Bayesian approach to estimation and testing in time-course microarray experiments. Stat Appl Genet Mol Biol 6: Article24
63. Angelini C, Cutillo L, De Canditiis D, Mutarelli M, Pensky M (2008) BATS: a Bayesian user-friendly software for analyzing time series microarray experiments. BMC Bioinformatics 9:415
64. Lian W, Jayapal K, Charaniya S, Mehra S, Glod F, Kyung Y-S, Sherman D, Hu W-S (2008) Genome-wide transcriptome analysis reveals that a pleiotropic antibiotic regulator, AfsS, modulates nutritional stress response in Streptomyces coelicolor A3(2). BMC Genomics 9:56
65. Gollub J, Sherlock G, Alan K, Brian O (2006) Clustering microarray data. Academic Press, London
66. Morrison DA, Ellis JT (2003) The design and analysis of microarray experiments: applications in parasitology. DNA Cell Biol 22:357
67. Boutros PC, Okey AB (2005) Unsupervised pattern recognition: an introduction to the whys and wherefores of clustering microarray data. Brief Bioinform 6:331
68. Jolliffe I (2005) Principal component analysis. Wiley, NY
69. Yeung KY, Ruzzo WL (2001) Principal component analysis for clustering gene expression data. Bioinformatics 17:763
70. Brunet J-P, Tamayo P, Golub TR, Mesirov JP (2004) Metagenes and molecular pattern discovery using matrix factorization. Proc Natl Acad Sci USA 101:4164
71. Lee DD, Seung HS (1999) Learning the parts of objects by non-negative matrix factorization. Nature 401:788
72. Schachtner R, Lutter D, Stadlthanner K, Lang EW, Schmitz G, Tome AM, Gomez Vilda P (2007) Routes to identify marker genes for microarray classification. In: Engineering in medicine and biology society, 2007 EMBS 2007 29th Annual International Conference of the IEEE
73. Aiba K, Sharov AA, Carter MG, Foroni C, Vescovi AL, Ko MSH (2006) Defining a developmental path to neural fate by global expression profiling of mouse embryonic stem cells and adult neural stem/progenitor cells. Stem Cells 24:889
74. Ulloa-Montoya F, Kidder B, Pauwelyn K, Chase L, Luttun A, Crabbe A, Geraerts M, Sharov A, Piao Y, Ko M, Hu W-S, Verfaillie C (2007) Comparative transcriptome analysis of embryonic and adult stem cells with extended and limited differentiation capacity. Genome Biol 8:R163
75. Liu W, Yuan K, Ye D (2008) Reducing microarray data via nonnegative matrix factorization for visualization and clustering analysis. J Biomed Inform 41:602
76. Han X (2008) Improving gene expression cancer molecular pattern discovery using nonnegative principal component analysis. Genome Inf 21:200
77. Frigyesi A, Hoglund M (2008) Non-negative matrix factorization for the analysis of complex gene expression data: identification of clinically relevant tumor subtypes. Cancer Inform 6:275

78. Sherlock G (2000) Analysis of large-scale gene expression data. Curr Opin Immunol 12:201
79. Nugent R, Meila M (2010) An overview of clustering applied to molecular biology. Humana Press, Clifton
80. Frades I, Matthiesen R (2010) Overview on techniques in cluster analysis. Bioinf Methods Clin Res 593:81
81. Anichini A, Scarito A, Molla A, Parmiani G, Mortarini R (2003) Differentiation of CD8+T cells from tumor-invaded and tumor-free lymph nodes of melanoma patients: role of common î³-chain cytokines. J Immunol 171:2134
82. Vega F, Coombes KR, Thomazy VA, Patel K, Lang W, Jones D (2006) Tissue-specific function of lymph node fibroblastic reticulum cells. Pathobiology 73:71
83. Ambrosi DJ, Tanasijevic B, Kaur A, Obergfell C, O'Neill RJ, Krueger W, Rasmussen TP (2007) Genome-wide reprogramming in hybrids of somatic cells and embryonic stem cells. Stem Cells 25:1104
84. Secco M, Moreira Y, Zucconi E, Vieira N, Jazedje T, Muotri A, Okamoto O, Verjovski-Almeida S, Zatz M (2009) Gene expression profile of mesenchymal stem cells from paired umbilical cord units: cord is different from blood. Stem Cell Rev R 5:387
85. Fortier JM, Payton JE, Cahan P, Ley TJ, Walter MJ, Graubert TA (2010) POU4F1 is associated with t(8;21) acute myeloid leukemia and contributes directly to its unique transcriptional signature. Leukemia 24:950
86. Ayache S, Panelli M, Byrne K, Slezak S, Leitman S, Marincola F, Stroncek D (2006) Comparison of proteomic profiles of serum, plasma, and modified media supplements used for cell culture and expansion. J Translational Med 4:40
87. Chong WPK, Goh LT, Reddy SG, Yusufi FNK, Lee DY, Wong NSC, Heng CK, Yap MGS, Ho YS (2009) Metabolomics profiling of extracellular metabolites in recombinant Chinese Hamster Ovary fed-batch culture. Rapid Commun Mass Spectrom 23:3763
88. De Bruyne V, Al-Mulla F, Pot B (2005) Methods for microarray data analysis. Humana Press, Clifton
89. Dopazo J, Zanders E, Dragoni I, Amphlett G, Falciani F (2001) Methods and approaches in the analysis of gene expression data. J Immunol Methods 250:93
90. Everitt BS (1974) Cluster analysis. Heinemann Educational [for] the Social Science Research Council, London
91. Do JH, Choi D-K (2008) Clustering approaches to identifying gene expression patterns from DNA microarray data. Mol Cells 25:279
92. Liu Y, Yang Y, Xu H, Dong X (2010) Implication of USP22 in the regulation of BMI-1, c-Myc, p16INK4a, p14ARF, and cyclin D2 expression in primary colorectal carcinomas. Diagn Mol Pathol 19:194
93. Way KJ, Dinh H, Keene MR, White KE, Clanchy FIL, Lusby P, Roiniotis J, Cook AD, Cassady AI, Curtis DJ, Hamilton JA (2009) The generation and properties of human macrophage populations from hemopoietic stem cells. J Leukoc Biol 85:766
94. Kohonen T (2001) Self-organizing maps. Springer, Berlin
95. Baker TK, Carfagna MA, Gao H, Dow ER, Li Q, Searfoss GH, Ryan TP (2001) Temporal gene expression analysis of monolayer cultured rat hepatocytes. Chem Res Toxicol 14:1218
96. Tamayo P, Slonim D, Mesirov J, Zhu Q, Kitareewan S, Dmitrovsky E, Lander ES, Golub TR (1999) Interpreting patterns of gene expression with self-organizing maps: Methods and application to hematopoietic differentiation. Proc Natl Acad Sci USA 96:2907
97. Pandey G, Yoshikawa K, Hirasawa T, Nagahisa K, Katakura Y, Furusawa C, Shimizu H, Shioya S (2007) Extracting the hidden features in saline osmotic tolerance in Saccharomyces cerevisiae from DNA microarray data using the self-organizing map: biosynthesis of amino acids. Appl Microb Biotechnol 75:415
98. Li W, You P, Wei Q, Li Y, Fu X, Ding X, Wang X, Hu Y (2007) Hepatic differentiation and transcriptional profile of the mouse liver epithelial progenitor cells (LEPCs) under the induction of sodium butyrate. Front Biosci 12:1691
99. Bezdek J (1981) Pattern Recognition with Fuzzy Objective Function Algorithms (Advanced Applications in Pattern Recognition). Springer, Berlin

100. Dembele D, Kastner P (2003) Fuzzy C-means method for clustering microarray data. Bioinformatics 19:973
101. Kim S, Lee J, Bae J (2006) Effect of data normalization on fuzzy clustering of DNA microarray data. BMC Bioinformatics 7:134
102. Schwammle V, Jensen ONJ (2010) A simple and fast method to determine the parameters for fuzzy c-means cluster analysis. Bioinformatics 26:2841
103. Czernicki T, Zegarska J, Paczek L, Cukrowska B, Grajokowska W, Zajaczkowska A, Brudzewski K, Ulaczyk J, Marchel A (2007) Gene expression profile as a prognostic factor in high-grade gliomas. Int J Oncol 30:55
104. Wang J, Bø T, Jonassen I, Myklebost O, Hovig E (2003) Tumor classification and marker gene prediction by feature selection and fuzzy c-means clustering using microarray data. BMC Bioinformatics 4:1
105. Tchagang A, Bui K, McGinnis T, Benos P (2009) Extracting biologically significant patterns from short time series gene expression data. BMC Bioinformatics 10:255
106. Luan Y, Li H (2003) Clustering of time-course gene expression data using a mixed-effects model with B-splines. Bioinformatics 19:474
107. Gaffney, S and P Smyth (2005) Joint probabilistic curve clustering and alignment. Adv Neural Inf Process Syst
108. De Hoon MJ, Imoto S, Miyano S (2002) Statistical analysis of a small set of time-ordered gene expression data using linear splines. Bioinformatics 18:1477
109. Peddada SD, Lobenhofer EK, Li L, Afshari CA, Weinberg CR, Umbach DM (2003) Gene selection and clustering for time-course and dose-response microarray experiments using order-restricted inference. Bioinformatics 19:834
110. Schliep A, Schonhuth A, Steinhoff C (2003) Using hidden Markov models to analyze gene expression time course data. Bioinformatics 19(1):255
111. Ramoni MF, Sebastiani P, Kohane IS (2002) Cluster analysis of gene expression dynamics. Proc Natl Acad Sci USA 99:9121
112. Moller-Levet CS, Cho KH, Wolkenhauer O (2003) Microarray data clustering based on temporal variation: FCV with TSD preclustering. Appl Bioinform 2:35
113. Ernst J, Bar-Joseph Z (2006) STEM: a tool for the analysis of short time series gene expression data. BMC Bioinformatics 7:191
114. Baker DA, Russell S (2009) Gene expression during Drosophila melanogaster egg development before and after reproductive diapause. BMC Genomics 10:242
115. Li D, Su Z, Dong J, Wang T (2009) An expression database for roots of the model legume Medicago truncatula under salt stress. BMC Genomics 10:517
116. Ozbudak E, Tassy O, Pourquie O (2010) Spatiotemporal compartmentalization of key physiological processes during muscle precursor differentiation. Proc Natl Acad Sci USA 107:4224
117. Cheng Y, Church GM (2000) Biclustering of expression data. Proc Int Conf Intell Syst Mol Biol 8:93
118. Madeira SC, Oliveira AL (2004) Biclustering algorithms for biological data analysis: a survey. IEEE/ACM Trans Comput Biol Bioinform 1:24
119. Wu C-J, Kasif S (2005) GEMS: a web server for biclustering analysis of expression data. Nucleic Acids Res 33:W596
120. Shamir R, Maron-Katz A, Tanay A, Linhart C, Steinfeld I, Sharan R, Shiloh Y, Elkon R (2005) EXPANDER—an integrative program suite for microarray data analysis. BMC Bioinformatics 6:232
121. Leung E, Bushel PR (2006) PAGE: phase-shifted analysis of gene expression. Bioinformatics 22:367
122. Goncalves JP, Madeira SC, Oliveira AL (2009) BiGGEsTS: integrated environment for biclustering analysis of time series gene expression data. BMC Res Notes 2:124
123. Cheng KO, Law NF, Siu WC, Lau TH (2007) BiVisu: software tool for bicluster detection and visualization. Bioinformatics 23:2342

124. Barkow S, Bleuler S, Prelic A, Zimmermann P, Zitzler E (2006) BicAT: a biclustering analysis toolbox. Bioinformatics 22:1282
125. Cover T, Hart P (1967) Nearest neighbor pattern classification. IEEE Trans Inf Theory 13:21
126. Tan P-N, Steinbach M, Kumar V (2005) Introduction to data mining. Addison-Wesley, Reading
127. Parry RM, Jones W, Stokes TH, Phan JH, Moffitt RA, Fang H, Shi L, Oberthuer A, Fischer M, Tong W, Wang MD (2010) k-Nearest neighbor models for microarray gene expression analysis and clinical outcome prediction. Pharmacogenomics J 10:292
128. Laiho P, Kokko A, Vanharanta S, Salovaara R, Sammalkorpi H, Jarvinen H, Mecklin JP, Karttunen TJ, Tuppurainen K, Davalos V, Schwartz S Jr, Arango D, Makinen MJ, Aaltonen LA (2006) Serrated carcinomas form a subclass of colorectal cancer with distinct molecular basis. Oncogene 26:312
129. Breiman L, Friedman J, Olshen R, Stone C (1984) Classification and regression trees. Wadsworth International Group, Belmont
130. Quinlan JR (1993) C4.5: Programs for machine learning. Morgan Kaufmann Publishers, Los Altos
131. Kingsford C, Salzberg SL (2008) What are decision trees? Nat Biotechnol 26:1011
132. Breiman L (2001) Random forests. Mach Learn 45:5
133. Tong W, Hong H, Fang H, Xie Q, Perkins R (2003) Decision forest: combining the predictions of multiple independent decision tree models. J Chem Inf Comput Sci 43:525
134. Li Y, Wang N, Perkins EJ, Zhang C, Gong P (2010) Identification and optimization of classifier genes from multi-class earthworm microarray dataset. PLoS One 5:e13715
135. Ihnen M, Wirtz RM, Kalogeras KT, Milde-Langosch K, Schmidt M, Witzel I, Eleftheraki AG, Papadimitriou C, Janicke F, Briassoulis E, Pectasides D, Rody A, Fountzilas G, Muller V (2010) Combination of osteopontin and activated leukocyte cell adhesion molecule as potent prognostic discriminators in HER2- and ER-negative breast cancer. Br J Cancer 103:1048
136. Minsky ML, Papert SA (1969) Perceptrons. MIT Press, Cambridge
137. Krogh A (2008) What are artificial neural networks? Nat Biotechnol 26:195
138. Khan J, Wei JS, Ringner M, Saal LH, Ladanyi M, Westermann F, Berthold F, Schwab M, Antonescu CR, Peterson C, Meltzer PS (2001) Classification and diagnostic prediction of cancers using gene expression profiling and artificial neural networks. Nat Med 7:673
139. Choi YL, Tsukasaki K, O'Neill MC, Yamada Y, Onimaru Y, Matsumoto K, Ohashi J, Yamashita Y, Tsutsumi S, Kaneda R, Takada S, Aburatani H, Kamihira S, Nakamura T, Tomonaga M, Mano H (2006) A genomic analysis of adult T-cell leukemia. Oncogene 26:1245
140. Boser BE, Guyon IM, Vapnik VN (1992) A training algorithm for optimal margin classifiers. Association for Computing Machinery, New York
141. Noble WS (2006) What is a support vector machine? Nat Biotechnol 24:1565
142. Furey TS, Cristianini N, Duffy N, Bednarski DW, Schummer ML, Haussler D (2000) Support vector machine classification and validation of cancer tissue samples using microarray expression data. Bioinformatics 16:906
143. Charaniya S, Karypis G, Hu W-S (2009) Mining transcriptome data for function–trait relationship of hyper productivity of recombinant antibody. Biotechnol Bioeng 102:1654
144. Gene Ontology. http://www.geneontology.org
145. Kyoto Encyclopaedia of Genes and Genomes. http://www.genome.jp/kegg/
146. GenMAPP. http://www.genmapp.org
147. Ingenuity. http://www.ingenuity.com/
148. MetaCore. http://www.genego.com/metacore.php
149. Subramanian A, Tamayo P, Mootha VK, Mukherjee S, Ebert BL, Gillette MA, Paulovich A, Pomeroy SL, Golub TR, Lander ES, Mesirov JP (2005) Gene set enrichment analysis: a knowledge-based approach for interpreting genome-wide expression profiles. Proc Natl Acad Sci USA 102:15545

150. Efron B, Tibshirani R (2007) On testing the significance of sets of genes. Ann Appl Stat 1:107
151. Dahlquist KD, Salomonis N, Vranizan K, Lawlor SC, Conklin BR (2002) GenMAPP, a new tool for viewing and analyzing microarray data on biological pathways. Nat Genet 31:19
152. Dahlquist KD (2002) Using GenMAPP and MAPPFinder to view microarray data on biological pathways and identify global trends in the data. Wiley, NY
153. Doniger S, Salomonis N, Dahlquist K, Vranizan K, Lawlor S, Conklin B (2003) MAPPFinder: using Gene Ontology and GenMAPP to create a global gene-expression profile from microarray data. Genome Biol 4:R7
154. Prickett D, Watson M (2009) Use of GenMAPP and MAPPFinder to analyse pathways involved in chickens infected with the protozoan parasite Eimeria. BMC Proc 3:S7
155. Yu X, Griffith WC, Hanspers K, Dillman JF, Ong H, Vredevoogd MA, Faustman EM (2006) A system-based approach to interpret dose- and time-dependent microarray data: quantitative integration of gene ontology analysis for risk assessment. Toxicol Sci 92:560
156. Mootha VK, Lindgren CM, Eriksson KF, Subramanian A, Sihag S, Lehar J, Puigserver P, Carlsson E, Ridderstrale M, Laurila E, Houstis N, Daly MJ, Patterson N, Mesirov JP, Golub TR, Tamayo P, Spiegelman B, Lander ES, Hirschhorn JN, Altshuler D, Groop LC (2003) PGC-1alpha-responsive genes involved in oxidative phosphorylation are coordinately downregulated in human diabetes. Nat Genet 34:267
157. Gene Set Enrichment Analysis. http://www.broadinstitute.org/gsea/index.jsp
158. Aryee DNT, Niedan S, Kauer M, Schwentner R, Bennani-Baiti IM, Ban J, Muehlbacher K, Kreppel M, Walker RL, Meltzer P, Poremba C, Kofler R, Kovar H (2010) Hypoxia modulates EWS-FLI1 transcriptional signature and enhances the malignant properties of ewing's sarcoma cells in vitro. Cancer Res 70:4015
159. Pemov A, Park C, Reilly K, Stewart D (2010) Evidence of perturbations of cell cycle and DNA repair pathways as a consequence of human and murine NF1-haploinsufficiency. BMC Genomics 11:194
160. Butte, A J and I S Kohane (2000) Mutual information relevance networks: functional genomic clustering using pairwise entropy measurements. Pac Symp Biocomput 418
161. Butte AJ, Tamayo P, Slonim D, Golub TR, Kohane IS (2000) Discovering functional relationships between RNA expression and chemotherapeutic susceptibility using relevance networks. Proc Natl Acad Sci USA 97:12182
162. Moriyama M, Hoshida Y, Otsuka M, Nishimura S, Kato N, Goto T, Taniguchi H, Shiratori Y, Seki N, Omata M (2003) Relevance network between chemosensitivity and transcriptome in human hepatoma cells. Mol Cancer Ther 2:199
163. Jiang W, Li X, Rao S, Wang L, Du L, Li C, Wu C, Wang H, Wang Y, Yang B (2008) Constructing disease-specific gene networks using pair-wise relevance metric: application to colon cancer identifies interleukin 8, desmin and enolase 1 as the central elements. BMC Syst Biol 2:72
164. Faith JJ, Hayete B, Thaden JT, Mogno I, Wierzbowski J, Cottarel G, Kasif S, Collins JJ, Gardner TS (2007) Large-scale mapping and validation of Escherichia coli transcriptional regulation from a compendium of expression profiles. PLoS Biol 5:e8
165. Taylor RC, Singhal M, Weller J, Khoshnevis S, Shi L, McDermott J (2009) A network inference workflow applied to virulence-related processes in Salmonella typhimurium. Ann N Y Acad Sci 1158:143
166. Basso K, Margolin AA, Stolovitzky G, Klein U, Dalla-Favera R, Califano A (2005) Reverse engineering of regulatory networks in human B cells. Nat Genet 37:382
167. Margolin AA, Nemenman I, Basso K, Wiggins C, Stolovitzky G, Dalla Favera R, Califano A (2006) ARACNE: an algorithm for the reconstruction of gene regulatory networks in a mammalian cellular context. BMC Bioinformatics 7(1):S7
168. Margolin AA, Wang K, Lim WK, Kustagi M, Nemenman I, Califano A (2006) Reverse engineering cellular networks. Nat Protoc 1:662
169. Basso K, Saito M, Sumazin P, Margolin AA, Wang K, Lim WK, Kitagawa Y, Schneider C, Alvarez MJ, Califano A, Dalla-Favera R (2010) Integrated biochemical and computational

approach identifies BCL6 direct target genes controlling multiple pathways in normal germinal center B cells. Blood 115:975

170. Nemenman I, Escola GS, Hlavacek WS, Unkefer PJ, Unkefer CJ, Wall ME (2007) Reconstruction of metabolic networks from high-throughput metabolite profiling data: in silico analysis of red blood cell metabolism. Ann N Y Acad Sci 1115:102
171. Castro-Melchor M, Charaniya S, Karypis G, Takano E, Hu W-S (2010) Genome-wide inference of regulatory networks in Streptomyces coelicolor. BMC Genomics 11:578
172. Taylor RC, Acquaah-Mensah G, Singhal M, Malhotra D, Biswal S (2008) Network inference algorithms elucidate Nrf2 regulation of mouse lung oxidative stress. PLoS Comput Biol 4:e1000166
173. Ciaccio MF, Wagner JP, Chuu CP, Lauffenburger DA, Jones RB (2010) Systems analysis of EGF receptor signaling dynamics with microwestern arrays. Nat Methods 7:148
174. Meyer PE, Kontos K, Lafitte F, Bontempi G (2007) Information-theoretic inference of large transcriptional regulatory networks. EURASIP J Bioinform Syst Biol 79879
175. Meyer PE, Lafitte F, Bontempi G (2008) minet: A R/Bioconductor package for inferring large transcriptional networks using mutual information. BMC Bioinformatics 9:461
176. Carey VJ, Gentry J, Whalen E, Gentleman R (2005) Network structures and algorithms in Bioconductor. Bioinformatics 21:135
177. Shannon P, Markiel A, Ozier O, Baliga NS, Wang JT, Ramage D, Amin N, Schwikowski B, Ideker T (2003) Cytoscape: a software environment for integrated models of biomolecular interaction networks. Genome Res 13:2498
178. Murphy K, Mian S (1999) Modelling gene expression data using dynamic Bayesian networks
179. Kim SY, Imoto S, Miyano S (2003) Inferring gene networks from time series microarray data using dynamic Bayesian networks. Brief Bioinform 4:228
180. Heckerman D (1998) A tutorial on learning with Bayesian networks. Kluwer Academic, Boston
181. Needham CJ, Bradford JR, Bulpitt AJ, Westhead DR (2006) Inference in Bayesian networks. Nat Biotechnol 24:51
182. Zhu J, Zhang B, Smith EN, Drees B, Brem RB, Kruglyak L, Bumgarner RE, Schadt EE (2008) Integrating large-scale functional genomic data to dissect the complexity of yeast regulatory networks. Nat Genet 40:854
183. Singh A, Elvitigala T, Cameron J, Ghosh B, Bhattacharyya-Pakrasi M, Pakrasi H (2010) Integrative analysis of large scale expression profiles reveals core transcriptional response and coordination between multiple cellular processes in a cyanobacterium. BMC Syst Biol 4:105
184. Zoppoli P, Morganella S, Ceccarelli M (2010) TimeDelay-ARACNE: reverse engineering of gene networks from time-course data by an information theoretic approach. BMC Bioinformatics 11:154
185. Zou M, Conzen SD (2005) A new dynamic Bayesian network (DBN) approach for identifying gene regulatory networks from time course microarray data. Bioinformatics 21:71
186. Perrin B-E, Ralaivola L, Mazurie A, Bottani S, Mallet J, d'Alche-Buc F (2003) Gene networks inference using dynamic Bayesian networks. Bioinformatics 19:ii138
187. Li P, Zhang C, Perkins E, Gong P, Deng Y (2007) Comparison of probabilistic Boolean network and dynamic Bayesian network approaches for inferring gene regulatory networks. BMC Bioinformatics 8:S13
188. Zou C, Feng J (2009) Granger causality vs dynamic Bayesian network inference: a comparative study. BMC Bioinformatics 10:122
189. Zhu J, Chen Y, Leonardson AS, Wang K, Lamb JR, Emilsson V, Schadt EE (2010) Characterizing dynamic changes in the human blood transcriptional network. PLoS Comput Biol 6:e1000671
190. Dojer N, Gambin A, Mizera A, Wilczynski B, Tiuryn J (2006) Applying dynamic Bayesian networks to perturbed gene expression data. BMC Bioinformatics 7:249

Adv Biochem Engin/Biotechnol (2012) 127: 71–108
DOI: 10.1007/10_2011_120

Published Online: 8 October 2011

Modeling Metabolic Networks for Mammalian Cell Systems: General Considerations, Modeling Strategies, and Available Tools

Ziomara P. Gerdtzen

Abstract Over the past decades, the availability of large amounts of information regarding cellular processes and reaction rates, along with increasing knowledge about the complex mechanisms involved in these processes, has changed the way we approach the understanding of cellular processes. We can no longer rely only on our intuition for interpreting experimental data and evaluating new hypotheses, as the information to analyze is becoming increasingly complex. The paradigm for the analysis of cellular systems has shifted from a focus on individual processes to comprehensive global mathematical descriptions that consider the interactions of metabolic, genomic, and signaling networks. Analysis and simulations are used to test our knowledge by refuting or validating new hypotheses regarding a complex system, which can result in predictive capabilities that lead to better experimental design. Different types of models can be used for this purpose, depending on the type and amount of information available for the specific system. Stoichiometric models are based on the metabolic structure of the system and allow explorations of steady state distributions in the network. Detailed kinetic models provide a description of the dynamics of the system, they involve a large number of reactions with varied kinetic characteristics and require a large number of parameters. Models based on statistical information provide a description of the system without information regarding structure and interactions of the networks involved. The development of detailed models for mammalian cell metabolism has only recently started to grow more strongly, due to the intrinsic complexities of mammalian systems, and the limited availability of experimental information and adequate modeling tools. In this work we review the strategies, tools, current advances, and recent models of mammalian cells, focusing mainly on metabolism, but discussing the methodology applied to other types of networks as well.

Keywords Mammalian cell models · Metabolic networks · Systems biology

---

Z. P. Gerdtzen (✉)
Department of Chemical Engineering and Biotechnology,
Millennium Institute for Cell Dynamics and Biotechnology:
a Centre for Systems Biology, University of Chile,
Beauchef 850, Santiago, Chile
e-mail: zgerdtze@ing.uchile.cl

Contents

1 Systems Biology Overview

Metabolic engineering aims at modifying an organism or a cell line with new or augmented pathways, for the production of specific compounds or the utilization of alternative substrates [1]. This can be achieved by altering the relative rates of some reactions, controlling the external conditions, or by introducing genetic modifications into the cell. The numerous applications of this methodology in the biotechnology and pharmaceutical industries require a deep understanding of the biological processes involved. There has been a growing interest in addressing the study of biochemical networks and cellular processes in a comprehensive way. This approach, often referred as a systems approach, arises as a promising option for a detailed analysis of these complex networks [2].

In a systems biology approach the aim is to examine the structure and dynamics of the different cellular functions instead of focusing on the characteristics of the individual cellular components. The complexity of cellular systems, especially of mammalian cells, cannot be understood as a detailed inventory of genes, proteins, and metabolites. A comprehensive understanding of the systems characteristics requires insight into its networks, components and interactions, dynamic behavior, control and regulation, and observable characteristics [3]. A systems approach assumes that the behavior of a system can be described as a direct consequence of the interactions of its components, and considers all the components and reactions

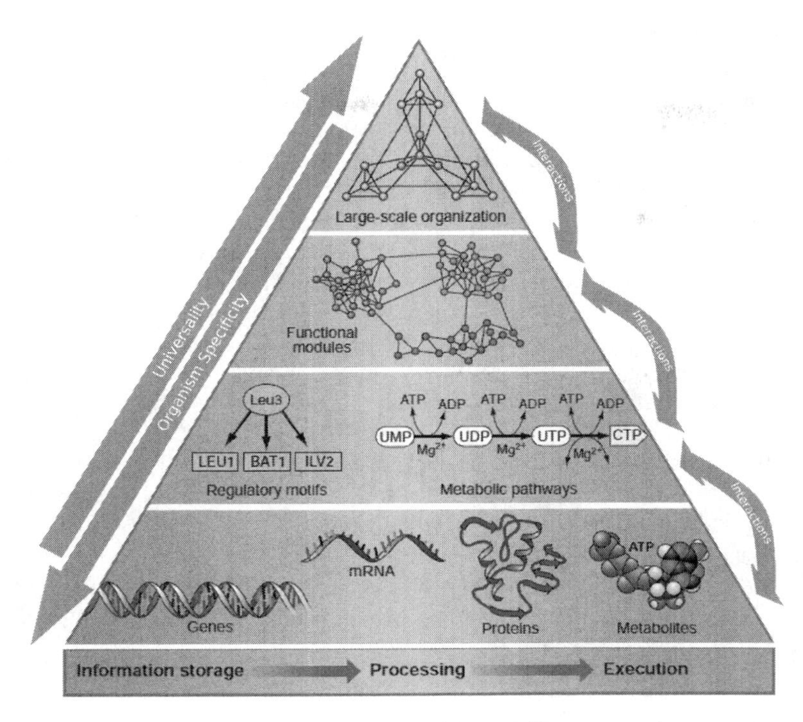

Fig. 1 A set of genes, metabolites, and proteins, with a specific copy number or concentration, defines the physiology of an organism (level 1). These components form genetic-regulatory motifs or metabolic pathways (level 2), which are the building blocks of functional modules (level 3). Nested modules generate a hierarchical architecture (level 4). Characteristics at many different levels can be conserved among organisms. Heterogeneity in physiology is likely due to different relative concentrations of these components. Adapted from [4] with permission

involved in a particular cellular process. Those reactions, ranging from DNA, RNA, protein synthesis and their regulation to biochemical conversions, form complex interaction networks. Modeling, computation, and experimental research data from metabolomics, genomics, and proteomics can be combined to generate a meaningful description of a cellular process or biochemical network.

This type of approach can be applied to identifying key control points in a metabolic network, limiting steps in a reaction pathway, and candidate genes for metabolic and cell engineering. This can be achieved by using large-scale screening techniques and a mathematical framework for processing this information. All these components store and control different levels of information: long-term in the genome, short-term in the proteome, and information retrieval and control in the transcriptome and metabolome [4]. These different levels of information are integrated in complex networks of interactions, in order to generate cellular functions and responses. This is illustrated in the complexity pyramid in Fig. 1. An organism can be described as a combination of a series of genes, metabolic networks, and regulatory networks, interacting with each other at different levels. Although the main individual component distribution may be unique for a given organism, there are a significant number of genes and proteins which are homologous and highly conserved among organisms. In addition, a large number of metabolites as well as the topological properties of cellular networks are commonly shared by different species. Hence, differences in cellular physiology are most likely due to different relative concentrations of these biological components as a result of disparate regulatory mechanisms. Thus, not only are the main components essential for characterizing an organism, but also the connections, interplay, and regulation of these components are essential in defining the final properties observed. The study of these complex networks and interactions is currently possible due to the advances in recent decades in analytical methods, genome sequencing, and high-throughput experimental methods and software development, which allow us to obtain comprehensive sets of data on a system's performance and gain insight into the specific molecules involved in each process. Integrative models provide system-level insights into the mechanisms by which each component interacts with the rest of the network, generating predictions and providing knowledge regarding perturbed networks. This methodology can be used to identify and study emergent properties of a system such as robustness and redundancy, in order to understand the underlying characteristics of the system. This system-level information has potential applications in medicine, agriculture, biofuels production, and many other areas [5].

2 The System: Mammalian Cells

Mammalian cells are complex compartmentalized cellular systems that perform a large number of biochemical reactions simultaneously, subject to a series of thermodynamic constraints. These cells have significant differences in characteristics with respect to other cellular systems such as bacteria or plants. Some of these differences are illustrated in Table 1.

The complexity of the intracellular environment in mammalian cells is the result of a large genome, with ten times more genes than bacteria and an even larger number of transcripts and proteins, as well as other elements such as lipids, sugars, and nucleotides. In this environment a large number of interactions can take place. This crowded intracellular environment facilitates the interactions between components located near to each other, reduces movement and transport of molecules by diffusion, and promotes protein folding and aggregation. There is a high level of internal spatial complexity in mammalian cells, since intracellular components and therefore component interactions are not distributed homogeneously inside the cell. The characteristics of this intracellular environment also change continuously as a result of cellular activity; e.g. the relative proportions of RNA, DNA, and protein are also related to cellular growth rate. In *Escherichia coli* the RNA-to-protein ratio increases with growth rate while the DNA-to-protein

Table 1 Characteristics of animal cells versus *E. coli*

Characteristic	Mammalian cell	*E. coli*
Diameter[a]	18–20 μm	0.8–2.0 μ m
Volume[b]	~ 10, 000 μm^3	~ 1 μm^3
Cell generation time[b]	20 h—non dividing	~ 30 min to h
Genome size[b]	3.0 10^9 bp	4.6 10^6 bp
Number of genes[c, d]	~ 25, 000–30, 000	3,200
RNA content[a, e]	~ 25 μg/10^6 cell	20–211 μ g/10^9 cell
RNA lifetime[b]	2–5 min	~ 10 min–10 h
DNA content[a, e]	~ 10 μ g/10^6 cell	7.6–18 μ g/10^9 cell
Protein content[a, e]	~ 250 μ g/10^6 cell	25–130 μ g/10^9 cell
Proteins/cell[b]	~ 410^{10}	~ 410^6
Diffusion time of protein across cell[b]	~ 100 s	~ 0.1 s
Carbohydrate content[a, e, f]	~ 150 μ g/10^6cell	4.4–26 μ g/10^9cell
Lipid content[a, e, f]	~ 120 μ g/10^6cell	3–17.3 μ g/10^9cell
Compartmentalization	Yes	No

[a][6], [b][7], [c][8], [d][9], [e][10], [f][11]

ratio decreases [6]. Hence, parameters related to protein and RNA synthesis rate should be estimated as a function of growth rate.

The first step in developing a mathematical model is identifying the components of the system to be modeled. Then, key interactions among these components as well as the directionality of these interactions need to be assessed. The characteristics of the model will depend on the nature and amount of information available on the system. Based on the interactions identified, a set of reactions is obtained which represents the interactions of each of the components in the network. From these reactions a mathematical description of the system is constructed. Some of the parameters required by the model may be fitted, some of them measured experimentally, and some obtained from the literature or databases. Mammalian cell physiology is well studied and documented for many cell lines, and sufficient information for constraining, refining and validating mathematical models for these cells can be obtained. Details can be found in the following section.

Metabolism is a dynamic process as organisms are continuously exposed to environmental changes to which they must react and adapt by triggering changes in intracellular conditions. Regarding metabolic networks, there is a vast amount of knowledge with respect to the reactions that take place within mammalian cells, which has become available in the last decade as a result of the development of *omics* (genomics, proteomics, metabolomics, transcriptomics, interactomics, etc). However, there are a large number of factors that affect the occurrence and the dynamics of intracellular processes and reactions. One key factor is macromolecular crowding. The very high total concentration of molecules such as proteins, nucleic acids, and complex sugars leads to a crowded and compact environment. This macromolecular crowding has energetic consequences for cellular processes as it can affect protein binding and hence enzymatic reaction rates, reduce diffusion rates, affect the formation of protein aggregates, and result in the confinement of macromolecules in small compartments [12]. Hence, many of the properties and parameters determined in vitro in dilute solutions, especially regarding kinetic parameters for enzymatic reactions, may not apply in the cellular biochemical network if the reaction rate is limited by substrate diffusion or by the stability of protein complexes. This issue is particularly significant in the processes of protein and nucleic acid synthesis as well as intermediary metabolism and cell signaling, which depend on variables affected by intracellular crowding such as non-covalent associations of molecules and conformational changes [13].

Integrated frameworks for the analysis of the underlying structure and properties of cellular processes have emerged as a result of the sequencing and characterization of genomes, the development of technologies for high-throughput identification of cellular components and interactions, the emergence of different modeling approaches, and the reduction of computational costs [9, 10]. For performing this type of analysis, the system's network must be assembled. The assembly process can be performed by collecting experimental data or by in-silico identification of binary interactions between components and known pathways. Proteomic and transcriptomic information can be accessed from public data repositories. For most organisms, information regading cellular components, cellular interactions, metabolic reactions, and pathways present in an organism, as well as genomic and kinetic data for organisms that have been sequenced, are freely available on public databases. Currently there is a large number of these resources focused on mammalian cells; some of them are listed in Table 2.

This information about the cell's genomic characteristics, metabolic and signaling pathways, and biological component interactions can be used for the assembly of an in-silico network that represents the system of interest. The development of a mathematical model provides the means for the organization of all the information available at different levels (genomic, proteomic, metabolomic) in a coherent structure that provides a framework for the qualitative and quantitative analysis of the integrated system, as illustrated in Fig. 2, in order to understand the system's structure and dynamics. A mathematical description allows the identification of the relative relevance and impact of each one of the components, identification of new properties that emerge from the topology of the system's integrated network, identification of key points that can be manipulated for modifying and modulating this complex system, and testing of hypothesis that cannot be verified experimentally with the current technology available.

3 Modeling Strategies

Systems biology is expected to provide deep insight into the underlying characteristics of metabolic, regulatory, and signaling networks of complex living organisms on the organism as a whole. A major goal of systems biology is to relate

Table 2 Web resources and databases for components and pathways for mammalian cells

Name	*Web address* and description	Reference
AfCS—Alliance for Cellular Signaling	http://www.afcs.org/index.html—protein–protein interactions and modifications. Public data-sets freely available	[14]
APID—Agile Protein Interaction Data Analyzer	bioinfow.dep.usal.es/apid—open-access web application where experimentally validated protein–protein interactions are unified	[15]
BioCyc	*biocyc.org*—collection of genome and metabolic pathways for a large number of organisms, including mammalian	[16]
BioGRID	*thebiogrid.org*—public repository with genetic and protein interaction data from model organisms and humans	[17]
BOND—Biomolecular Object Network Databank	*bond.unleashedinformatics.com*—comprehensive information on small molecule and protein interactions	[18]
BRENDA	http://www.brenda-enzymes.org—freely available comprehensive collection of biochemical, molecular and metabolic information on all classified enzymes, based on primary literature	[19]
DIP—Database of Interacting Proteins	*dip.doe-mbi.ucla.edu*—experimentally determined protein–protein interactions	[20]
GeneNet	wwwmgs.bionet.nsc.ru—integrates databases on gene network components and data processing tools for structure and function of DNA, RNA, and proteins	[21]
GenMAPP	http://www.genmapp.org—computer application for analysis and visualization of genome-scale data for rapid interpretation.	[22]
HPID—Human Protein Interaction Database	wilab.inha.ac.kr/hpid—integrates information from other databases on proteins and their interactions. Predicts potential interactions between human proteins	[23]
HPRD—Human Protein Reference Database	http://www.hprd.org—platform for integration of information on post-translational modifications, interaction networks and disease association for proteins in the human proteome	[24]
IBIS—Inferred Biomolecular Interactions Server	http://www.ncbi.nlm.nih.gov/Structure/ibis/ibis.cgi—protein interactions, both experimentally determined and inferred by homology	[25]
iHOP—Information Hyperlinked over Proteins	http://www.ihop-net.org—protein association network built by literature mining	[26]
Ii2D	*ophid.utoronto.ca*—database with known, experimental, and predicted PPIs for five model organisms and human	[27]
IntAct	http://www.ebi.ac.uk/intact—open-source database system and analysis tools for protein interaction data	[28]

(continued)

Table 2 (continued)

Name	*Web address* and description	Reference
KDBI—Kinetic Data of Bio-molecular Interaction	xin.cz3.nus.edu.sg/group/kdbi/kdbi.asp—experimentally determined kinetic data of interactions between proteins, RNA, DNA, and ligands, or reaction events	[29]
KEGG BRITE	http://www.genome.jp/kegg/brite—contains functional hierarchies and binary relationships between biological objects	[30]
KEGG PATHWAY	http://www.genome.jp/kegg/pathway.html—collection of pathway maps for a large number of organisms	[30]
LOCATE	*locate.imb.uq.edu.au*—curated database describing membrane organization and subcellular localization of proteins from mouse and human protein sequence sets	[31]
Mammalian Degradome Database	*degradome.uniovi.es*—proteases expressed in human, chimpanzee, mouse, and rat	[32]
MINT—Molecular INTeraction database	*mint.bio.uniroma2.it/mint*—contains experimentally verified protein–protein interactions	[33]
MIPS—Mammalian Protein–Protein Interaction Database	*mips.helmholtz-muenchen.de/proj/ppi*—collection of manually curated high-quality PPI data from the scientific literature	[34]
MetaCyc	*metacyc.org*—experimentally determined metabolic pathways and enzymes	[35]
Predictome	*visant.bu.ed*—database of predicted functional associations among genes and proteins	[36]
PRIME—PRotein Interaction and Molecular information databasE	*prime.ontology.ims.u-tokyo.ac.jp*—contains integrated gene/protein information automatically extracted based on natural language processing	[37]
PSIbase	*psibase.kobic.re.kr*—molecular interaction database	[38]
Sigmoid	http://www.sigmoid.org—database of models of cellular signaling and metabolic pathways	
Signaling Gateway Molecule Pages	http://www.signaling-gateway.org/molecule—information on proteins involved in cellular signaling	[39]
SIMPATHWAY—Smart Integrated Molecular Pathway	http://www.helios-bioscience.com/technologies-molecular.php—commercial molecular interaction database. Includes an integrated intracellular signal transduction network, query and data representation software	
STRING	*string.embl.de*—database of known and predicted protein interactions	[40]
ResNet Mammalian Database	http://www.ariadnegenomics.com/products/databases—commercial database of molecular interactions for human, rat, and mouse	

PPI protein–protein interaction

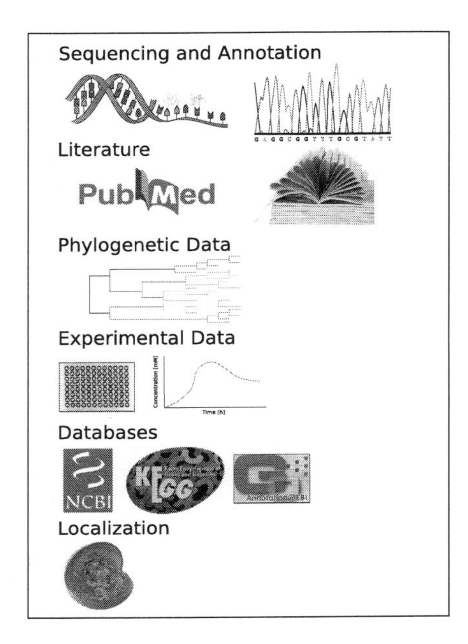

Fig. 2 Several data sources provide the information required to define metabolic networks. A comprehensive list of the cellular components is found in the genome's sequence. Literature provides information on biochemical characterization of enzymes. Physiological data are required to validate the reconstruction. Phylogenetic data can be useful to infer characteristics of organisms that are not well studied. Databases contain a vast amount of data about gene function and associated metabolic activities. Cellular localization is crucial when multiple compartments are considered. Adapted from [41] with permission

genomic, proteomic, and metabolomic data to cell physiology and the observed cellular phenotype. This requires the identification of the components of the networks involved, their interactions in the system, and the incorporation of this information into a mathematical model. Information obtained from high-throughput methods can span several levels, from network components, component interactions, spatial distribution of components, and potential adaptive changes the network can undergo [41]. Network components correspond to genes with functional annotation, or functionally related proteins and their interactions. The interactions between these components can be associated with a biological context for the network. The final levels of information correspond to the location of the network components in the different cellular compartments and the identification of thermodynamic properties that define directionality of the reactions in the network.

For the detailed analysis of a cellular system, a mathematical representation of the system is required. There are a number of different representations of metabolism that can be used, which vary in the level of detail, the level of available information, and the type of predictive capacity of the model. It is essential to define beforehand the purpose and scope of the model; if it is to understand the integration of each component of the system or to predict the system's behavior in response to external perturbations, to what degree: qualitative or quantitative, and what are the assumptions and restrictions to be considered. Accurately identifying biochemical components in a metabolic network is crucial for the accuracy of subsequent predictions of network properties generated through a mathematical model. The modeling strategy to be followed depends on the final objective of the analysis, the characteristics of the experimental data, and the available details of the structure of the network. Other approaches for the development of qualitative and quantitative models applying other methodologies such as graph theory, topology analysis, regulatory motifs, and module interactions, as well as the use of deterministic, stochastic, and hybrid models, for the development of predictive models for mammalian cells are reviewed in [42]. General strategies for model development are discussed in the following section.

3.1 Physicochemical Models

Physicochemical models are able to describe the different kinds of molecular transformations that occur in biological systems, such as association, translocation, and modification through reactions. These models use standard tools in biochemistry, chemistry, and physics and consider information regarding the system's properties, behavior, and observable characteristics. Given the nature of this type of model, each of the equations in the model can be associated to a specific process in the system and hence each of the parameters can be assigned a physical interpretation.

For well-studied organisms, a high level of detail is available in terms of the components that form a pathway and how they interact. In this case, a mechanistic type of model can be constructed to capture the properties of the system. First, a detailed list of cellular components and their interactions needs to be assembled. These interactions are translated into reactions and then into equations, as illustrated in Fig. 3. In this case the mathematical model is only a translation of the physical description of the system into equations. A mechanistic model can be used for exploring hypotheses in cases where the interactions amongst the different components in the system are too intricate to allow an intuitive evaluation of the system's response. Even though mechanistic models require making assumptions regarding how interactions take place, this type of model greatly expedites the process of hypothesis testing.

The type of analysis performed on the model will depend on the characteristics of the biological knowledge available and the modeling objectives. Once the initial conditions for the dynamic representation of a system are specified, responses to a variety of external and internal perturbations can be predicted. These initial conditions can be obtained from literature or experimentally, e.g. from representative measurements of the intracellular and extracellular cellular environment under steady-state conditions. For a steady-state analysis only the network's structure is required and no information on rate constants is needed. For instance, flux balance

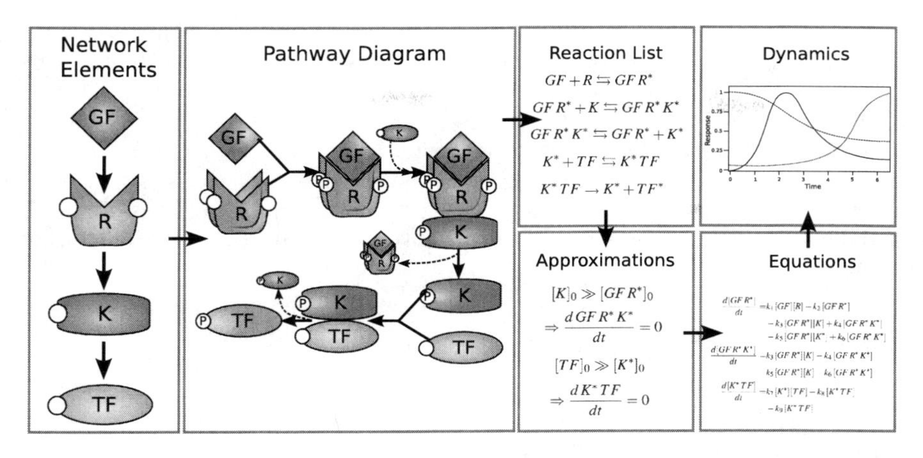

Fig. 3 Stages for the development of a mechanistic model for a signaling network: identify elements, generate diagram, translate into reactions, assumptions and approximations, differential equations

analysis has been used to predict the metabolic switch that *E. coli* undergoes for different nutritional conditions, using only the metabolic network's structure. Results for this theoretical analysis have also been validated experimentally [43].

Simulation, stability, and sensitivity analysis can be performed to test the system's dynamic response to perturbations, if information on rate constants is available. In this case each differential equation represents the production and consumption rates of a particular species or component in the network. Reaction rates are often nonlinear functions and can be expressed in a simplified equation, either empirical or mechanistic. The effect of the variables involved in a reaction can be represented in different ways: by a mass-action form which considers all the mechanistic steps of the reaction, in a rate-law form as in Michaelis–Menten models and Hill kinetics, or by using power law and lin-log approximations or thermokinetic considerations [44, 45]. An extensive review on the formulation and analysis of kinetic models can be found in [46].

Once a detailed mechanistic model is available, specific characteristics of the system can be assessed. Bifurcation analysis can be used to evaluate the existence of multiple steady states that the system can achieve for the same extracellular conditions, given a different culture history or parameter values [47]. For instance, cell cycle regulation was studied for different cell types including mammalian cells, using a generic model of eukaryotic cell cycle control and bifurcation theory [48]. The conditions required for a balance between cell cycle progression and overall cell growth, as well as for cell homeostasis were identified. Other characteristics such as robustness, the ability of the system to maintain its characteristics despite perturbations and changes in its environment, can be assessed by observation of the properties exhibited by the system. These includes the system's adaptation to environmental changes, its sensitivity to parameter changes, and the slow loss of function upon damage. This can be achieved by the utilization of different strategies such as system control strategies and separation of components into modules with redundancy of these modules, as well as an intrinsic structure stability analysis associated to the network's design. Cloutier et al. [49] studied the application of these strategies for the control of energy metabolism, using a generic mathematical model of glycolysis and oxidative phosphorylation. This study resulted in the identification of essential control components (proportional, derivative, integral) and structures (feedback, feed-forward) in energy metabolism.

3.2 Statistical Models and Tools

When data are abundant but intricate or difficult to manage, and there is a scarce amount of information available about the structure of the system (i.e. the system's components, interactions, and regulation), experimental data can be used to obtain some insight into the system using data-driven statistical modeling.

In this case, given the lack of information about the system's structure, no assumptions can be made regarding the underlying mechanisms involved. Hence, the objective is not to represent the characteristics of the processes that occur in the system but to mathematically identify relevant variables or dimensions in the system and extract them from the data space. This permits an intuitive understanding of data and reveals interesting unexpected information about the system's characteristics by introducing a new quantitative perspective that could not be obtained from data alone. Statistical tools are useful in functionally connecting different layers of cellular information, filling the gaps in our knowledge of kinetic and regulatory phenomena. The main limitation of these models is that they are restricted in their predictive power and biological representativity by the scope of the data they are based on and the dimensions identified as relevant.

For instance, network reconstruction involves infering the characteristics of the network from observed gene expression and metabolic data. In the work of Ma and Zeng [50], the metabolic networks of 80 sequenced organisms are reconstructed in silico from genome data and a bioreaction database. A recent work presents a method that integrates multiple inference methods and experiments using multi-objective optimization [51]. This method was applied for modeling *E. coli* acid stress and in-vivo tumour development. For networks with a more limited amount of experimental data and knowledge available, the use of an integral additive model for yeast cell cycle has been proposed [52]. A comprehensive protocol describing the necessary steps to build a high-quality genome-scale metabolic reconstruction was developed by Thiele and Palsson, including instructions for all stages of the reconstruction process, the use of genome annotation, phylogenetically close organisms, and gap filling, as well as available resources and a confidence scoring system [53].

A common methodology for the initial analysis of large data sets is data clustering. This methodology allows the identification of natural organization patterns into groups based on similarities among data. Cluster analysis identifies subgroups in multivariate data, in such a way that objects in the same cluster resemble each other, while objects belonging to different clusters are dissimilar. In this case similarity is defined as a distance in the n-dimensional data space, e.g. Euclidean distance. Based on this distance, a proximity matrix gives pair distances or similarities between data points, following either a partitional or an agglomerative algorithm for grouping data. It has been applied, for instance, to transcriptional microarray data to identify groups of genes that behave similarly and hence may be co-regulated [54].

Principal component analysis (PCA) can be used for data condensation and identification of a reduced coherent set of variables. With this methodology, a large number of independent variables can be systematically condensed in fewer dimensions that capture the data variance. The original dataset can be projected onto a reduced dimension space formed by combinations of the original variables providing an additional physical interpretation of the data [55]. From PCA, principal component regression (PCR) can be performed to predict the relationship of one variable with a set of inputs, e.g. correlating a given phenotype with a certain metabolic profile.

Artificial neural networks are multi-layered statistical models that represent the variation in a complex response variable to a collection of input variables. This is a supervised learning algorithm that attempts to approximate the description of a phenomenon by mathematical models deduced from observation of cases using regression arguments. The reader is referred to De Iorio et al. [56] for a more detailed review of these methods, and a discussion on data preprocessing, extraction, and discrimination procedures.

3.3 *Modeling Stages*

The following stages for model construction are identified in the scheme proposed by Wiechert et al. [57]: (i) definition of the elementary units of the system; (ii) characterization of connectivity and interactions between units; (iii) association of a biochemical rate equation to each interaction. These stages in model construction are followed by (iv) model validation with experimental data and model interrogation for emergent properties and predictions.

3.3.1 Requirements, Construction, and Initial Analysis

Prior to modeling, the assumptions that will be considered for describing the system must be stated. The amount and type of experimental data available must be assessed to obtain parameter values and operation range, as well as initial conditions for the simulation. For instance, the values of a representative cellular state can be used as the initial point for the simulation. Alternatively, if the system is to be perturbed, the natural steady state of the model can be used as an initial condition.

In order to develop a mathematical model for a biological system or process, the individual components that are to be considered in the systems structure must be defined for the level of detail outlined in the scope of the model. These components are the variables that characterize the state of the system. The relevant biological interactions among components must then be identified. For these two steps, besides experimental data, information available on the resources like the ones listed in Table 2 can be of use. As a result, all the available knowledge regarding a specific system can be translated into a list of components and interactions, or reactants and reactions. Putting together all these elements for the reconstruction of the network's structure also requires expert knowledge of the available data, in order to reduce errors and inconsistencies from the literature.

A general form of a dynamic mass balance for metabolites j inside a cell can be represented by:

$$\frac{dC_j}{dt} = Sr(C,p) \tag{1}$$

where C is the concentration vector ($n \times 1$), C_j is the concentration of metabolite j, r is the vector of metabolic reaction rates ($m \times 1$) which is a function of the metabolite's concentration vector ($n \times 1$) and a parameter vector p, and S is the stoichiometric matrix that contains the stoichiometric coefficients which relate the n species in the m reactions that form the network. This generates a system of n differential equations and m unknown time-dependent fluxes.

Once the general characteristics of the interactions among components, such as directionality and stoichiometry, are formalized, initial exploration of the system's qualitative properties can be performed. Since not all species in the system are independent, the stoichiometric matrix is not full rank, and since metabolites often participate in multiple reactions, the number of reactions is greater than the number of metabolites. Therefore $rank(S) < n < m$. This is due to constraints associated with structural conservations such as conserved moieties (mainly enzymes and cofactor). Therefore the model can be reduced to

$$\begin{aligned} C^d &= L_0 C^d + T \\ \frac{dC^i}{dt} &= S_R r(C^i, C^d, p) \end{aligned} \tag{2}$$

where the superscript refers to dependent d and independent i species. S_R corresponds to a reduced stoichiometric matrix and L_0 is the link matrix that relates dependent and independent variables in conservation constrains up to a total in vector T [58].

The stoichiometric matrix (S), consisting of n rows and m columns, can provide a large amount of information regarding the structure of the system and its properties, independent of the parameters and expressions for reaction rates and other biomolecular interactions. The stoichiometric matrix S can provide good insight into the basic topological characteristics of the network, such as highly connected nodes and shortest metabolic pathways, through simple analysis. Stoichiometric matrices are generally sparse as only a few compounds participate in each reaction. The elementary topological properties are determined from their nonzero entries. The sum of all nonzero entries in a column j corresponds to the number of compounds involved in the reaction j. The sum of all nonzero entries in a row indicates the number of reactions in which a compound i is involved, and it is a measure of the node's connectivity. A highly connected node is expected to have a greater influence on the network behavior [59].

The *Left null space* L of S is formed by a set of $n-$rank(S) linearly independent row vectors such that $LS = 0$. These vectors correspond to all the conservation relationships, or time invariants, that a network contains, e.g. mass conservation of atomic elements, molecular subunits, or chemical moieties.

The *Null space* N of S is formed by a set of $m-$rank(S) linearly independent column vectors that contains all the steady-state flux through the network represented by $S, Sr(C,p) = 0$. The analysis of $N(S)$ allows exploration of the steady-state behavior of the system. This is of particular interest in an initial analysis since this permits assessment of the quality of the model through the verification of compliance with thermodynamic constrains for equilibrium and the correspondence of the allowable steady states with the characteristic equilibrium states of the experimental system.

An assembled network structure will evolve as new experimental data becomes available, the components used for network assembly are reassessed, and the genome annotation is updated. High-throughput data such as transcriptomic, proteomic, and metabolomic information provides a context that allows the evaluation of inconsistencies between the model simulation results and experimental data. In this way the model can be used as a tool to determine the coherence of data sets from different types and sources in the context of a specific biological system and additional hypothesis can be generated that drive experimental discovery.

Computer-assisted methods for automated reconstruction have emerged in the past few years based on genome sequence data available on public databases. Only a few of these efforts are focused on mammalian cells, specifically human [60] and mouse [61]. Since the first attempts to model cell metabolism, there have been important developments in genetic tools, computer hardware, and high-throughput analysis techniques. These advances have provided more information than ever before regarding the cellular state and cell metabolism. Recently, a resource for the high-throughput generation and analysis of genomic models, Model SEED, was developed [62]. This methodology allows the production of detailed and quantitative predictions of organism behavior. Biomass reactions are also included, along with stoichiometric coefficients. If data for obtaining these coefficients are not available, a set of rules for obtaining approximate coefficients for each reaction is applied. In addition, a minimal set of reactions that must be added to each model is identified through an optimization algorithm.

Several automated reconstructions have been generated with Pathway Tools and are available through BioCyc [16]. Pathway Tools is a program for automated network reconstruction using metabolic reactions that are associated with Enzyme Commission numbers and/or enzyme names from one-dimensional genome annotation. It uses known metabolic pathways to evaluate reactions and pathways in a reconstruction [63]. Defined pathways are evaluated and scored by the program and included in the reconstruction based on the number of enzymes in a pathway that are found in the annotation. Pathway Tools also completes the pathway by including missing reactions if a significant fraction of the other enzymes in the pathway are present [64]. Other available software tools and databases available for model implementation are listed in Table 3.

Automated reconstructions for metabolic networks are often not suitable for modeling since additional details and manual curation are required in order to obtain a higher quality assembled network and get meaningful results. Reconstruction efforts have successfully targeted many organisms, including *E. coli* and yeast [65–67], and genome-scale reconstructions for over 30 organisms have been published [53]. A high-quality network would include specific biochemical and physiological data regarding reaction reversibility, cofactor usage, transport reactions, cellular compartments, and biomass composition. Hence, only well-studied organisms for which this information is available are considered suitable for modeling. The model for *Mus musculus* developed by Sheikh et al. [61] was based not only on annotated genomic data and pathway databases, but also on currently available biochemical and physiological information. This is one of the first attempts to collect and characterize the metabolic network of a mammalian cell in an automated manner. It considers compartmentalization between the cytosol and mitochondria, transport between compartments, and over 1,200 reactions. The level of compartmentalization considered implies a number of assumptions about the system. A structured model implies a variety of species located in different compartments, each of them with its one characteristic time-scale. Transport between compartments can generate spatial heterogeneity in the intracellular environment.

4 Calibration and Revision

In order to perform a detailed dynamic analysis of the system, each of the reaction rates and interactions must be formally described. For this, the kinetic parameters such as half saturation constants, turnover rates, and inhibition constants associated with the expressions for the reaction rates r_k must be identified. Experimental data and literature resources such as BRENDA and others listed in Table 2, are key at this point to determine the value of these parameters. To estimate model parameters by regression on empirical data, large amounts of high-quality experimental data are needed. Alternatively, an exploration of the parametric space can be performed using literature values as a reference starting point. If the model complexity is such that it cannot be solved numerically, additional

Table 3 Modeling resources and databases for systems biology research in mammalian cells

Name	*Web address*and description	Reference
BioModels	*biomodels.net*—model collection in SML format. Includes data resources and publication references. Supports model visualization	[131]
CellDesigner	http://www.celldesigner.org—diagram editor for drawing gene-regulatory and biochemical networks. Models are stored in SBML format	[132]
Cellerator	http://www.cellerator.org—mathematical package for automatic generation of differential equations of biochemical networks	[133]
CellNetAnalyzer	http://www.mpi-magdeburg.mpg.de/projects/cna/cna.html—Matlab-based software package for structural and functional analysis of networks based on their topology	[134]
CellWare	http://www.bii.a-star.edu.sg/achievements/applications/cellware—grid-based tool for modeling, simulation, parameter estimation, and optimization	[135]
COBRA	*systemsbiology.ucsd.edu/downloads/COBRAToolbox*— Matlab package for quantitative prediction of cellular behavior using a constraint-based approach	[136]
COPASI	http://www.copasi.org—supports models in the SBML standard. Uses ODEs or Gillespie's stochastic simulation algorithms	[137]
DOQCS	*doqcs.ncbs.res.in*—database of quantitative cellular signaling. Repository of models of signaling pathways. It includes reaction schemes, concentrations, and rate constants, as well as annotations on the models	[138]
E-Cell	http://www.e-cell.org—software platform for modeling, simulation and analysis of complex systems	[139]
JDesigner	http://www.sys-bio.org—visual network design tool for systems biology	[129]
Metatool	*pinguin.biologie.uni-jena.de/bioinformatik/networks*—program for calculating elementary modes, compatible with octave and Matlab. Distributed with CellNetAnalyzer	[140]
OptFlux	http://www.optflux.org—tools for in-silico metabolic engineering	[141]
Pathway analyser	*sourceforge.net/projects/pathwayanalyser*—software for flux-based analyses and simulations on SBML models	
SBRT	http://www.ieu.uzh.ch/wagner/software/SBRT—systems biology research tool. Software platform for analyzing stoichiometric networks	[142]
SNA	http://www.bioinformatics.org/project/?group_id=546—toolbox for steady-state analysis of metabolic networks	[143]
STOCKS	http://www.sysbio.pl/stocks—software for stochastic simulation of biochemical processes with Gillespie algorithm. Supports SBML	[144]
Virtual cell	http://www.vcell.org—web-based computational environment for modeling and simulation of cell biology	[145]
WebCell	*webcell.kaist.ac.kr*—integrated simulation environment for the analysis of cellular networks over the web	[146]
YANAsquare	*yana.bioapps.biozentrum.uni-wuerzburg.de*—software package for the analysis of metabolic networks. It incorporates database extraction and visualization tools	[147]

assumptions will be required in order to simplify the model to allow the problem to be solved numerically.

After the network reconstruction process, which is based on genomic and biochemical evidence, network gaps may have occurred as a result of missing components that have not been identified or have not been reported in literature. These gaps can be identified through the verification of the capacity of the network to produce essential components such as amino acids, and the verification of growth capabilities and requirements. The resulting phenotypic behavior after genetic perturbation under given growth conditions can be used to assess the capacity of a metabolic network. The addition of reactions required to fill the gaps is supported by empirical observation or by the identification of putative genes in a related organism. If the gap occurs due to genomic annotation information, these could be revised. If the gaps are the consequence of biochemical data, this indicates incomplete metabolic knowledge and should be explored further to improve the model. In addition, computational tools such as flux-coupling analysis [68] or Pathway Tools [63] can be used for identifying dead-end metabolites isolated from the rest of the network. A key point is that network functionalities obtained from the model must be in correspondence with observable phenotypic characteristics.

5 Validation and Testing

Results obtained from simulations must be compared to biological, biochemical, and physiological information. At this point expert interpretation of the results and manual error correction must be performed, in order to verify that the model equations accurately represent the system's behavior observed experimentally. The model should also evaluated for its ability to perform predictions that can be verified experimentally. Other tests are also performed such as robustness, sensitivity to parameter values, bistability, and evaluation of simplicity versus accuracy of the model. Data used for validation should be different from data used for parameter calibration. If data availability is limited, the bootstrap method can be applied to combine data sets in order to generate different groups of data for calibration and validation [69].

6 Limitations and Challenges

One of the main considerations to be taken into account is the applicability of the model beyond the region considered valid under the assumptions used for its construction. The model is valid for capturing and predicting the system's behavior under the assumptions, conditions, and ranges of the experimental data used to define the system structure, and calibrating its parameters. Although an empirical model could perform well under conditions not previously considered, a case-by-case analysis is required when extrapolating to other conditions for prediction, due the fact that some changes may occur when modifying operational conditions. This is something to consider if simulation is going to be performed to predict the behavior of the system under conditions that cannot be achieved experimentally.

One of the main challenges in model development is associated with their construction. The quality and hence the results obtained from modeling a metabolic network are directly dependent on the quality of the model development process. Hence, the development of efficient methods for collecting information, specifically for mammalian cells, and translating it into network structures is crucial for the further progress of the modeling field in systems biology applied to mammalian cells. The ability to gather and analyze good quality data is also crucial, as multidimensional experimental data is essential for model construction, calibration, and validation. Another important point lies in the preliminary phase of model construction. A more detailed model will have a broader scope, with added complexity which makes numerical solution more difficult. A compromise between scope and level of detail must be achieved.

The development of mathematical models for mammalian cells is of great importance for the characterization of intracellular networks and to drive experimental design and improve process development. Models can be used, for instance, for the identification of genes whose function remains unknown, identification of molecular cell components that have not been determined or characterization of kinetic properties still unknown. In addition, mathematical models which have been developed using a systems approach permit testing for knowledge completeness and assist in experimental design, synthetic network design and devising strategies for system control. The latter requires methods of analyzing large mathematical models, model reduction techniques, and an improvement in the understanding of control mechanism for intracellular networks.

7 Mathematical Descriptions of Cell Metabolism

Mathematical models of mammalian cell metabolism allow the organization of the information available for a system, in order to simulate and predict the behavior of the system for the optimization of production conditions, testing of hypothesis, data integration, and process design and control. The rapid development of sequencing projects has set a challenge to incorporate this information, along with the phenotypic information, into detailed genome-based models or mechanistic metabolic models. Once the structure of the network is defined, a topological analysis can be performed to identify pathways, redundancies in the network, and connectivity properties. Once a detailed model is available, simulation of metabolism can be used to decipher the system's phenotypic response for a given structure or genotype under a set of environmental conditions. Two main modeling approaches are generally applied for modeling metabolic networks. A mechanistic kinetic approach or a constraint-based stoichiometric approach, which are reviewed in detail in [70, 71]. An overview of these approaches and their applications to mammalian cell modeling is given in the following section.

7.1 Steady-State Models

The simplest way to mathematically describe a metabolic network, Metabolic Flux Balance Analysis, is based on the fundamental law of mass conservation [72]. This methodology provides a framework that is independent of reaction kinetics and thermodynamics. This method gives information on the steady-state rates of a reaction network, but does not provide any insight into their dynamics. The first step in metabolic flux analysis is the determination of the metabolic network structure and the definition of its stoichiometry. Mass balances are then applied to the intracellular metabolites involved in the network by considering metabolic demands. Steady-state values for the accumulation rates of extracellular metabolites are experimentally determined and used to calculate the intracellular fluxes, considering a stoichiometric model for the main intracellular reactions under a pseudo-steady-state assumption [58].

The main input of these types of models is measured extracellular fluxes, and the output is the net flux map of the cell including estimated steady-state rates for each reaction. Metabolic flux analysis (MFA) provides a measure of the influence of the various pathways on the overall cellular functions and metabolic processes. The results obtained from these balances depend strongly on the reactions included in the network and the stoichiometry constraints. Several systemic properties can be identified from the stoichiometric matrix, such as metabolite connectivity and systemic reactions [73]. Metabolite connectivity refers to the number of reactions a given metabolite participates in. Systemic reactions represent the overall or dominant types of chemical transformation in a given network [59].

Considering the general form of a dynamic mass balance intracellular metabolites in Eq. 1

$$\frac{dC_j}{dt} = Sr(C, p)$$

and assuming a steady state for the intracellular metabolite concentrations, the mass balances become:

$$Sr = 0 \tag{3}$$

This assumption is based on the fact that the characteristic time of metabolite concentration changes is much smaller that the cellular process changes [58]. To complement this analysis, steady-state flux distributions can be estimated or measured by ^{13}C labeling [74].

If the number of metabolite rates is greater than the number of equations ($N > R$) the system of equations defined in Eq. 3 is underdetermined. In this case many solutions for the reaction rates$r_i$satisfy the balance equations. Additional constraints need to be introduced in order to uniquely determine the values of the reaction rates. This can be done by adding mass balance equations, making assumptions on the system, or obtaining additional information on the internal fluxes. The unconstrained system considers all possible behaviors of the system, defining a large feasible solution space. By imposing constraints such as physicochemical constrains (thermodynamic or conservation restrictions) or biological constraints (external environment, enzyme capacity, regulatory constraints) a smaller allowable solution space is defined [73]. However, the addition of constraints reduces the allowable solution space, but usually not to a single point (exact solution). For an underdetermined system, a solution can also be obtained by using linear programming (LP) to optimize a particular network function such as growth rate or biomass production, following the methodology known as flux balance analysis (FBA) [75, 76]. An alternative approach based on FBA is the minimization of metabolic adjustment (MOMA), which attempts to determine more realistic flux distributions by calculating intermediate profiles that are intermediate between the base case and the perturbed system, based on the hypothesis that perturbed metabolic fluxes undergo a minimal redistribution [77].

If the number of measured fluxes is grater than the number of unknown fluxes, the system is overdetermined. In this case a regression analysis method must be used in order to minimize the error between the calculated fluxes from the mass balance equations and the measured extracellular fluxes [72]. The additional information available can be used to improve the values obtained for the calculated fluxes and the measured ones [58]. As a result, an exact solution or a range of allowable solutions can be found. This approach is limited by the availability of measurable fluxes and the uncertainty of experimental measurements. The advantage of this approach lies in the fact that it is a relatively simple and easy to understand method, which can give an estimate of the relative order of magnitude of the reactions in a pathway. It also permits the exploration of the metabolic capabilities of an organism without specific biochemical data. Its predictive power when modifications on the fluxes and different growth conditions are introduced has been shown by Varma and Palsson for *E. coli* [76]. However, one must consider that the calculated flux distribution is extremely sensitive to errors in the measurement of the extracellular rates.

Among the many applications of the Metabolic Flux Balance methodology are studies of metabolic physiology [78], simulation and interpretation of experimental data, metabolic engineering, bioprocess design and monitoring [79], and determination of the production capabilities of a given strain [76]. The availability of genomic information makes this a useful method for understanding the genotype–phenotype relationship. Different gene expression levels can be associated with different steady-state values for the reaction rates, making this a good tool for analyzing, interpreting, and predicting the possible phenotypic response of a given strain, based on its genotype [72, 80]. This method gives valuable information about the net flux distribution at steady state, but provides no insight into the dynamics of the process. This approach is also limited by the assumption of optimal behavior of the organism. The integration of flux, concentration and kinetic variables in a single unified framework allows an increase in the quantitative predictive capacity of flux balance analysis. In Fleming et al. [81], experimental and theoretical bounds on thermodynamic and kinetic variables are included to ensure the thermodynamic as well as biochemical feasibility of the predicted steady-state fluxes in *E. coli*. Computational resources for metabolic flux and flux balance analysis are listed in Table 3. A hybrid strategy was proposed by Carinhas et al. [82] to link estimated metabolic fluxes with measured productivities in an insect cell line. This tool was shown to be useful for metabolic identification and quantification in incomplete or ill-defined metabolic networks such as those with complex products. The ill-defined part of the network is substituted by a statistical sub-model based on empirical data.

A detailed discussion of the metabolism of mammalian cells in culture and a review of the methods for flux analysis and their application to mammalian cells can be found in Bonarius et al. [83]. This text also covers isotope distribution modeling, redox balances and formulation of objective functions for linear optimization techniques for flux analysis. The flux network and confidence intervals for each flux in a carcinoma cell line was determined by Metallo et al. [84] using ^{13}C labeled glucose and glutamine tracers. A more recent article by Queket al. discusses the advances in conventional and constraint-based metabolic flux analysis for the study of the metabolic phenotype of mammalian cells, the use of tracer analysis for network model validation and the advances in these two fields towards large-scale ^{13}C metabolic flux analysis for mammalian cells [85]. Examples for cultures of hybridoma and CHO cells are presented.

There are only a few works associated with metabolic models of mammalian cells. This can be attributed to the complexity of mammalian cell metabolism, reduced availability of information about the system, and difficulties in measuring in-vivo metabolic fluxes in mammalian cells. In Llaneras et al. [86], a method is presented to compute the ranges of possible values for non-calculable flux, resulting in a flux region. The method was applied to the CHO cell metabolic network proposed by Provost et al. [87], which describes the metabolism concerned only with glucose and glutamine. MFA has also been used in CHO cells to investigate cellular metabolism in cultures containing glucose and galactose as a carbon source [88, 89].

In other cell lines, MFA has been applied to HEK293 cells, to determine the best conditions to perform adenovirus infection and determine metabolic changes upon infection [90, 91]. Different culture conditions at low glutamine concentrations have been compared using a simple metabolic network for HEK293 cells [92]. Zupke et al. [93] applied a stoichiometric balance to estimate intracellular fluxes and to study energy metabolism of hybridoma cells. Flux estimates were validated with labelling experiments, with good agreement [93]. Xie and Wang applied material balance analysis to an hybridoma cell stoichiometric reaction network. The roles of essential and non-essential amino acids, together with the metabolism of glucose and lactate as well as ATP production rates, were assessed [94, 95].

A reconstruction of the cellular metabolic network of *Mus musculus* was presented recently by Sheikh et al. [61]. This reconstruction is based on annotated genomic data, pathway databases, and currently available biochemical and physiological information from the Kyoto Encyclopedia of Genes and Genomes (KEGG). This is the first genomic reconstruction of a mammalian metabolic network. The reconstruction captures carbon, energy, and nitrogen metabolism in a compartmentalized setting, including transport reactions between the compartments and the extracellular medium. It considers 872 internal metabolites and 1,220 reactions. As part of an initial in-silico analysis, metabolic flux analysis of the reconstructed network was performed based on the mass balances for all metabolites and a pseudo-steady-state assumption on metabolite concentrations. Since the system is underdetermined, a solution was found by linear programming considering three objective functions: maximization of cell growth, minimization of substrate uptake rate, and maximization of production of monoclonal antibody. The model predicts growth, lactate, and ammonia production given glucose, oxygen, and glutamine uptake, but it fails to predict alanine production, illustrating the limitations of the model. This model was improved by Selvarasu et al. [96] to include biomass and monoclonal antibody (mAb) synthesis as well as updated lipid, amino acid, and nucleotide metabolic pathways. The resulting model is capable of producing alanine, aspartate, and glutamate. The model was used for in-silico analysis of a fed-batch hybridoma culture, allowing the study of the physiological and metabolic states of hybridoma cells during fed-batch culture. As a result of the analysis, the utilization of feed media without glutamate and alanine, maintaining a low glucose concentration during culture, and controlled glutamine concentration were identified as strategies for increased cell density and mAb productivity. The genome-scale mouse metabolic model was further improved and expanded by the same group, including additional information on gene–protein-reaction association, and improved network connectivity through lipid, amino acid, carbohydrate, and nucleotide biosynthetic pathways, based on integrated biochemical and genomic data of *Mus musculus* [97]. The improved model was used for additional studies using constraints-based flux analysis, and the evaluation of the structural and functional characteristics of mouse metabolism.

Metabolic flux analysis of CHO cells was recently performed by Zamorano et al. [98] using a detailed metabolic network involving 100 reactions which considers the main pathways of mammalian cell metabolism. The model was used to assess the efficiency of flux analysis when using a small set of extracellular measurements without using isotopic tracer methods. This contribution showed that for this underdetermined mass balance system, narrow intervals are found for most fluxes [98]. A comparison of flux estimates from MFA with a simplified model and ^{13}C glucose and 2D-NMR spectroscopy was performed by Goudar et al. [99] for CHO cells in perfusion culture. In the reduced space, there is good agreement between measurements and estimations in the glycolytic, TCA cycle, and oxidative phosphorylation fluxes. These results corroborate the fact that MFA provides a good representation of cellular metabolism and can be used for process development and metabolism characterization, which is of great importance especially for CHO cells, as they are widely used in the pharmaceutical industry for recombinant protein production. A stoichiometric model for CHO cells was used by Sengupta et al. [100] in combination with steady-state isotopomer balancing to evaluate flux distribution in the late non-growth phase. This model allowed the evaluation of redox balances in the system and comparison of flux distribution for growth and non-growth phases. Ahn et al. [101] studied CHO cells at the growth phase and early stationary phase in fed-batch culture, using isotopic tracers and mass spectrometry and a compartmentalized metabolic network model of CHO cell metabolism. Significant changes in metabolic fluxes were identified as the culture progressed. At the exponential phase, there was a high flux of glycolysis from glucose to lactate. At the stationary phase, the flux map was characterized by a reduced flux of glycolysis and net lactate uptake, with similar fluxes of pyruvate dehydrogenase and TCA cycle compared with the exponential phase. The key influence of the oxidative pentose phosphate pathway and anaplerosis fluxes was identified. In a recent work, Famili's group presented a validated computational metabolic modeling platform based on a genome-scale metabolic model for CHO cell metabolism [102]. This platform has been applied to process design and novel media formulation, improving product titers and reducing byproduct accumulation. This illustrates how MFA and mathematical models, coupled with experimental techniques, can help understand and elucidate the changes that cells undergo in culture, and improve culture productivity.

7.2 Dynamic Models

Kinetic models result from the combination of stoichiometric equations derived from mass balances and kinetic expressions for the reaction rates. They have been employed for a long time in chemical reaction systems, which are relatively simple and well defined. Biological systems, however, are highly regulated and complex, so a detailed knowledge not only of the kinetics but also of the regulation and interaction of the components involved is required for simulation of the system's behavior. In addition, for the development of a mechanistic kinetic model, rate equations and their associated parameters are required for all reactions. In a system where this type of detailed knowledge is available, elaborated mechanistic models can be developed. Several approaches to developing dynamic cellular models have been formulated, based on the rate equations for the individual reactions that form the network. Van Riel utilized a combination of metabolic flux analysis and Michaelis–Menten enzyme kinetics to model the central nitrogen metabolism of *S. cerevisiae* [103]. Van Dien and Keasling [104] developed a dynamic model of the *E. coli* phosphate-starvation response. The model includes phosphate transport, detection at the cell surface, signal transduction cascade, and mRNA translation. This model integrates a large amount of biological information regarding this system and it was applied to the overexpression of heterologous genes. A structured extension of this model was also developed, incorporating differences in gene regulation among cells of the population and the heterogeneity of their response, as well as ATP synthesis and utilization [105]. This is a good example of the capability of detailed kinetic models to explore the dynamic response of cellular and metabolic processes.

To develop a dynamic mechanistic model, an accurate estimation of the mechanistic equation parameters is required, keeping in mind that in-vitro obtained kinetic parameters might not be valid in the invivo situation. Rizzi developed a compartmentalized model of glycolysis in *S. cerevisiae* based on invitro determined mechanistic rate equations with invivo estimated parameters. Metabolic flux analysis was used to obtain steady state fluxes from glycolysis and the TCA cycle, assuming the macromolecular composition for a given growth rate was known [106]. This kind of dynamic formulation requires solving a large number of coupled ordinary differential equations simultaneously. It also involves a large number of parameters. The main advantage of kinetic models is that they capture the dynamics of the metabolic processes, allow their simulation, and, since they may consider regulatory effects, they are more likely to predict the system response under different conditions. Regulation of enzyme activity can be included easily in the model via augmented kinetic expressions for the reaction rates.

The main difficulty associated with the development of kinetic models for mammalian cells has been that large amount of data are required to parameterize large-scale models, making them experimentally intractable. One of the first large-scale models published was the red blood cell model developed by Jamshidi et al. [107]. This model is based on mass balances for metabolites in the red blood cell and consists of kinetic expressions for 35 enzymes, six transport channels, sodium and potassium leak fluxes, and the Na^+K^+ ATPase pump. The kinetic expressions for enzymatic reactions used in this model were obtained from the literature. Changes in metabolite concentrations and enzyme fluxes were simulated over time.

Detailed enzyme kinetic data such as kinetic parameters are scarce and their acquisition is costly, prone to error, and not always feasible under physiological conditions. In spite of that, models based on this type of data are convenient for the characterization of the kinetic mechanism of enzyme-catalyzed reactions. A random sampling methodology of kinetic parameters has been proposed to bypass the need to characterize detailed kinetic parameters. This technique has proved to be well suited to handling uncertainty in parameters for metabolic kinetic model construction [108, 109].

Different aspects of mammalian systems, such as transcriptional profiles, biochemical reactions in the cell or in a bioreactor, cell growth, and population changes, can be described by a set of differential equation using their reaction kinetic and thermodynamic properties. The time-dependent behavior of the system can be obtained by solving the resulting system of differential equations, provided with a set of initial conditions. However, in this case changes of state are assumed to be continuous and deterministic, and describe the average properties of a population. Population models as well as single cell models and their attributes are carefully described in Sidoli et al. [110]. A quantitative comparison between a number of unstructured hybridoma cell growth and death models which consider glucose or glutamine as limiting substrates was presented by Portner et al. [111].

However, this approach cannot be applied to all systems. In a population, the behavior of individuals has a random component, which is averaged out in a large population. As the population size becomes small, continuous functions do not accurately describe the system's behavior. When reactants that are few in number or are in a restricted location such that the perfectly stirred assumption often used in single cell models cannot be applied, reactions cannot be described by mass action laws. Position and thermal motion effects make reactants interact with one another in a probabilistic manner which is referred to as a stochastic reaction. Cellular processes such as binding of transcriptional regulators to specific sites for gene expression regulation and many cell signaling processes are stochastic phenomena. In that case stochastic descriptions are more suitable, and could be coupled with kinetic models in a hybrid model that considers both deterministic and stochastic reactions to generate a biologically realistic representation.

Boolean models are a discrete modeling approach formed by a system of interconnected binary elements. The interactions between elements are governed by logical rules and each element in the system can only be in one of two states (On or Off). The state of the network at any time t can be evaluated as the state of all the elements. The output state of any element at any time $t + 1$ can be computed from the input states at time t. Therefore, for any given initial condition, the successive states of the network can be computed. To calculate each successive state, the elements can be updated sequentially or randomly, which may alter the final steady-state solution obtained. An application of this approach to a complex protein interaction network is presented by Faure et al. [112]. A generic model of mammalian cell cycle control was translated into a logical framework and implemented as a boolean model using a regulatory graph, identifying the logical parameters of the system and specifying the updating assumptions. Different strategies and assumptions were applied to explore its dynamic properties and assess its asymptotic behavior.

Petri nets are special graphs with two types of nodes, called places and transitions. Places, represented as circles, usually describe passive elements of the system such as conditions, states, or compounds, and can take discrete or continuous values. Transitions on the other hand, represented as boxes, describe active elements of the system, such as events, actions, and reactions. They can also take discrete or continuous values and fire at discrete intervals of time defined by the parameter assigned to them. Arcs connect nodes of different types and describe the causal relation between active and passive elements [113]. Petri nets have been used to represent molecular interactions and mechanisms of signaling pathways [114]. In addition, a simulation method for determining some of the characteristics of the interactions is proposed in this work, and the overall strategy is illustrated in the apoptosis signaling pathways. In Peleg et al. [115], Petri net formalisms and tools are applied to the modeling and simulation of a variety of biological systems. For illustration purposes, three case studies which encompass most of the features of biological systems are tested: a cellular-level process, molecular-level functions involved in a protein translation process, and a molecular-level interaction process.

For mammalian cells, many of the available dynamic models correspond to macroscopic bioreaction models rather than detailed mechanistic physicochemical models [116–118]. Provost et al. [119] designed an unstructured dynamic bioreaction model for CHO cells in culture based on measurements of extracellular species. This model is aimed at the design of on-line algorithms for process monitoring, control, and optimization. The model was validated experimentally and takes into account the metabolic changes that cells undergo in culture due to substrate availability. An unstructured, unsegregated, deterministic model for describing mAb production in hybridoma cells was developed by Jang and Barford [120]. The model considers the anabolism of cellular macromolecules and the dependence of protein concentration on gene expression. Kontoravdi et al. [121] presented a single unstructured model structure for describing the cell growth kinetics and metabolism of HEK293 and CHO cells, by applying minor changes in the model for the two cell lines such as growth rate expressions and dependence of cell death on metabolite concentrations. The network considered is consistent with the information available in literature sources and the pathways available in KEGG. Model simulation results are in good agreement with experimental data [121].

Hybrid models have been introduced in an attempt to simulate whole-cell mammalian cell dynamics which combine steady state flux analysis and kinetic rate expressions. These models have to deal with the complexity of describing all extracellular reactions with concentration-dependent rate equations and maintain sufficient constraints to allow flux calculations. A dynamic metabolic model for hybridoma cells was constructed by Gao et al. by combining stoichiometric and dynamic mass balances, and using Monod-based kinetic expressions. To reduce the size and complexity of the system, significant intracellular fluxes were identified using MFA, and distinct characteristics were assigned to the different phases of the culture. The model is capable of predicting glucose and glutamine and the key amino acid concentrations as well as product formation [122]. This model was taken by Baughman et al. [123] to illustrate an optimization-based technique for the estimation of kinetic parameter values in dynamic models of cell metabolism, based on algebraic approximations of continuous differential equations. Reported results in [123] show an improvement in the agreement of experimental and simulated results and the ability to solve the unsimplified variant of the model.

Recently, a very detailed kinetic model for CHO cells was developed by Nolan and Lee [124], based on a network of reactions collected from the KEGG database for the hamster analog mouse. Linear pathways were collapsed for simplification while preserving stoichiometric relationships between metabolites. The model was compartmentalized into the cytosol and mitochondria for NADH transport. Cytosolic enzyme-catalyzed reactions were described with kinetic rate expressions, where metabolite uptake rates are controlled by several intracellular reactions, rather than a single transporter. The effect of temperature and the redox state of the cell was also accounted for. In Nolan and Lee [125] the effects of temperature shift, seed density, specific productivity, and metabolite concentrations on viable cell density (VCD), antibody, lactate, asparagine, and the redox state were studied by considering a detailed kinetic model of CHO cell metabolism. A metabolic network similar to the one in previous works was used, with a novel framework for simulating the dynamics of metabolic and biosynthetic pathways, where rate expressions are used to calculate pseudo-steady-state flux distributions and extracellular metabolite concentrations at discrete time points for fed-batch cultures. The model provides time profiles for all metabolites in the reactor and successfully predicts the effects of several process perturbations on cell growth and product titer.

8 Modeling Resources

Dynamic models of metabolic networks involve large sets of highly nonlinear differential equations, which are solved numerically due to their complexity. These equations can be solved using any pertinent algorithm implemented in any numerical computing and programming language such as Mathematica, Scilab or Matlab. Hence the availability of suitable software is crucial for the development of systems biology research. Many simulation software packages have been developed to provide specific analysis and computing packages similar to the ones used in engineering and other disciplines, but without a universal standard to enable integration of these resources.

8.1 Systems Biology Markup Language (SBML) and Systems Biology Workbench (SBW)

The Systems Biology Markup Language (SBML) arose as a result of the observation that there is a large number of software tools that researchers use for different needs, which generate results that are not easily compatible and cannot be shared. SMBL provides a standard that enables models to be exchanged between software tools, and allows simulation and analysis results in systems biology to be shared through a common model exchange language. SBML uses a machine-readable format for representing computational models in systems biology and can be used as a common exchange format for transferring, for instance, computational models of biochemical reactions between different software tools [126, 127]. For the model to be useful it should be sufficiently characterized and accompanied by a minimum amount of information, as outlined in Minimum Information Requested In the Annotation of biochemical Models (MIRIAM) [128].

The systems biology workbench (SBW) is a software framework built on SBML that permits sharing computational resources and applications associated with different programming languages and running on different platforms in a simple, high-performance, open-source software infrastructure [129].

There is a large number of software tools and databases freely available on the web that can simplify the implementation of such models or provide information regarding structure and parameters of the network. These resources are listed in Table 3. Other resources and tools for systems biology (genome information and analysis, transcriptome and proteome databases, metabolic profiling and metabolic control analysis, metabolic and regulatory information databases, and software for computational systems biology and simulation) are reviewed in [130].

8.2 Metabolic Engineering Tools

Currently there are many tools available for the analysis of stoichiometric networks, such as the COBRA Toolbox, CellNetAnalyzer, PathwayAnalyser [148], Metatool [140], YANAsquare, SNA and OptFlux. Some of these programs require a specific programming environment, such as Matlab or Mathematica, as indicated in Table 3.

CellNetAnalyzer is a comprehensive software tool that runs in the Matlab environment, and allows performance of analyses for metabolic engineering such as FBA on metabolic, regulatory, and signalling networks in a user-friendly environment [134]. COBRA is a Matlab package for performing flux and pathway analysis, either with or without experimental data. In addition to standard pathway analysis, SBRT includes other capabilities such as data analysis tools [136]. Other tools for stoichiometric analysis are YANAsquare and SNA which are focused mainly on elementary flux mode analysis. YANAsquare provides a framework for rapid network assembly, visualization, and elementary flux mode analysis [147]. SNA is a Mathematica toolbox for elementary flux mode analysis. It also supports steady-state analysis by linear programming [143]. These two applications lack algorithms for the identification of potential metabolic engineering targets and model visualization tools.

OptFlux is a software designed to support in-silico metabolic engineering. It incorporates the identification of metabolic engineering targets for strain development and allows phenotype simulation and flux space determination of stoichiometric metabolic models using FBA, MOMA, MFA and elementary flux mode analysis, and it also includes visualization tools [141]. The main characteristics of this software which make it interesting for metabolic engineering applications are that it is open-source, user-friendly, modular, and compatible with SBML.

8.3 Diagrammatic Simulation Tools

To date there are a large number of mathematical packages for simulating cellular networks that take the user interface to an visual level. In this format, models are represented as network diagrams drawn on a canvas, with connections associated with the reactions between components. The user only requires knowledge of the components of the network, interactions, rate expressions, and parameters. Knowledge of programming, differential equations and numerical integration methods is not required as the diagram generated is converted seamlessly into a mathematical representation for simulation. Examples of such tools include CellDesigner, Cellware, JDesigner and Virtual Cell (see Table 3). Some of these tools are reviewed and compared in [149].

JDesigner is an open-source network design tool which allows the drawing of biochemical networks and exporting them in SBML format. It also acts as a simulator via SBW [129]. Cellware provides a modeling and simulation environment for models of multiple levels such as gene regulatory networks, signal transduction, and metabolic pathways. It includes stochastic and deterministic algorithms based on ordinary differential equation solvers, and hybrid algorithms for a flexible architecture. It also provides parameter estimation and optimization tools in a user-friendly intuitive environment [135].

CellDesigner is a modeling and simulation tool that allows visualization, modeling, and simulation of gene-regulatory and biochemical networks. The main characteristics that define the utility of this tool for creating and transferring models are its comprehensive graphical representation of network models and the fact that it is based on SBML, which facilitates model transfer. It is integrated with SBW-enabled simulation and analysis modules, and with the SBML ODE Solver library [150].

8.4 Other Simulation Tools

Models of drug metabolism are an essential element used by pharmaceutical companies in preclinical pharmacokinetic studies and to maximize the efficiency of lead compound selection, facilitating drug discovery while reducing the need for animal testing. In this direction, a biosimulation tool based on yeast as a model organism was proposed by Pieper et al. [151]. This approach is restricted to drugs metabolized by enzymes in the central carbon metabolism.

Pharmaceutical as well as specialized companies provide simulation tools and services for cellular as well as whole-organ systems. Entelos developed PhysioLab, a platform of mathematical models that integrates genomic, proteomic, physiological, environmental, and behavioral data to represent the physiology of a disease and its response to treatment. This tool allows the optimization of trial design by simulation on virtual patients, and has been applied to develop a model for diabetes [152]. In order to predict the toxicity of potential drug candidates, Strand Life Sciences developed a virtual liver system based on mathematical modeling of the kinetics of essential biochemical pathways involved in liver homeostasis. On a cellular scale, Genomatica provides an integrated metabolic engineering and bio-process development platform, which combines predictive computational modeling with experimental lab technology to design and test highly-optimized organisms in order to accelerate process development and production.

For general cellular model development, Virtual Cell is a computational environment for analysis, modeling, and simulation of cell biology that allows the development of a full range of models. These models can range from simple descriptions used for evaluating hypotheses or interpreting experimental data, to models used to test the predicted behavior of complex, highly non-linear systems. It integrates molecular mechanisms, such as reaction kinetics, diffusion, flow, membrane transport, etc., and allows the development of spatially explicit models. Components can be uniquely identified and annotated following the MIRIAM standard. It features a web-based Java interface to specify compartmentalized topology and geometry, molecular characteristics, and relevant interaction parameters. The biological description obtained is automatically translated into a corresponding system of ordinary or partial differential equations to be solved numerically. For model simulation, both deterministic and stochastic algorithms are supported. For simulations of combined reaction, diffusion, and advection in complex geometries, a partial differential equation solver is available [145].

9 Final Remarks

The progress in experimental biology from the molecular level to the systems level has modified the way we understand biological systems and cellular processes, leading to knowledge integration from different fields. This paradigm change in understanding of these complex systems has also modified the way we address modeling of mammalian cell systems. Detailed quantitative models are required to capture the mechanisms of the underlying processes involved from a physico-chemical point of view, in order to increase our understanding of these systems. In spite of the current accomplishments, systems biology still has challenges to face. For instance, in the recombinant protein production field, since the molecular and physiological factors involved in high protein production in mammalian cell lines are not well understood, identification of differentially expressed genes in high producers could enrich the development of models in order to improve heterologous protein production by providing an understanding of the regulatory layer behind the observed effects. The understanding of these elements on a system-wide level is of great importance, as recombinant protein production has recognized effects on the host organism.

Nowadays information is becoming available at a faster rate, allowing the development of detailed predictive models of mammalian cells. Therefore, since the metabolic network modeling field has grown significantly in the past decade, and will continue to do so, it is expected that in the next decade the knowledge of specific genetic characteristics and the analysis and modeling of empirical data will continue to help predict the response and behavior of mammalian cell systems for application in production processes as well as in the medical field.

On the other hand, mathematical models can assist in strain design and complement metabolic methods in order to increase productivity. However, many of the models currently available have limitations in their description of the system due to the scope of the solutions and availability and accuracy of the data used on their construction. Further improvements are needed to overcome these problems and to obtain a global view of mammalian cell systems, in order to provide a better understanding of their metabolic responses due to perturbation and during protein production. In order to address new biological applications, the further development of dynamic and genome-based models for mammalian cells will require abundant high-quality experimental data, and new software, algorithms, and visualization techniques.

Acknowledgements The financial support of CONICYT, FONDECYT Initiation Grant 11090268 is gratefully acknowledged.

References

1. Schuler M, Kargi F (1992) Bioprocess engineering: basic concepts. Prentice-Hall Inc., New Jersey
2. Ideker T, Galitski T, Hood L (2001) A new approach to decoding life: systems biology. Annu Rev Genomics Hum Genet 2:343–372
3. Kitano H (2002) Systems biology: a brief overview. Science 295(5560):1662–1664
4. Oltvai ZN, Barabasi AL (2002) Systems biology. Life's complexity pyramid. Science 298(5594):763–764
5. Hood L (2003) Systems biology: integrating technology, biology, and computation. Mech Ageing Dev 124(1):9–16
6. Hu W.S. (2009) Cellular bioprocess technology: an advanced course in bioprocessing. University of Minnesota, Minnesota
7. Alon U (2007) An introduction to systems biology: design principles of biological circuits. Math Comput Biol, Chapman & Hall, London
8. Waterston RH, Lindblad-Toh K et al (2002) Initial sequencing and comparative analysis of the mouse genome. Nature 420(6915):520–562
9. Blattner FR, Plunkett 3rd G et al (1997) The complete genome sequence of *Escherichia coli* K-12. Science 277(5331):1453–1462
10. Bremer H, Dennis PP (1996) Modulation of chemical composition and other parameters of the cell by growth rate. In: *Escherichia coli* and *Salmonella*: Cellular and Molecular Biology. Neidhart FC et al. (ed), ASM Press, Washington DC, 1553–1569
11. Watson J (1988) Molecular biology of the gene. 4th edn. Benjamin/Cummings Pub. Co, Menlo Park, California
12. Zimmerman SB, Minton AP (1993) Macromolecular crowding: biochemical, biophysical, and physiological consequences. Annu Rev Biophys Biomol Struct 22:27–65

13. Ellis RJ, Minton AP (2003) Cell biology: join the crowd. Nature 425(6953):27–28
14. Li J, Ning Y et al (2002) The molecule pages database. Nature 420(6916):716–717
15. Prieto C, DeLas Rivas J (2006) APID: agile protein interaction dataanalyzer. Nucl Acids Res 34(Web Server issue):W298–W302
16. Karp PD, Ouzounis CA et al (2005) Expansion of the BioCyc collection of pathway/genome databases to 160 genomes. Nucleic Acids Res 33(19):6083–6089
17. Stark C, Breitkreutz BJ et al (2011) The BioGRID interaction database: 2011 update. Nucl Acids Res 39(Database issue):D698–D704
18. Willis RC, Hogue CW (2006) Searching, viewing, and visualizing data in the biomolecular interaction network database (BIND). Curr Protoc Bioinform Chap 8:Unit 8.9
19. Chang A, Scheer M, Grote A et al (2009) BRENDA, AMENDA and FRENDA the enzyme information system: new content and tools in 2009. Nucleic Acids Res 37(Database issue): D588–D592
20. Salwinski L, Miller CS et al (2004) The database of interacting proteins: 2004 update. Nucl Acids Res 32(Database issue):D449–D451
21. Kolchanov NA, Nedosekina EA et al (2002) GeneNet database: description and modeling of gene networks. In Silico Biol 2(2):97–110
22. Salomonis N, Hanspers K et al (2007) GenMAPP 2: new features and resources for pathway analysis. BMC Bioinform 8:217
23. Han K, Park B et al (2004) HPID: the human protein interaction database. Bioinformatics 20(15):2466–2470
24. Keshava Prasad TS, Goel R et al (2009) Human protein reference database—2009 update. Nucl Acids Res 37(Database issue):D767–D772
25. Shoemaker ba, Zhang d et al (2010) Inferred biomolecular interaction server—a web server to analyze and predict protein interacting partners and binding sites. Nucl Acids Res 38(Database issue):D518–D524
26. Hoffmann R, Valencia A (2004) A gene network for navigating the literature. Nat Genet 36(7):664
27. Brown KR, Jurisica I (2005) Online predicted human interaction database. Bioinformatics 21(9):2076–2082
28. Aranda B, Achuthan P et al (2010) The IntAct molecular interaction database in 2010. Nucl Acids Res 38(Database issue):D525–D531
29. Kumar P, Han BC et al (2009) Update of KDBI: Kinetic data of bio-molecular interaction database. Nucl Acids Res 37(Database issue):D636–D641
30. Kanehisa M, Araki M et al (2008) KEGG for linking genomes to life and the environment. Nucl Acids Res 36(Database issue):D480–D444
31. Sprenger J, Fink JL et al (2008) LOCATE: a mammalian protein subcellular localization database. Nucl Acids Res 36(Database issue):D230–D233
32. Quesada V, Ordóñez GR et al (2009) The degradome database: mammalian proteases and diseases of proteolysis. Nucl Acids Res 37(Database issue):D239–D243
33. Ceol A, Chatr Aryamontri A et al (2010) MINT, the molecular interaction database: 2009 update. Nucl Acids Res 38(Database issue):D532–D539
34. Pagel P, Kovac S et al (2005) The MIPS mammalian protein–protein interaction database. Bioinformatics 21(6):832–834
35. Caspi R, Altman T et al (2010) The MetaCyc database of metabolic pathways and enzymes and the BioCyc collection of pathway/genome databases. Nucl Acids Res 38(Database issue): D473–D479
36. Hu Z, Snitkin ES, DeLisi C (2008) VisANT: an integrative framework for networks in systems biology. Brief Bioinform 9(4):317–325
37. Koike A, Takagi T (2005) PRIME: automatically extracted protein interactions and molecular information database. In Silico Biol 5(1):9–20
38. Gong S, Yoon G et al (2005) PSIbase: a database of Protein Structural Interactome map (PSIMAP). Bioinformatics 21(10):2541–2543

39. Saunders B, Lyon S et al (2008) The molecule pages database. Nucl Acids Res 36(Database issue):D700–D706
40. Szklarczyk D, Franceschini A et al (2011) The STRING database in 2011: functional interaction networks of proteins, globally integrated and scored. Nucl Acids Res 39(Database issue):D561–D568
41. Reed JL, Famili I et al (2006) Towards multidimensional genome annotation. Nat Rev Genet 7(2):130–141
42. Ma'ayan A, Blitzer RD, Iyengar R (2005) Toward predictive models of mammalian cells. Annu Rev Biophys Biomol Struct 34:319–349
43. Edwards JS, Ibarra RU, Palsson BO (2001) In silico predictions of*Escherichia coli*metabolic capabilities are consistent with experimental data. Nat Biotechnol 19(2):125–130
44. Voit EO, Savageau MA (1987) Accuracy of alternative representations for integrated biochemical systems. Biochemistry 26(21):6869–6880
45. Heijnen JJ (2005) Approximative kinetic formats used in metabolic network modeling. Biotechnol Bioeng 91(5):534–545
46. Steuer R, Junker BH (2008) Computational models of metabolism:stability and regulation in metabolic networks. In: Advances in Chemical Physics, vol 142: Wiley, New York
47. Tyson JJ, Chen K, Novak B (2001) Network dynamics and cell physiology. Nat Rev Mol Cell Biol 2(12):908–916
48. Csikasz-Nagy A, Battogtokh D et al (2006) Analysis of a generic model of eukaryotic cell-cycle regulation. Biophys J 90(12):4361–4379
49. Cloutier M, Wellstead P (2010) The control systems structures of energy metabolism. J R Soc Interf 7(45):651–665
50. Ma H, Zeng AP (2003) Reconstruction of metabolic networks from genome data and analysis of their global structure for various organisms. Bioinformatics 19(2):270–277
51. Gupta R, Stincone A, Antczak P, others, Falciani F (2011) A computational framework for gene regulatory network inference that combines multiple methods and datasets. BMC Syst Biol 5:52
52. Novikov E, Barillot E (2008) Regulatory network reconstruction using an integral additive model with flexible kernel functions. BMC Syst Biol 2:8
53. Thiele I, Palsson BO (2010) A protocol for generating a high-quality genome-scale metabolic reconstruction. Nat Protoc 5(1):93–121
54. Zhao Y, Karypis G (2005) Data clustering in life sciences. Mol Biotechnol 31(1):55–80
55. Yeung KY, Ruzzo WL (2001) Principal component analysis for clustering gene expression data. Bioinformatics 17(9):763–774
56. De Iorio M, Ebbels TMD, Stephens DA (2007) Statistical techniques. Metabolic profiling. In: Handbook of Statistical Genetics, 3rd edn. 347–370. Wiley-Interscience, New York
57. Wiechert W, Takors R (2004) Validation of metabolic models: concepts, tools, and problems. In: Metabolic engineering in the post genomic era. Horizon Biosci, UK, pp 277–320
58. Stephanopoulos G, Aristidou A, Nielsen J (1998) Metabolic engineering: principles and methodologies. 4th edn. Academic Press, Boston
59. Palsson BO (2006) Systems biology: properties of reconstructed networks. 4th edn. Cambridge University Press, New York
60. Romero P, Wagg J et al (2005) Computational prediction of human metabolic pathways from the complete human genome. Genome Biol 6(1):R2
61. Sheikh K, Forster J, Nielsen LK (2005) Modeling hybridoma cell metabolism using a generic genome-scale metabolic model of Mus musculus. Biotechnol Prog 21(1):112–121
62. Henry CS, DeJongh M et al (2010) High-throughput generation, optimization and analysis of genome-scale metabolic models. Nat Biotechnol 28(9):977–982
63. Karp PD, Paley S, Romero P (2002) The pathway tools software. Bioinformatics 18(1): S225–S232
64. Karp PD, Paley S et al (2010) Pathway tools version 13.0: integrated software for pathway/ genome informatics and systems biology. Brief Bioinform 11(1):40–79

65. Forster J, Famili I, Fu P et al (2003) Genome-scale reconstruction of the Saccharomyces cerevisiae metabolic network. Genome Res 13(2):244–253
66. Aho T, Almusa H, Matilainen J et al (2010) Reconstruction and validation of RefRec: a global model for the yeast molecular interaction network. PLoS One 5(5):e10662
67. Chung BK, Selvarasu S, Andrea C et al (2010) Genome-scale metabolic reconstruction and in silico analysis of methylotrophic yeast Pichia pastoris for strain improvement. Microb Cell Fact 9:50
68. Burgard AP, Nikolaev EV et al (2004) Flux coupling analysis of genome-scale metabolic network reconstructions. Genome Res 14(2):301–312
69. Efron B, Tibshirani R (1994) An Introduction to the Bootstrap. 4th edn. Chapman & Hall/CRC, London
70. Wiechert W (2002) Modeling and simulation: tools for metabolic engineering. J Biotechnol 94(1):37–63
71. Covert MW, Famili I, Palsson BO (2003) Identifying constraints that govern cell behavior: a key to converting conceptual to computational models in biology? Biotechnol Bioeng 84(7):763–772
72. Edwards JS, Ramakrishna R et al (1999) Metabolic flux balance analysis, volume metabolic engineering. Marcel Deker, enter text here
73. Price ND, Reed JL, Palsson BO (2004) Genome-scale models of microbial cells: evaluating the consequences of constraints. Nat Rev Microbiol 2(11):886–897
74. Yang TH, Heinzle E, Wittmann C (2005) Theoretical aspects of 13C metabolic flux analysis with sole quantification of carbon dioxide labeling. Comput Biol Chem 29(2):121–133
75. Reed LJ (1981) Regulation of mammalian pyruvate dehydrogenase complex by a phosphorylation-dephosphorylation cycle. Curr Top Cell Regul 18:95–106
76. Varma A, Palsson BO (1994) Stoichiometric flux balance models quantitatively predict growth and metabolic by-product secretion in wild-type *Escherichia coli* W3110. Appl Environ Microbiol 60(10):3724–3731
77. Segre D, Vitkup D, Church GM (2002) Analysis of optimality in natural and perturbed metabolic networks. Proc Natl Acad Sci USA 99(23):15112–15117
78. Follstad BD, Balcarcel RR et al (1999) Metabolic flux analysis of hybridoma continuous culture steady state multiplicity. Biotechnol Bioeng 63(6):675–683
79. Lee K, Berthiaume F et al (1999) Metabolic flux analysis: a powerful tool for monitoring tissue function. Tissue Eng 5(4):347–368
80. Schilling CH, Edwards JS, Palsson BO (1999) Toward metabolic phenomics: analysis of genomic data using flux balances. Biotechnol Prog 15(3):288–295
81. Fleming RM, Thiele I et al (2010) Integrated stoichiometric, thermodynamic and kinetic modelling of steady state metabolism. J Theor Biol 264(3):683–692
82. Carinhas N, Bernal V et al (2011) Hybrid metabolic flux analysis: combining stoichiometric and statistical constraints to model the formation of complex recombinant products. BMC Syst Biol 5:34
83. Bonarius HP, De Goolier CD, Tramper J (2000) Flux analysis of mammalian cell culture: methods and applications. In: Encyclopedia of Cell Technology, Wiley, New York, pp 726–735
84. Metallo CM, Walther JL, Stephanopoulos G (2009) Evaluation of 13C isotopic tracers for metabolic flux analysis in mammalian cells. J Biotechnol 144(3):167–174
85. Quek LE, Dietmair S et al (2010) Metabolic flux analysis in mammalian cell culture. Metab Eng 12(2):161–171
86. Llaneras F, Pico J (2007) A procedure for the estimation over time of metabolic fluxes in scenarios where measurements are uncertain and/or insufficient. BMC Bioinform 8:421
87. Provost A, Bastin G (2004) Dynamic metabolic modelling under the balanced growth condition. J Process Control 2004(14):717–728
88. Altamirano C, Illanes A et al (2006) Considerations on the lactate consumption by CHO cells in the presence of galactose. J Biotechnol 125(4):547–556
89. Wilkens CA, Altamirano C, Gerdtzen ZP (2011) Comparative metabolic analysis of lactate for CHO cells in glucose and galactose. Biotechnol Bioprocess Eng 16

90. Henry O, Perrier M, Kamen A (2005) Metabolic flux analysis of HEK-293 cells in perfusion cultures for the production of adenoviral vectors. Metab Eng 7(5–6):467–476
91. Martinez V, Gerdtzen ZP et al (2010) Viral vectors for the treatment of alcoholism: use of metabolic flux analysis for cell cultivation and vector production. Metab Eng 12(2):129–137
92. Nadeau I, Jacob D et al (2000) 293SF metabolic flux analysis during cell growth and infection with an adenoviral vector. Biotechnol Prog 16(5):872–884
93. Zupke C, Stephanopoulos G (1995) Intracellular flux analysis in hybridomas using mass balances and in vitro (13)C NMR. Biotechnol Bioeng 45(4):292–303
94. Xie L, Wang DI (1996) Material balance studies on animal cell metabolism using a stoichiometrically based reaction network. Biotechnol Bioeng 52(5):579–590
95. Xie L, Wang DI (1996) Energy metabolism and ATP balance in animal cell cultivation using a stoichiometrically based reaction network. Biotechnol Bioeng 52(5):591–601
96. Selvarasu S, Wong VV et al (2009) Elucidation of metabolism in hybridoma cells grown in fed-batch culture by genome-scale modeling. Biotechnol Bioeng 102(5):1494–1504
97. Selvarasu S, Karimi IA et al (2010) Genome-scale modeling and in silico analysis of mouse cell metabolic network. Mol Biosyst 6(1):152–161
98. Zamorano F, Wouwer AV, Bastin G (2010) A detailed metabolic flux analysis of an underdetermined network of CHO cells. J Biotechnol 150(4):497–508
99. Goudar C, Biener R et al (2010) Metabolic flux analysis of CHO cells in perfusion culture by metabolite balancing and 2D [13C, 1H] COSY NMR spectroscopy. Metab Eng 12(2): 138–149
100. Sengupta N, Rose ST, Morgan JA (2011) Metabolic flux analysis of CHO cell metabolism in the late non-growth phase. Biotechnol Bioeng 108(1):82–92
101. Ahn WS, Antoniewicz MR (2011) Metabolic flux analysis of CHO cells at growth and non-growth phases using isotopic tracers and mass spectrometry. Metab Eng 13:598–609
102. Feist A, Rosenbloom J et al (2011) Improving mammalian cell line protein production using a metabolic model-based approach. Eur Soc Animal Cell Technol, vol 22: ESACT meeting
103. Riel NAV, Giuseppin ML et al (1998) A structured, minimal parameter model of the central nitrogen metabolism in *Saccharomyces cerevisiae*: the prediction of the behavior of mutants. J Theor Biol 191(4):397–414
104. Van Dien SJ, Keasling JD (1998) A dynamic model of the *Escherichia coli* phosphate-starvation response. J Theor Biol 190(1):37–49
105. Van Dien SJ, Keasling JD (1999) Effect of polyphosphate metabolism on the *Escherichia coli* phosphate-starvation response. Biotechnol Prog 15(4):587–593
106. Rizzi M, Baltes M et al (1997) In vivo analysis of metabolic dynamics in *Saccharomyces cerevisiae*: II mathematical model. Biotechnol Bioeng 55(4):592–608
107. Jamshidi N, Edwards JS et al (2001) Dynamic simulation of the human red blood cell metabolic network. Bioinformatics 17(3):286–287
108. Wang L, Birol I, Hatzimanikatis V (2004) Metabolic control analysis under uncertainty: framework development and case studies. Biophys J 87(6):3750–3763
109. Schellenberger J, Palsson BO (2009) Use of randomized sampling for analysis of metabolic networks. J Biol Chem 284(9):5457–5461
110. Sidoli FR, Mantalaris A, Asprey SP (2004) Modelling of mammalian cells and cell culture processes. Cytotechnology 44(1–2):27–46
111. Portner R, Schafer T (1996) Modelling hybridoma cell growth and metabolism—a comparison of selected models and data. J Biotechnol 49(1–3):119–35
112. Faure A, Naldi A et al (2006) Dynamical analysis of a generic Boolean model for the control of the mammalian cell cycle. Bioinformatics 22(14):e124–e131
113. Junker BH, Schreiber F (2008) Analysis of biological networks. 4th edn. Wiley, New York
114. Li C, Ge QW et al (2007) Modelling and simulation of signal transductions in an apoptosis pathway by using timed Petri nets. J Biosci 32(1):113–127
115. Peleg M, Rubin D, Altman RB (2005) Using Petri Net tools to study properties and dynamics of biological systems. J Am Med Inform Assoc 12(2):181–199

116. Batt BC, Kompala DS (1989) A structured kinetic modeling framework for the dynamics of hybridoma growth and monoclonal antibody production in continuous suspension cultures. Biotechnol Bioeng 34(4):515–531
117. Cazzador L, Mariani L (1993) Growth and production modeling in hybridoma continuous cultures. Biotechnol Bioeng 42(11):1322–1330
118. Sanderson CS, Barton GW, Barford JP (1995) Optimisation of animal cell culture media using dynamic simulation. Comput Chem Eng 19:681–686
119. Provost A, Bastin G et al (2006) Metabolic design of macroscopic bioreaction models: application to Chinese hamster ovary cells. Bioprocess Biosyst Eng 29(5–6):349–366
120. Jang JD, Barford JP (2000) An unstructured kinetic model of macromolecular metabolism in batch and fed-batch cultures of hybridoma cells producing monoclonal antibody. Biochem Eng J 4:153–168
121. Kontoravdi C, Wong D et al (2007) Modeling amino acid metabolism in mammalian cells-toward the development of a model library. Biotechnol Prog 23(6):1261–1269
122. Gao J, Gorenflo VM et al (2007) Dynamic metabolic modeling for a MAB bioprocess. Biotechnol Prog 23(1):168–181
123. Baughman AC, Huang X et al (2010) On the dynamic modeling of mammalian cell metabolism and mAb production. Comput Chem Eng 34:210–222
124. Nolan R, Lee K (2009) Modeling the dynamics of cellular networks. In: Methods in bioengineering: systems analysis of biological networks, Artech House Publishers, Boston
125. Nolan RP, Lee K (2011) Dynamic model of CHO cell metabolism. Metab Eng 13(1): 108–124
126. Dada JO, Spasic I et al (2010) SBRML: a markup language for associating systems biology data with models. Bioinformatics 26(7):932–938
127. Hucka M, Finney A et al (2003) The systems biology markup language (SBML): a medium for representation and exchange of biochemical network models. Bioinformatics 19(4): 524–531
128. Le Novere N, Finney A et al (2005) Minimum information requested in the annotation of biochemical models (MIRIAM). Nat Biotechnol 23(12):1509–1515
129. Sauro HM, Hucka M et al (2003) Next generation simulation tools: the systems biology workbench and BioSPICE integration. OMICS 7(4):355–372
130. Kim JS, Yun H et al (2006) Resources for systems biology research. J Microbiol Biotechnol 16(6):832–848
131. Li C, Courtot M et al (2010) BioModels.net Web Services, a free and integrated toolkit for computational modelling software. Brief Bioinform 11(3):270–277
132. Kitano H, Funahashi A et al (2005) Using process diagrams for the graphical representation of biological networks. Nat Biotechnol 23(8):961–966
133. Shapiro BE, Levchenko A et al (2003) Cellerator: extending a computer algebra system to include biochemical arrows for signal transduction simulations. Bioinformatics 19(5): 677–678
134. Klamt S, Saez-Rodriguez J, Gilles ED (2007) Structural and functional analysis of cellular networks with CellNetAnalyzer. BMC Syst Biol 1:2
135. Dhar PK, Meng TC et al (2005) Grid cellware: the first grid-enabled tool for modelling and simulating cellular processes. Bioinformatics 21(7):1284–1287
136. Becker SA, Feist AM et al (2007) Quantitative prediction of cellular metabolism with constraint-based models: the COBRA toolbox. Nat Protoc 2(3):727–738
137. Hoops S, Sahle S et al (2006) COPASI—a complex pathway simulator. Bioinformatics 22(24):3067–3074
138. Sivakumaran S, Hariharaputran S et al (2003) The database of quantitative cellular signaling: management and analysis of chemical kinetic models of signaling networks. Bioinformatics 19(3):408–415
139. Takahashi K, Ishikawa N et al (2003) E-cell 2: multi-platform E-cell simulation system. Bioinformatics 19(13):1727–1729

140. von Kamp A, Schuster S (2006) Metatool 5.0: fast and flexible elementary modes analysis. Bioinformatics 22(15):1930–1931
141. Rocha I, Maia P et al (2010) OptFlux: an open-source software platform for in silico metabolic engineering. BMC Syst Biol 4:45
142. Wright J, Wagner A (2008) The systems biology research tool: evolvable open-source software. BMC Syst Biol 2:55
143. Urbanczik R (2006) SNA—a toolbox for the stoichiometric analysis of metabolic networks. BMC Bioinform 7:129
144. Kierzek AM (2002) STOCKS: STOChastic kinetic simulations of biochemical systems with Gillespie algorithm. Bioinformatics 18(3):470–481
145. Moraru II, Schaff JC et al (2008) Virtual cell modelling and simulation software environment. IET Syst Biol 2(5):352–362
146. Lee DY, Yun C et al (2006) Webcell: a web-based environment for kinetic modeling and dynamic simulation of cellular networks. Bioinformatics 22(9):1150–1151
147. Schwarz R, Liang C et al (2007) Integrated network reconstruction, visualization and analysis using YANAsquare. BMC Bioinform 8:313
148. Raman K, Chandra N et al (2008) PathwayAnalyser: a systems biology tool for flux analysis of metabolic pathways. Nature Precedings. doi:10.1038/npre.2008.1868.1
149. Alves R, Antunes F, Salvador A (2006) Tools for kinetic modeling of biochemical networks. Nat Biotechnol 24(6):667–672
150. Funahashi A, Matsuoka Y et al (2008) Celldesigner 3.5: a versatile modeling tool for biochemical networks. Proc IEEE 96(8):1254–1265
151. Pieper I, Wechler K et al (2009) Biosimulation of drug metabolism—a yeast based model. Eur J Pharm Sci 36(1):157–170
152. Shoda L, Kreuwel H et al (2010) The Type 1 diabetes physiolab platform: a validated physiologically based mathematical model of pathogenesis in the non-obese diabetic mouse. Clin Exp Immunol 161(2):250–267

Adv Biochem Engin/Biotechnol (2012) 127: 109–132
DOI: 10.1007/10_2011_99

Published Online: 24 March 2011

Metabolic Flux Analysis in Systems Biology of Mammalian Cells

Jens Niklas and Elmar Heinzle

Abstract Reaction rates or metabolic fluxes reflect the integrated phenotype of genome, transcriptome and proteome interactions, including regulation at all levels of the cellular hierarchy. Different methods have been developed in the past to analyse intracellular fluxes. However, compartmentation of mammalian cells, varying utilisation of multiple substrates, reversibility of metabolite uptake and production, unbalanced growth behaviour and adaptation of cells to changing environment during cultivation are just some reasons that make metabolic flux analysis (MFA) in mammalian cell culture more challenging compared to microorganisms. In this article MFA using the metabolite balancing methodology and the advantages and disadvantages of ^{13}C MFA in mammalian cell systems are reviewed. Application examples of MFA in the optimisation of cell culture processes for the production of biopharmaceuticals are presented with a focus on the metabolism of the main industrial workhorse. Another area in which mammalian cell culture plays a key role is in medical and toxicological research. It is shown that MFA can be used to understand pathophysiological mechanisms and can assist in understanding effects of drugs or other compounds on cellular metabolism.

Keywords Biologics · Cell factories · Human cells · ^{13}C isotope labelling · Metabolite balancing · Production · Therapeutic proteins

Contents

J. Niklas · E. Heinzle (✉)
Biochemical Engineering Institute, Saarland University,
Campus A 1.5, 66123 Saarbrücken, Germany
e-mail: e.heinzle@mx.uni-saarland.de

1 Introduction

Systems biology has become an increasingly exciting field in biological science. An increasing number of research institutions and groups are focusing on this field, and various public and private funding agencies and companies are substantially supporting research in this field. The major reasons for this substantial investment are the anticipated increase in understanding the functioning of biological systems that is expected to strongly support the creation of new and improved therapies, drugs as well as biotechnological processes [1, 2]. The enormous booming of this area is heavily driven by the breathtaking progress in molecular biology combined with the development of large-scale experiments whose data collection and analysis only became possible with the new bioinformatic methods and tools. Bioinformatics has made it possible to store relevant data on databases that are quickly accessible by everybody at any time on a global scale. In parallel, techniques of mechanistic modelling are evolving that permit structured development of mathematical models and their solution on a larger scale. It is primarily these models that are essential in the process of conceptual clarification and that allow testing power of prediction.

Mammalian cells are model organisms that are used to help understanding diseases like cancer or neurological diseases [3] and identifying suitable drug targets. Mammalian cells, particularly human cells, are of increasing relevance for testing the metabolism and toxicity of drug candidates [4–6]. Another major application of mammalian cells is in the production of biopharmaceuticals [7], especially of proteins such as antibodies, but also of vaccines and viral carriers for gene therapy [6, 8, 9]. Here a solid systems understanding will assist in improving product quality, e.g. correct human glycosylation and product titers, e.g. by systems-supported media design or model-driven genetic modifications. Engineering of producing cells will be supported by a new discipline, synthetic biology [7], that helps designing new biological systems or elements thereof, e.g. new promoters, switches or sensors [10, 11].

In order to understand and improve production processes or to understand toxic effects, methods are needed that can describe the different phenotypes of the cells under different conditions. Metabolic fluxes or intracellular reaction rates represent an endpoint of a metabolic network reflecting all types of network events including regulation at different levels, i.e. gene, protein and metabolic interactions [12, 13], making it a very powerful method for systems biology research (Fig. 1).

The analysis of metabolic fluxes has been employed extensively in the past to understand, design and optimise a number of cell types and biological processes [12, 14, 15]. Metabolic flux analysis (MFA) provides a quantitative description of

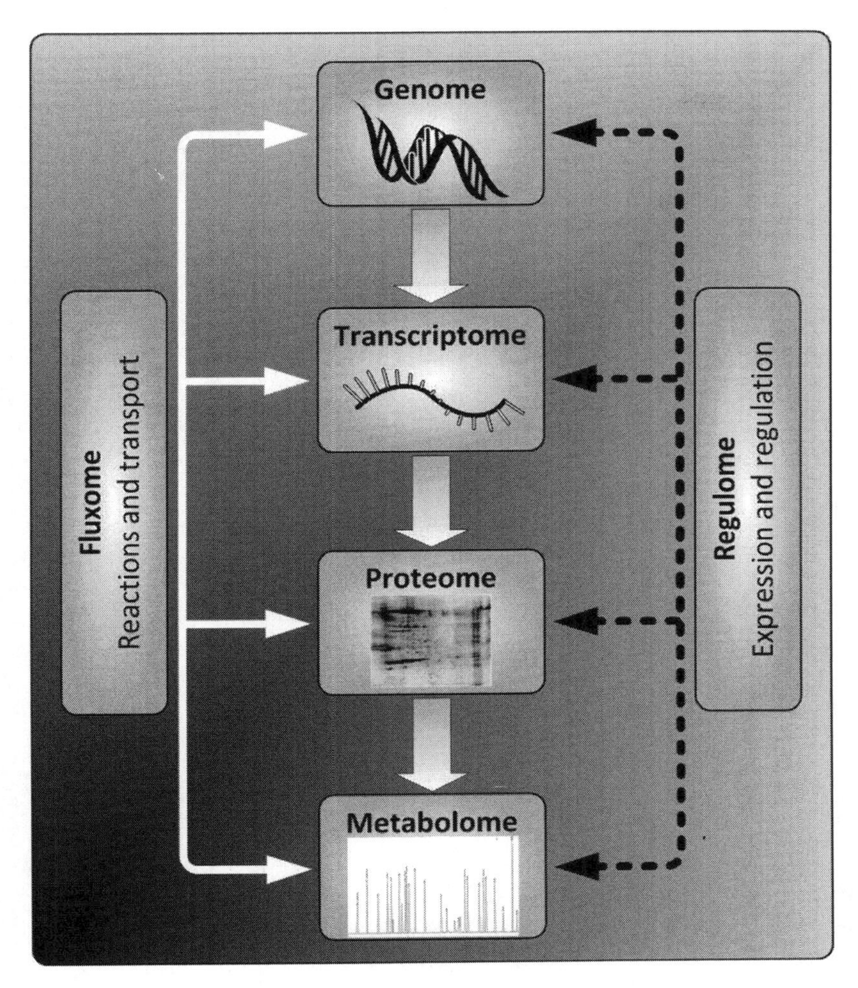

Fig. 1 Interactions on the various levels of the cellular hierarchy

in vivo intracellular reaction rates in metabolic networks describing activities of intracellular enzymes and whole pathways [16]. Fluxes can be quantified either by using metabolite balancing, often also called flux balance analysis (FBA), or by ^{13}C MFA, which involves the use of ^{13}C isotope-labelled substrates [17]. Table 1 presents methodological milestones that contributed to the development and improvement of MFA methods.

Flux analysis using metabolite balancing for microorganisms was described as early as 1978 [18]. It has been and still is the most commonly applied method for the analysis of the metabolism of mammalian cells [5, 15, 19, 20]. The accuracy of flux estimates can be improved compared to pure metabolite balancing by using ^{13}C tracers in metabolic flux studies and later incorporation of the resulting

Table 1 Methodological milestones in MFA method development

Flux estimation method	Year of introduction	Principle	Applications	References
Metabolite balancing	1978	Measure substrate and product conversion rates	Prokaryotes, yeast, Aspergillus, Penicillium, mammalian cells	[18]
^{13}C labelling, atom mapping matrices	1994	Measure fractional labelling of individual metabolite carbon atoms or of average using NMR or GC-MS	Various	[21]
^{13}C labelling, isotopomers, isotopomer mapping matrices	1997	Systematic description of carbon isotopes in any network	Whole network isotopomer flux estimations in prokaryotes and eukaryotes	[22]
Local flux split ratio estimation	1997	Calculate local flux split ratios based on local mass isotopomer balances	Local calculations, flux constraints combined with metabolite balancing	[24, 25]
Link to mass isotopomers	1999	Correction for natural isotopes using matrix approach	Whole network isotopomer models and mass spectrometric analysis	[29]
Cumomers	1999	Explicit solution of large isotopomer networks	Whole isotopomer networks	[30]
Flux screening on a microtiter scale	2003	Miniaturised cultivation methods combined with mass spectrometry	Investigation of mutant libraries	[26, 44, 45]
EMUs-elementary metabolite units	2007	Systematic simplification of isotopomer network for improved estimation	Particularly useful for dynamic isotopomer models	[31]
FRET metabolite nanosensors	2009	FRET sensors detect conformational change of reporter proteins caused by binding of a metabolic ligand, e.g. glucose	Dynamic measurement of individual metabolite concentration in cellular compartments	[32]

labelling information stored in the metabolites into the flux calculation. Different mathematical methods that have been developed in the past have significantly contributed to the applicability of ^{13}C flux analysis. Zupke and Stephanopoulos introduced the concept of atom mapping matrices (AMM) for the modelling of isotope distributions in metabolic networks [21]. A following important advancement was the introduction of isotopomer mapping matrices (IMM) by

Schmidt et al. [22], which allowed using the complete information of the isotopomer distributions of metabolites and the elegant use of matrices for solving complex isotopomer models. Another method is based on local isotopomer balances and allows the estimation of local flux split ratios [23–25]. This method is generally very suitable for high throughput because of its easy calculation. Flux split ratios can be used directly and interpreted biologically, or can serve as additional constraints for FBA. Metabolic flux screening on a miniaturised scale using mass spectrometry was shown to be a promising method for high-throughput analysis of cellular phenotypes [26]. Whereas earlier labelling patterns were mostly analysed by NMR [27, 28], in the late 1990s GC-MS was proposed to be a possible competitive technique, and mass isotopomer distributions obtained from GC-MS measurements can be sufficient for detailed analysis of metabolic fluxes [29]. In 1999 Wiechert et al. provided an elegant procedure for solving isotopomer balances and introduced the concept of cumulative isotopomers (cumomers) [30]. Another significant improvement for flux calculation was presented by Antoniewicz et al. [31]. They introduced an efficient decomposition algorithm that identifies the minimum amount of information needed to simulate the isotopic labelling in a reaction network. This so-called elementary metabolite unit (EMU) framework significantly reduces the computation time needed for flux estimation. Another milestone especially for studying compartmented systems is the introduction of specific fluorescence resonance energy transfer (FRET) techniques [32]. The determination of individual departmental concentrations by using FRET nanosensors that can be combined with ^{13}C flux analysis might allow studying even fluxes between compartments and different cells, providing deeper insights into metabolic compartmentation in eukaryotic cells in future studies.

As can be seen in the literature, MFA was mostly applied in biotechnology to increase the productivity of producing microorganisms [33–35]. In mammalian cells, it is still less established owing to their higher complexity, particularly concerning compartmentation, complex medium requirements and unbalanced growth behaviour. However, there have been a number of interesting studies applying MFA in the past, mainly in the areas of cell culture technology and toxicology/medicine [5, 36–39]. These studies are mostly performed under the assumption of metabolic steady state, which is a prerequisite for stationary MFA. In mammalian cells the question remains of whether a true steady state and balanced growth behaviour can be achieved. Especially in batch cultures, in which the medium composition is changing and the cellular metabolism has to adapt to a changing environment, true metabolic steady state will not be achieved. Mammalian cells change their metabolism as a response to different substrate concentrations, e.g. low and high glucose concentrations [40]. There are different possibilities to deal with this challenge of usually unbalanced growth. In most cases mean fluxes are calculated for the exponential growth phase of the cells, and during this phase metabolic steady state is assumed, which is often a fair assumption [5, 20]. Another possibility is to examine metabolism and flux distributions over time using dynamic approaches [41, 42]. If specific short time effects on the metabolism were to be investigated, e.g. effects of certain compounds, transient ^{13}C flux analysis would be an interesting but still very sophisticated method [37]. For reliable ^{13}C metabolic profiling in mammalian cell culture processes, which usually takes several days, a special medium design could be an option to come close to metabolic and isotopic steady state allowing stationary ^{13}C flux analysis [43]. In this article different methods for MFA in mammalian cells will be reviewed and then application examples will be presented.

2 Theoretical Aspects: Methods for MFA

Different MFA methods are available as discussed before. A typical workflow of a stationary ^{13}C MFA is depicted in Fig. 2. If metabolite balancing is applied without the use of any tracers, only the measurement of labelling patterns would be excluded from the workflow. Another main difference between metabolite balancing methodology and ^{13}C MFA is computation. Flux calculations can be straight forward using metabolite balancing as described in Sect. 2.1. In ^{13}C MFA (Sect. 2.2) metabolite balancing is extended by carbon isotopomer balances, resulting in a nonlinear least squares problem. This can be solved for example by using efficient numerical optimisation techniques [46, 47].

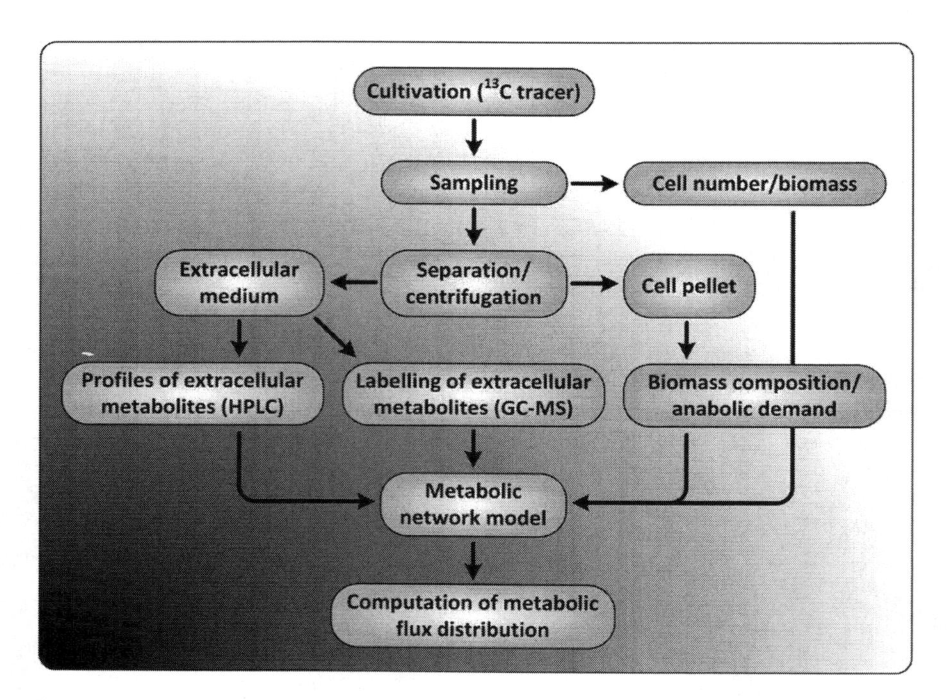

Fig. 2 Exemplary workflow of an experiment for metabolic flux analysis

2.1 Stoichiometric Models and Metabolite Balancing in Mammalian Cells

The metabolite balancing or flux balancing technique is the MFA method that has been applied most often for the analysis of animal cell metabolism. The stoichiometric models used for flux balancing can also be applied for in silico prediction of network characteristics (e.g. maximal yields, optimal pathways, minimum substrate requirements) [48–50] or prediction of optimal genetic modifications using different algorithms [51–55]. The importance of these targeted optimisation approaches is rapidly increasing, which is also caused by an increasing availability of genomic information as well as genome-scale models of different mammalian species [56–58]. The general metabolite balancing methodology is depicted in Fig. 3.

The first step is to set up a network that is describing the part of the metabolism that should be investigated. Metabolic network models of the central metabolism of mammalian cells have been described and applied in a number of studies [5, 15, 20, 38, 49]. As an example a model of the human central metabolism is presented in Fig. 4. An important database that can be used to set up metabolic networks is the Kyoto Encyclopedia of Genes and Genomes (KEGG) pathway database (http://www.kegg.com). If a metabolic network consists of N fluxes and M internal metabolites, it has $F = N - M$ degrees of freedom, meaning that F fluxes have to be measured to determine all remaining fluxes [59]. The calculation of metabolic fluxes in an overdetermined network (more measurements than necessary) results in a set of calculated fluxes that represents a least squares solution. In this case it can also be checked if measurements are consistent, meaning that the balance around each metabolite is zero. The network is underdetermined if not enough measurements are available, which can also be seen by calculating the rank of the stoichiometric matrix. If the rank is lower than the number of internal metabolites, the network is underdetermined. In this case the network has to be simplified or specific fluxes must be assumed a priori. If insufficient fluxes are measured, this can sometimes be easily solved by measuring extra rates. Other options exist to calculate fluxes in underdetermined parts of the network, which would extend the metabolite balancing method. Thermodynamic constraints can be used, indicating that certain reactions do not take place or can only proceed in one direction [60]. Expression data or measurements of in vitro enzyme activities can be used to exclude specific reactions in the network [49, 61–63]. Another possibility is to use specific objective functions such as for example maximising energy production to find flux distributions that optimise the applied objective function [64, 65]. The probably best possibility is the use of labelled substrates (^{13}C tracers) and analysis of resulting labelling patterns in metabolites providing additional information about cellular metabolism. The labelling of specific metabolites can be used for example to get information about fluxes at branch points like the glycolysis/pentose phosphate pathway (PPP) split [66]. ^{13}C MFA is further explained in the next paragraph.

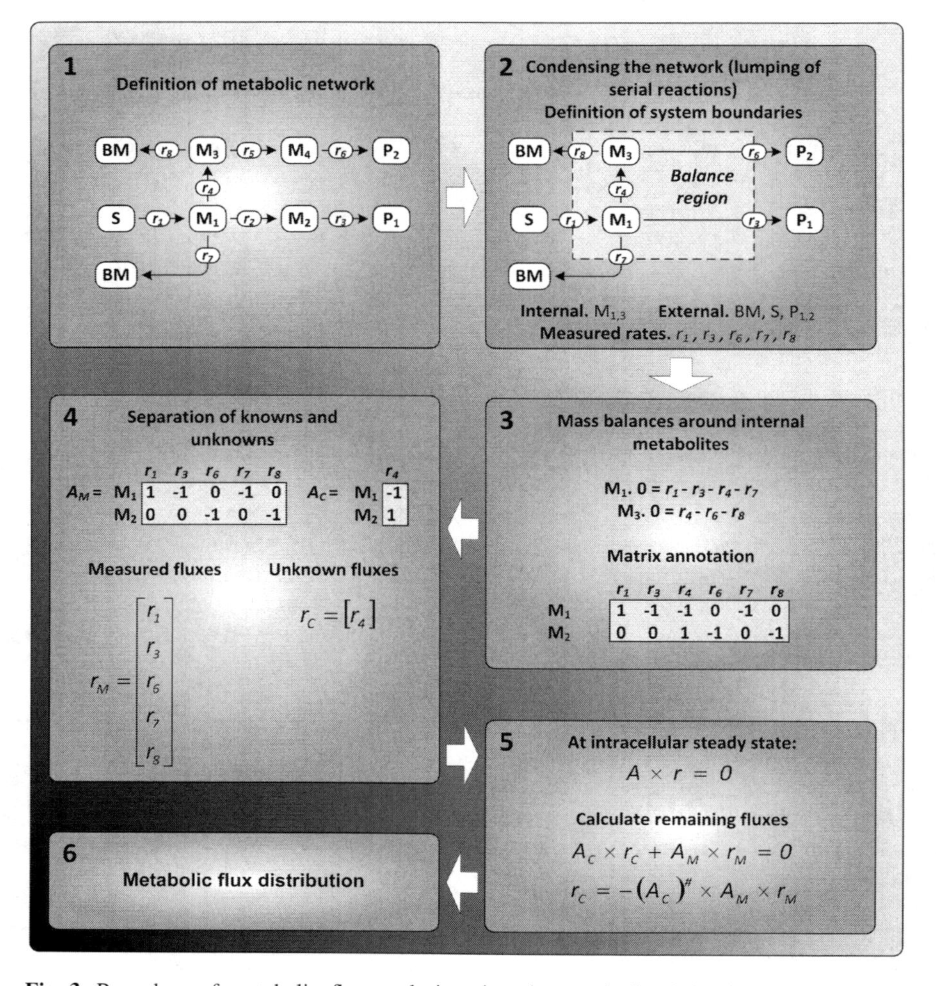

Fig. 3 Procedure of metabolic flux analysis using the metabolite balancing technique. *BM* biomass, *M* metabolite, *S* substrate, *P* product, *r* reaction rates, *A* stoichiometric matrix, $(A)^{\#}$ pseudo inverse of the matrix

The metabolic network model in Fig. 4 for flux balancing in human cells and general aspects of modelling metabolic networks in mammalian cells will be explained in this section. The metabolic network can be divided into the following parts:

Central energy metabolism The main pathways of the energy metabolism are represented, i.e. glycolysis, oxidative decarboxylation, TCA-cycle, electron transport chain and oxidative phosphorylation. Since it is not known for some reactions in the model whether NADH or NADPH take part, and due to possible activity of nicotinamide nucleotide transhydrogenases, NADH and NADPH were lumped together. The excess of NAD(P)H and $FADH_2$ considering their

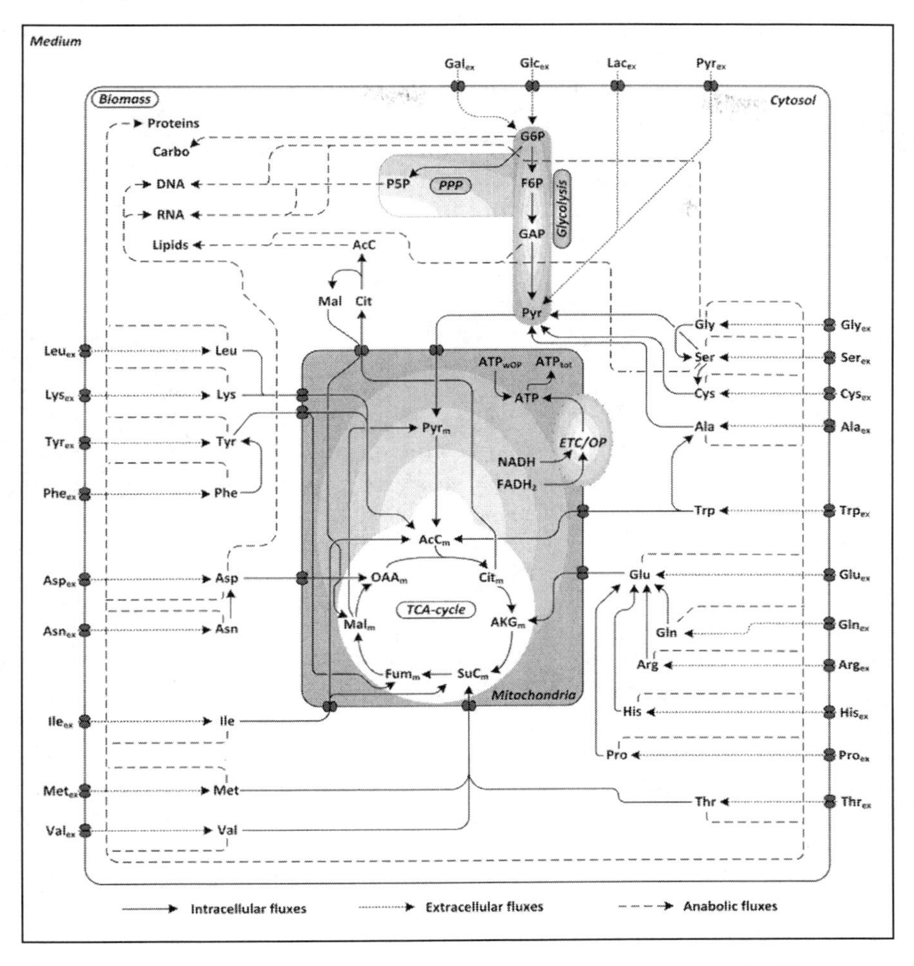

Fig. 4 Exemplary stoichiometric metabolic network model of a human cell. *ETC* electron transport chain, *OP* oxidative phosphorylation, *PPP* pentose phosphate pathway, *TCA* tricarboxylic acid, *AcC* acetyl coenzyme A, *AKG* α-ketoglutarate, *ATP* adenosine triphosphate, ATP_{tot} total ATP, ATP_{wOP} ATP without oxidative phosphorylation, *Carbo* carbohydrates, *Cit* citrate, *F6P* fructose 6-phosphate, $FADH_2$ flavin adenine dinucleotide, *Fum* fumarate, *G6P* glucose 6-phosphate, *GAP* glyceraldehyde 3-phosphate, *Gal* galactose, *Glc* glucose, *Lac* lactate, *Mal* malate, *NADH* nicotinamide adenine dinucleotide, *OAA* oxaloacetate, *P5P* pentose 5-phosphate, *Pyr* pyruvate, *SuC* succinyl coenzyme A, standard abbreviations for amino acids. Indices: *m* mitochondrial, *ex* extracellular

consumption and production in all reactions permits estimating the total ATP excess. However, the P/O ratio, which is the amount of ATP formed per NADH oxidised, is usually not exactly known. In the literature different P/O values were assumed or estimated [49, 67–69]. The calculated ATP excess in the presented model example (Fig. 4) represents an estimate of the amount of ATP that is needed in the cell, e.g. for maintenance and transport reactions, but also for so-called substrate or futile cycles [70].

Pentose phosphate pathway The PPP consists of an oxidative and a nonoxidative branch. The split between glycolysis and PPP and the reversible reactions of the non-oxidative part cannot be resolved by metabolite balancing alone. Therefore, PPP is usually assumed to be responsible only for nucleic acid synthesis in pure metabolite balancing studies, and the nonoxidative part is neglected. However, there are possibilities for obtaining some information concerning the glycolysis/PPP split by using additional ^{13}C tracer experiments [66], which can be included in the metabolic flux calculation.

Amino acid metabolism The metabolism of proteinogenic amino acids is usually modelled by selected degradation pathways. Where several pathways are possible, the most probable and suitable pathway can be chosen, or additional experiments (e.g. enzyme activity measurements) can be performed to evaluate this. Since metabolite balancing can only be used to calculate net fluxes, there are no data concerning reversibility as for example in the synthesis and degradation of non-essential amino acids such as alanine and glutamate. The degradation flux of essential amino acids should usually be close to zero or higher, but never below zero. Values below zero would indicate that there are errors in the metabolite measurement or the applied anabolic demand.

Further reactions The reactions catalysed by the enzymes malic enzyme (cytosolic and mitochondrial), phosphoenolpyruvate carboxykinase and pyruvate carboxylase represent parallel or reversible pathways and also cannot be distinguished by pure metabolite balancing. Again this can be solved by assuming activities for some of these enzymes or by taking data from other experiments or lumping all these reactions together into one reaction representing the sum of all these fluxes converting malate/oxaloacetate to pyruvate, as was done in the model in Fig. 4.

Synthesis of biomass Fluxes to biomass can be represented as five fluxes to the major macromolecules of the cell, namely proteins, carbohydrates, DNA, RNA and lipids. Hereby the lipid fraction also contains the cholesterol part of the cells. These anabolic fluxes are calculated using the specific anabolic demand of the cells, which is derived from the biomass composition of the cells. In most flux studies the biomass composition is assumed constant. The macromolecular composition (Table 2) and amino acid composition (Table 3) of Hep G2 cells are shown as an example. This was applied in a metabolite balancing study in which

Table 2 Macromolecular composition in Hep G2 cells [5]

	Macromolecule fraction (%)
DNA	3.9
RNA	2.4
Carbohydrates	3.4
Lipids	18.0
Proteins	61.4
Rest/ash	10.9

Table 3 Amino acid composition of total cellular protein in Hep G2 cells [5]

	Amino acid fraction (%)
Alanine	8.5
Arginine	4.7
Aspartate/asparagine	10.6
Cysteine	2.6
Glutamate/glutamine	12.3
Glycine	12.7
Histidine	1.4
Isoleucine	2.5
Leucine	7.2
Lysine	12
Methionine	1.3
Phenylalanine	2.8
Proline	4.6
Serine	6.6
Threonine	3.7
Tryptophan	0.8
Tyrosine	2.6
Valine	3.3

constant composition was assumed [5]. However, in detailed studies concerning cellular biomass dynamics, it was shown that this composition and also the total biomass of the cells (e.g. dry weight) can vary during cultivation or at different growth conditions [71, 72]. The methods that are usually used to determine cellular macromolecules are, however, time-consuming and require relatively large quantities of sample material, making it usually not possible to determine the biomass composition in every flux analysis experiment, which would, strictly speaking, be required.

2.2 ^{13}C *Metabolic Flux Analysis*

In mammalian cells, relevant information can be obtained already by the metabolite-balancing methodology since the number of measurable uptake and production fluxes of metabolites is large. However, there are several cases in which the balancing technique is insufficient. Particularly certain circular pathways, reversible reactions and alternative pathways cannot be resolved (Fig. 5). Most important, underdetermined parts in the metabolic network of mammalian cells are typically the PPP split, the anaplerotic/gluconeogenic fluxes around pyruvate/phosphoenolpyruvate/malate/oxaloacetate and reversibility of uptake and production of substrate metabolites. Specific metabolic compartmentations, such as for example different intracellular pools of metabolites as suggested for pyruvate [73], are other parts that cannot be resolved just by balancing.

In some situations it might be possible to use well-defined constraints to solve some underdetermined parts. Mass balances of cofactors, e.g. ATP and NAD(P)H,

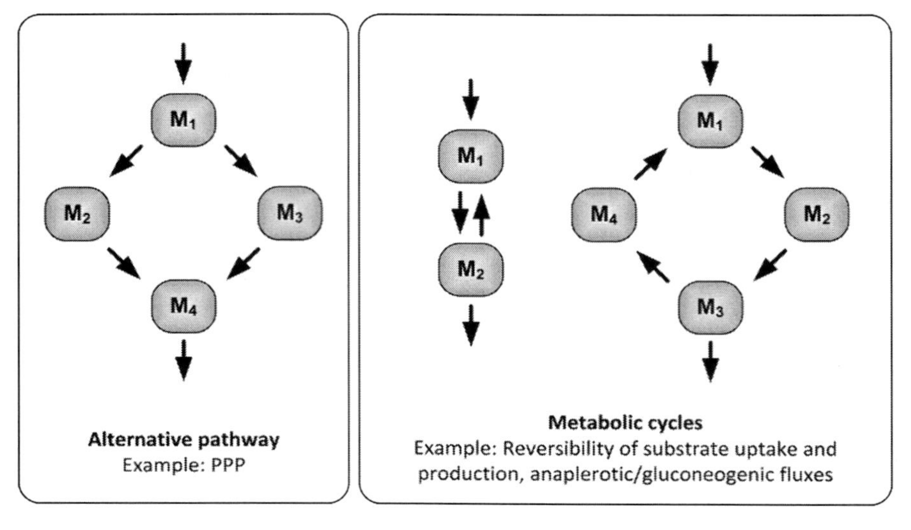

Fig. 5 Cases in which the metabolite-balancing technique is limited and examples in the metabolism

irreversibility of reactions or specific objective functions have been proposed and reviewed as additional constraints [74]. However, balancing of the energy metabolites [ATP, NAD(P)H] does not seem generally applicable since this would require knowing all energy-producing and -consuming reactions as well as all conversion reactions between the energy metabolites [75]. In addition the P/O ratio can vary and can usually not be determined precisely [76], substrate or futile cycles might impair results, e.g. in the anaplerosis [77], and for some reactions it is just not known which co-metabolite, NADH or NADPH is used. For example malic enzyme and isocitrate dehydrogenase enzymes can utilise NADH or NADPH (http://www.kegg.com). In case of PPP split, it would be possible to get some estimates about its activity by balancing NADPH. However, transhydrogenation reactions can occur in the cells, transferring reducing equivalents between NADPH and NADH, which would falsify PPP estimates. In a study comparing results obtained by metabolite balancing with those from ^{13}C MFA, discrepancies were found concerning PPP split [78].

All the shortcomings and limitations of the metabolite balancing method can be overcome by getting more information through application of isotopically labelled ^{13}C tracers. These tracers are ^{13}C labelled substrates that are taken up by the cell, and the labelled carbon atoms are distributed through the cellular metabolism in a clearly defined way (Fig. 6). Depending on the activity of enzymes and metabolic pathways, this will result in specific labelling patterns in metabolites. These chemically identical compounds with different isotope composition are referred to as isotopomers. As depicted in Fig. 6, labelling in extracellular and intracellular metabolites can be measured, but also the labelling in macromolecule building blocks, e.g. amino acids in proteins provide equivalent information. Extracellular metabolites can be easily measured since these are directly present in the medium,

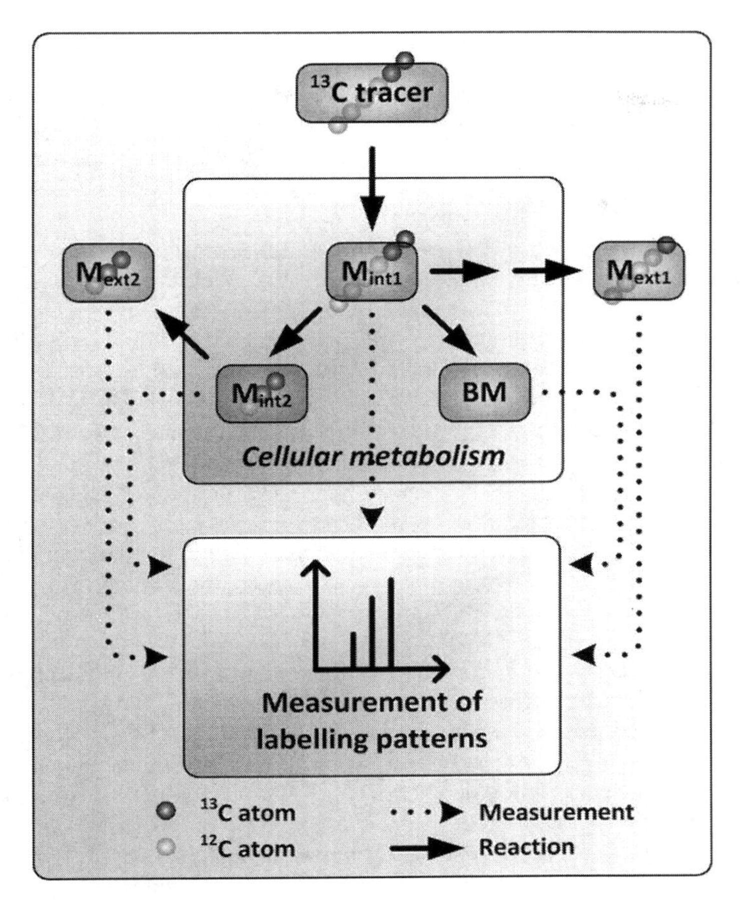

Fig. 6 ^{13}C tracer experiment. The labelling of the tracer is distributed through the metabolism, resulting in specifically labelled intracellular metabolites (M_{int}), extracellular metabolites (M_{ext}) and biomass (*BM*) components. Measurement of labelling patterns can be performed directly in intracellular or extracellular metabolites, but also in macromolecule building blocks of biomass constituents

usually at high abundance. However, measurement of intracellular metabolites requires reliable quenching of the metabolism and appropriate extraction methods. This is fairly established for adherent cells [79, 80], but still much more complicated for suspension cells [81]. Measurement of the labelling in monomers of macromolecules, as is often done in studies on prokaryotes [33], is usually not suitable for flux analysis in mammalian cells. This is because mammalian cells generally have much lower growth rates than microorganisms and therefore slow macromolecule turnover and slow labelling incorporation.

The measurement of labelling patterns in metabolites can be performed by nuclear magnetic resonance (NMR) measurements, gas chromatography mass spectrometry (GC-MS), liquid chromatography mass spectrometry (LC-MS), [16, 25, 29, 82–87], matrix-assisted laser desorption ionisation-time of flight mass spectrometry (MALDI-ToF MS) [88, 89], capillary electrophoresis MS [90] or membrane inlet MS [91]. Recently it was demonstrated that gas chromatography-combustion-isotope ratio mass spectrometry (GC–C–IRMS) is another interesting method for labelling quantification in ^{13}C MFA with a low labelling degree of tracer substrate, which is interesting for performing MFA in larger scale cultivations like industrial pilot scale fermentations [92, 93].

Compared to NMR, MS seems to be more attractive, which is mainly due to higher sensitivity and rapid data accumulation [94]. Certain potential problems of MS that would impair flux analysis, like isotope effects or naturally occurring isotopes particularly in atoms other than carbon, can be solved efficiently using specific correction methods [29, 92, 95–99].

For ^{13}C MFA the carbon atom transitions in the metabolic network of the cell have to be modelled. The concept of AMM, which is a systematic formulation of atom transfers [21], was further expanded by the IMM concept [22]. This allowed to calculate mass and NMR spectra directly from isotopomer abundance [100]. The introduction of cumomers [30] and later EMUs [31] improved computation of fluxes. The computational part in ^{13}C MFA can be performed by using available software packages [101–103]. Further description of ^{13}C MFA can also be found in several review articles [13, 16, 29, 45, 75, 100].

^{13}C MFA can be further divided into stationary and dynamic approaches, which both have been applied successfully in mammalian cells [80, 104, 105]. Metabolic steady state is a prerequisite for stationary ^{13}C MFA. This is still a challenge, especially for suspension cells used in industrial production. Appropriate medium design can be an option also enabling detailed ^{13}C MFA in suspension cells during exponential growth [43]. For adherent cells the problem of instationarity can be nicely overcome using transient ^{13}C MFA as demonstrated on hepatic cells [37, 79, 80].

The tracers that are mostly applied in ^{13}C MFA on mammalian cells are different glucose and glutamine tracers since these metabolites are the main substrates of mammalian cells [106]. Depending on the question of the study and the applied metabolic network structure, different tracers or combinations of tracers will be best suitable [105].

3 Application of MFA in Systems Biology of Mammalian Cells

3.1 Application of MFA in Optimisation of Cell Culture Processes

Mammalian cells are extensively applied for the production of vaccines [9] and therapeutic proteins requiring specific post-translational glycosylation [8, 107].

The number of biopharmaceuticals available for treatment of severe diseases as well as the quantity produced is steadily growing [108]. Therapeutic proteins are mainly expressed in Chinese hamster ovary (CHO) cells, but also other cell lines are commonly employed, such as murine myeloma, hybridoma, baby hamster kidney (BHK) or human embryo kidney cells (HEK-293) [8, 109]. Newly engineered human cell lines also represent very promising production systems, for example the cell lines AGE1.HN [42] and PER.C6 [110]. Vaccine production is conventionally carried out in embryonated chicken eggs. However, especially in the last decade several cell culture-derived vaccines have been established [9].

Much effort has been made to optimise cell culture processes to increase productivity and product quality. This includes on the one hand optimisation of the cultivation and on the other hand targeted engineering of the cell. Different cellular pathways that are associated with superior characteristics concerning cell growth and production were engineered. This includes central metabolism, protein synthesis and secretion, protein glycosylation, post-translational modifications, cell cycle control and apoptosis [107, 111, 112].

Optimisation of the metabolism of the producer cell is mandatory for different reasons. On the one hand efficient energy metabolism is important. Particularly the production of recombinant proteins requires a great deal of energy. On the other hand final product titres are directly dependent on integral viable cell density and lifespan of the culture, which can be increased when the substrate usage of the cell is very efficient. In an optimum scenario this would mean that the cell does not accumulate toxic waste products, which would decrease the lifespan of the culture, and substrates are taken up just to fulfil the cellular demand.

Analysis of the metabolism and metabolic flux studies has significantly contributed in the past to understanding and optimising the metabolism of mammalian cells. In this section we will review metabolic flux studies in hybridoma and myeloma cells, CHO cells and cell lines that are applied for vaccine production.

3.1.1 MFA in Hybridoma and Myeloma Cells

Hybridoma and myeloma cells are widely applied for the production of monoclonal antibodies [109]. In many studies MFA was used to understand the metabolism of these cells. Savinell and Palsson [64, 65] applied linear optimisation theory to understand the influences of fluxes on overall cell behaviour and to analyse limitations. They concluded that neither antibody production nor maintenance demand for ATP limited cell growth. Medium design represents one of the most important issues in optimising cell culture processes, which is also reflected by many metabolic flux studies in this area. Xie and Wang [113, 114] presented a balancing approach to design culture media for fed batch cultures that integrated substrates, products, pH, osmolarity and cell growth. They also estimated stoichiometric ATP production in batch and fed batch cultures of hybridoma cells [69]. Another interesting study focussed on regulation of fluxes in the central metabolism of myeloma cells [63]. By determining fluxes and enzyme activities in the central metabolism, the regulation of metabolic fluxes could be shown to occur mainly through modulation of enzyme activity. Determination of metabolic fluxes for multiple steady states in hybridoma continuous culture indicated that the performance could be improved by inducing specific cellular metabolic shifts, leading to favourable flux distribution [115]. Europa et al. [116] performed fed batch cultivations that were then switched to continuous mode. This approach enabled reaching a more desirable steady state with higher concentrations of cells and product. Additionally from analyses of hybridoma cells at different physiological states it was reported that the amino acid metabolism is very important for reducing lactate production [117]. In an earlier metabolite balancing study using additional constraints to resolve underdetermined parts in the metabolic network, Bonarius et al. [19] found that around 90% of the glucose was channelled through PPP, which was very surprising. In a following study using $^{13}CO_2$ mass spectrometry in combination with ^{13}C lactate NMR spectroscopy and metabolite balancing, it was found that just 20% is channelled through PPP [118]. This shows that different methods can yield very different results. As mentioned before, cofactor balance constraints must be used very carefully. Especially for estimation of PPP, constraints based on ^{13}C labelling might be much more realistic since the labelling measurement is usually very accurate and carbon transition in the reactions is exactly determined. Another interesting aspect that was analysed by Bonarius et al. [119] using MFA was the cellular response to oxidative and reductive stress. They reported that particularly dehydrogenase reactions producing NAD(P)H were decreased under oxygen limitation. In a recent study it was reported that antibody production in hybridoma cells could be increased by enhancing specific fluxes through the addition of specific metabolites to the medium [120]. Fluxes between malate and pyruvate were increased by the addition of the intermediates pyruvate, malate and citrate, resulting in increased ATP and antibody production. This is a very nice example showing how metabolic flux data can be used to improve the metabolic phenotype in a cell culture process.

Another important step is the construction of mathematical models that can be used to predict growth, metabolism and product formation during the cultivation. Dorka et al. [121] utilised an approach based on MFA to model batch and fed batch cultures of hybridoma cells. Genome-scale modelling and in silico simulations for fed batch cultures of hybridoma cells were recently carried out, suggesting that in the future the applied methodology might serve as a valuable tool for targeted optimisation [122].

3.1.2 MFA in CHO Cells

CHO cells [123] represent the main workhorse for the industrial production of biopharmaceuticals [109]. In several studies the metabolism of these cells under different conditions was analysed. CHO cells are often cultured in media supplemented with specific hydrolysates that contain many peptides, which makes MFA more complicated. Nyberg et al. [124] reported that these potential substrates must be balanced for accurate metabolic flux estimates. In another study, metabolic effects on the glycosylation of recombinant protein were reported. Particularly glutamine limitation seemed to influence glycosylation remarkably [125]. Altamirano et al. [126–128] published several results on medium design and favourable fed-batch strategies for CHO. MFA was performed particularly to understand lactate consumption in CHO cells grown on galactose. It was found that lactate was not used as a fuel in the TCA cycle [126]. A very important aspect in mammalian cell culture processes is the optimisation of the bioreactor operation. Goudar et al. [129] presented an approach for quasi real-time estimation of metabolic fluxes. Cellular physiology and metabolism can be monitored by combining on-line and off-line data to calculate metabolic fluxes. This methodology can help in optimising the cultivation process. Recently the same group analysed metabolic fluxes of CHO cells in perfusion culture by applying metabolite balancing and 2D-NMR spectroscopy [104]. Flux data obtained by metabolite balancing were in this case in good agreement with flux information from 2D-NMR spectroscopy. Recently a study was published in which MFA was performed for the late non-growth phase in CHO cultivation [130]. In this case a combination of metabolite balancing and isotopomer analysis was used. The most surprising finding in this study was that almost all of the consumed glucose was channelled through PPP. This result is in contrast with several publications [39, 66, 80, 104, 105, 131] that determined different flux distributions in the growth phase of CHO or other cells.

3.1.3 MFA in Cell Lines for Production of Vaccines and Viral Vectors

Another promising and important application of mammalian cell culture is the production of vaccines [9, 132]. A number of cell lines were identified to be suitable for high-yield vaccine manufacturing [133], such as for example madin darby canine kidney (MDCK) cells [134], HEK-293 cells [135] or specifically engineered cell lines such as PER.C6 or AGE1.CR [110, 136]. Optimisation of cell culture processes for vaccine production is still mainly done by trial and error. Detailed metabolic studies might help substantially to understand cell culture-based vaccine production [137] and enable targeted optimisation.

Wahl et al. investigated an influenza vaccine production process in MDCK cells using a segregated growth model for distinct growth phases in the batch process. Comparison of observed metabolic fluxes with theoretical minimum requirements revealed large optimisation potential for this process [49]. Flux analysis of MDCK in glutamine-containing media and media in which glutamine was replaced by pyruvate was presented in another publication [20]. Ammonia and lactate release were remarkably reduced in a high-pyruvate medium without further dramatic changes in the central metabolism.

Henry and co-workers [138] showed that MFA can provide a basis to develop a feeding strategy for perfusion cultures of HEK-293 cells for production of adenovirus vectors. Martinez et al. [139] compared metabolic states of HEK-293 cells during growth and adenovirus production to optimise media according to the cellular demand. Higher cell densities and increased adenovirus production were achieved.

3.2 MFA in Medical Research

MFA was mostly applied in the past by bioengineers to understand phenotypes of producer cells. But another very important field in which MFA can contribute substantially is the medical sector.

Defects in mitochondrial function contribute to many physiological diseases. Ramakrishna et al. [140] stated that FBA of mitochondrial energy metabolism might be a useful methodology to characterise the pathophysiology of mitochondrial diseases. The strength of metabolic flux data as mentioned in the beginning of this review and shown in Fig. 1 is the integration of all interactions at different levels of the cellular hierarchy. Therefore, specific flux patterns that reflect a certain physiological response might be very nice indicators for specific diseases or genetic defects.

Lee et al. [141] proposed that MFA could be a very useful tool for tissue engineering. By applying MFA, it is possible to obtain a very comprehensive view of the metabolic state, and flux estimates under different conditions can be used to monitor and optimise tissue function. Forbes and co-workers [36, 142] described an interesting method that uses isotopomer path tracing to quantify fluxes in metabolic models containing reversible reactions and applied MFA to analyse the effects of estradiol on breast cancer cells. Metabolic fluxes were calculated from extracellular fluxes and isotope enrichment data generated by NMR. They observed that breast cancer cells are dependent on PPP and glutamine consumption for estradiol-stimulated biosynthesis and concluded that these pathways might be possible targets for estrogen-independent breast cancer therapy.

Brain function and especially physiological and pathophysiological regulation of neural metabolism were investigated in several metabolic studies. Zwingmann et al. [143] investigated glial metabolism using ^{13}C-NMR. They concluded that the observed metabolic flexibility of astrocytes might buffer the brain tissue against extracellular cytotoxic stimuli and metabolic impairments. In another study the coupling between metabolic pathways of astrocytes and neurons was modelled and investigated by FBA [144]. By using the reconstructed model, effects of hypoxia could be fairly well predicted. This shows nicely how stoichiometric models can be used in medical metabolic flux modelling. Teixeira et al. [145] investigated the metabolism of astrocytes by combining ^{13}C-NMR spectroscopy and MFA. In a following study the method was applied to analyse metabolic alterations induced by ischaemia in astrocytes [146].

3.3 MFA in Toxicology

The analysis of effects of drugs and chemicals on cellular metabolism is another very promising application of MFA and highly relevant for toxicological research. Toxicity of drugs is one of the leading causes of attrition at all stages of the drug development process [147] and is mainly detected late in the pipeline where also most of the costs are incurred [148]. Identification of toxicity early in the drug development process would save much money. Metabonomics is a system approach for studying in vivo metabolic profiles and has emerged in the last decade as a very powerful technique for studying drug toxicity, disease processes and gene function [149–151]. MFA, which provides potentially more information than metabolic profiling, is however still not routinely used in toxicity studies, which might be mainly due to its presently low throughput. Some studies have focussed on developing and adapting MFA methods and setups for high-throughput screening [23, 26, 45, 152–154]. Balcarcel et al. [154] presented a method called High-Throughput Metabolic Screening that can be used for faster screening of the overall activity of metabolic pathways in mammalian cells. In a recent study it was shown that MFA can be applied in a high-throughput setup to analyse subtoxic drug effects [5]. Several changes in the metabolism of Hep G2 cells could be detected upon exposure to subtoxic drug levels. In the future it might be possible to use MFA in a high-throughput format to detect specific metabolic signatures or flux patterns that are associated with specific toxicity mechanisms. Other studies analysed effects of different compounds on cellular metabolism without focussing on high-throughput application of the applied methods. Srivastava and co-workers [38] applied MFA to identify the toxicity mechanism of free fatty acids and metabolic changes in Hep G2 cells. They observed that free fatty acid toxicity is associated with the limitation of cysteine import causing reduced glutathione synthesis. Another very promising MFA method was implemented by Maier et al. [37] to quantify statin effects on hepatic cholesterol synthesis. Transient ^{13}C flux analysis was applied to study effects of atorvastatin at a therapeutic concentration.

Summarising this section, it can be seen that there have been some interesting applications of MFA to identify toxicity mechanisms and metabolic effects of compounds. However, these methods must be applicable in high-throughput format to be attractive for larger scale toxicity screening. Additionally the analysis has to be more detailed, calling for improved analytical methods. Specific flux patterns and signatures for different toxicity mechanisms must be identified and clearly defined in the future, which would lead to a better understanding of toxicity at the metabolome level; then it might be possible to elucidate possible side effects of compounds early in the drug development process.

4 Conclusion and Future Perspectives

MFA is a very important method for understanding the metabolism of mammalian cells under various conditions. The acquired knowledge can be used to optimise cell culture media and cellular phenotypes, to define favourable feeding strategies as well as to understand mechanisms of toxicity and diseases. In suspension cells that are employed in industrial production, mainly MFA using metabolite balancing was applied. ^{13}C MFA cannot be directly applied in industrially relevant processes since metabolic steady state is not reached. As presented in some studies, continuous cultures are an option to enable detailed stationary ^{13}C flux studies. Dynamic methods are very promising tools to describe the dynamic and adaptive behaviour of the cells during batch and fed-batch processes, but they are still fairly complex and laborious. Therefore they are not yet widely applied. ^{13}C MFA is relatively established in tissues and adherent cells where the flux experiment can be performed at a short time scale and metabolic and isotopic steady state may be reached very fast. Transient ^{13}C flux analysis represents a very interesting method that may in many cases solve the problem of changing metabolism during cultivation of mammalian cells. The application examples presented in this review indicate that MFA can contribute significantly in many areas in which the metabolism of mammalian cells is of interest. However, MFA method development has to be intensified in the future to enable broader application in mammalian cell culture and to permit robust and realistic studies. Compartmentation of the metabolism is an issue that is often only rudimentarily considered in metabolic flux studies but is very important, especially for ^{13}C MFA.

Acknowledgments We thank Malina Orsini for valuable help.

References

1. Alberghina L, Westerhoff HV (eds) (2005) Systems biology—definitions and perspectives. Springer, Heidelberg
2. Choi S (2007) Introduction to systems biology. Humana, Totowa
3. Villoslada P, Steinman L, Baranzini SE (2009) Ann Neurol 65:124–139
4. Beckers S, Noor F, Muller-Vieira U, Mayer M, Strigun A, Heinzle E (2010) Toxicol In Vitro 24:686–694
5. Niklas J, Noor F, Heinzle E (2009) Toxicol Appl Pharmacol 240:327–336
6. Noor F, Niklas J, Muller-Vieira U, Heinzle E (2009) Toxicol Appl Pharmacol 237:221–231
7. O'Callaghan PM, James DC (2008) Brief Funct Genomic Proteomic 7:95–110
8. Wurm FM (2004) Nat Biotechnol 22:1393–1398
9. Genzel Y, Reichl U (2009) Expert Rev Vaccines 8:1681–1692
10. Weber W, Fussenegger M (2007) Curr Opin Biotechnol 18:399–410
11. Koide T, Pang WL, Baliga NS (2009) Nat Rev Microbiol 7:297–305
12. Niklas J, Schneider K, Heinzle E (2010) Curr Opin Biotechnol 21:63–69
13. Sauer U (2006) Mol Syst Biol 2:62
14. Boghigian BA, Seth G, Kiss R, Pfeifer BA (2010) Metab Eng 12:81–95

15. Quek LE, Dietmair S, Kromer JO, Nielsen LK (2010) Metab Eng 12:161–171
16. Wittmann C (2007) Microb Cell Fact 6:6
17. Wiechert W, de Graaf AA (1996) Adv Biochem Eng Biotechnol 54:109–154
18. Aiba S, Matsuoka M (1978) Eur J Appl Microbiol Biotechnol 5:247–261
19. Bonarius HP, Hatzimanikatis V, Meesters KP, de Gooijer CD, Schmid G, Tramper J (1996) Biotechnol Bioeng 50:299–318
20. Sidorenko Y, Wahl A, Dauner M, Genzel Y, Reichl U (2008) Biotechnol Prog 24:311–320
21. Zupke C, Stephanopoulos G (1994) Biotechnol Prog 10:489–498
22. Schmidt K, Carlsen M, Nielsen J, Villadsen J (1997) Biotechnol Bioeng 55:831–840
23. Fischer E, Zamboni N, Sauer U (2004) Anal Biochem 325:308–316
24. Nanchen A, Fuhrer T, Sauer U (2007) Methods Mol Biol 358:177–197
25. Sauer U, Hatzimanikatis V, Bailey JE, Hochuli M, Szyperski T, Wuthrich K (1997) Nat Biotechnol 15:448–452
26. Velagapudi VR, Wittmann C, Schneider K, Heinzle E (2007) J Biotechnol 132:395–404
27. Sonntag K, Eggeling L, De Graaf AA, Sahm H (1993) Eur J Biochem 213:1325–1331
28. Zupke C, Stephanopoulos G (1995) Biotechnol Bioeng 45:292–303
29. Wittmann C, Heinzle E (1999) Biotechnol Bioeng 62:739–750
30. Wiechert W, Mollney M, Isermann N, Wurzel M, de Graaf AA (1999) Biotechnol Bioeng 66:69–85
31. Antoniewicz MR, Kelleher JK, Stephanopoulos G (2007) Metab Eng 9:68–86
32. Niittylae T, Chaudhuri B, Sauer U, Frommer WB (2009) Methods Mol Biol 553:355–372
33. Becker J, Klopprogge C, Zelder O, Heinzle E, Wittmann C (2005) Appl Environ Microbiol 71:8587–8596
34. Kiefer P, Heinzle E, Zelder O, Wittmann C (2004) Appl Environ Microbiol 70:229–239
35. Wittmann C, Heinzle E (2002) Appl Environ Microbiol 68:5843–5859
36. Forbes NS, Meadows AL, Clark DS, Blanch HW (2006) Metab Eng 8:639–652
37. Maier K, Hofmann U, Bauer A, Niebel A, Vacun G, Reuss M, Mauch K (2009) Metab Eng 11:292–309
38. Srivastava S, Chan C (2008) Biotechnol Bioeng 99:399–410
39. Vo TD, Palsson BO (2006) Biotechnol Bioeng 95:972–983
40. Sanfeliu A, Paredes C, Cairo JJ, Godia F (1997) Enzyme Microb Technol 21:421–428
41. Llaneras F, Pico J (2007) BMC Bioinformatics 8:421
42. Niklas J, Schräder E, Sandig V, Noll T, Heinzle E (2011) Bioprocess Biosyst Eng. doi: 10.1007/s00449-010-0502-y
43. Deshpande R, Yang TH, Heinzle E (2009) Biotechnol J 4:247–263
44. Blank LM, Lehmbeck F, Sauer U (2005) FEMS Yeast Res 5:545–558
45. Sauer U (2004) Curr Opin Biotechnol 15:58–63
46. Yang TH, Frick O, Heinzle E (2008) BMC Syst Biol 2:29
47. Weitzel M, Wiechert W, Noh K (2007) BMC Bioinformatics 8:315
48. Fong SS, Burgard AP, Herring CD, Knight EM, Blattner FR, Maranas CD, Palsson BO (2005) Biotechnol Bioeng 91:643–648
49. Wahl A, Sidorenko Y, Dauner M, Genzel Y, Reichl U (2008) Biotechnol Bioeng 101:135–152
50. Kromer JO, Wittmann C, Schroder H, Heinzle E (2006) Metab Eng 8:353–369
51. Melzer G, Esfandabadi ME, Franco-Lara E, Wittmann C (2009) BMC Syst Biol 3:120
52. Patil KR, Rocha I, Forster J, Nielsen J (2005) BMC Bioinformatics 6:308
53. Segre D, Vitkup D, Church GM (2002) Proc Natl Acad Sci USA 99:15112–15117
54. Suthers PF, Burgard AP, Dasika MS, Nowroozi F, Van Dien S, Keasling JD, Maranas CD (2007) Metab Eng 9:387–405
55. Trinh CT, Wlaschin A, Srienc F (2009) Appl Microbiol Biotechnol 81:813–826
56. Duarte NC, Becker SA, Jamshidi N, Thiele I, Mo ML, Vo TD, Srivas R, Palsson BO (2007) Proc Natl Acad Sci USA 104:1777–1782
57. Quek LE, Nielsen LK (2008) Genome Inf 21:89–100
58. Selvarasu S, Karimi IA, Ghim GH, Lee DY (2010) Mol Biosyst 6:152–161

59. Nielsen J (2003) J Bacteriol 185:7031–7035
60. Beard DA, Babson E, Curtis E, Qian H (2004) J Theor Biol 228:327–333
61. Akesson M, Forster J, Nielsen J (2004) Metab Eng 6:285–293
62. Korke R, Gatti Mde L, Lau AL, Lim JW, Seow TK, Chung MC, Hu WS (2004) J Biotechnol 107:1–17
63. Vriezen N, van Dijken JP (1998) Biotechnol Bioeng 59:28–39
64. Savinell JM, Palsson BO (1992) J Theor Biol 154:455–473
65. Savinell JM, Palsson BO (1992) J Theor Biol 154:421–454
66. Lee WN, Boros LG, Puigjaner J, Bassilian S, Lim S, Cascante M (1998) Am J Physiol 274:E843–E851
67. Martens DE (2007) In: Al-Rubeai M, Fussenegger M (eds) Systems biology, vol 1, Springer, Berlin, pp. 275–299
68. Miller WM, Wilke CR, Blanch HW (1987) J Cell Physiol 132:524–530
69. Xie L, Wang DI (1996) Biotechnol Bioeng 52:591–601
70. Russell JB (2007) J Mol Microbiol Biotechnol 13:1–11
71. Jang JD, Barford JP (2000) Cytotechnology 32:229–242
72. Nielsen LK, Reid S, Greenfield PF (1997) Biotechnol Bioeng 56:372–379
73. Zwingmann C, Richter-Landsberg C, Leibfritz D (2001) Glia 34:200–212
74. Bonarius HPJ, Schmidt G, Tramper J (1997) Trends Biotechnol 15:308–314
75. Wiechert W (2001) Metab Eng 3:195–206
76. Sauer U, Bailey JE (1999) Biotechnol Bioeng 64:750–754
77. Petersen S, de Graaf AA, Eggeling L, Mollney M, Wiechert W, Sahm H (2000) J Biol Chem 275:35932–35941
78. Schmidt K, Marx A, de Graaf AA, Wiechert W, Sahm H, Nielsen J, Villadsen J (1998) Biotechnol Bioeng 58:254–257
79. Hofmann U, Maier K, Niebel A, Vacun G, Reuss M, Mauch K (2008) Biotechnol Bioeng 100:344–354
80. Maier K, Hofmann U, Reuss M, Mauch K (2008) Biotechnol Bioeng 100:355–370
81. Dietmair S, Timmins NE, Gray PP, Nielsen LK, Kromer JO (2010) Anal Biochem 404:155–164
82. Des Rosiers C, Lloyd S, Comte B, Chatham JC (2004) Metab Eng 6:44–58
83. Kelleher JK (2001) Metab Eng 3:100–110
84. Wittmann C, Hans M, Heinzle E (2002) Anal Biochem 307:379–382
85. Christensen B, Nielsen J (1999) Metab Eng 1:282–290
86. Maaheimo H, Fiaux J, Cakar ZP, Bailey JE, Sauer U, Szyperski T (2001) Eur J Biochem 268:2464–2479
87. Matsuda F, Morino K, Miyashita M, Miyagawa H (2003) Plant Cell Physiol 44:510–517
88. Wittmann C, Heinzle E (2001) Eur J Biochem 268:2441–2455
89. Wittmann C, Heinzle E (2001) Biotechnol Bioeng 72:642–647
90. Toya Y, Ishii N, Hirasawa T, Naba M, Hirai K, Sugawara K, Igarashi S, Shimizu K, Tomita M, Soga T (2007) J Chromatogr A 1159:134–141
91. Yang TH, Wittmann C, Heinzle E (2006) Metab Eng 8:417–431
92. Heinzle E, Yuan Y, Kumar S, Wittmann C, Gehre M, Richnow HH, Wehrung P, Adam P, Albrecht P (2008) Anal Biochem 380:202–210
93. Yuan Y, Hoon Yang T, Heinzle E (2010) Metab Eng 12:392–400
94. Wittmann C (2002) Adv Biochem Eng Biotechnol 74:39–64
95. Moseley HN (2010) BMC Bioinformatics 11:139
96. Wahl SA, Dauner M, Wiechert W (2004) Biotechnol Bioeng 85:259–268
97. Yang TH, Bolten CJ, Coppi MV, Sun J, Heinzle E (2009) Anal Biochem 388:192–203
98. van Winden WA, Wittmann C, Heinzle E, Heijnen JJ (2002) Biotechnol Bioeng 80:477–479
99. Dauner M, Sauer U (2000) Biotechnol Prog 16:642–649
100. Wittmann C, Heinzle E (2008) In: Burkovski A (ed) Corynebacteria: genomics and molecular biology. Caister Academic Press, Norfolk
101. Quek LE, Wittmann C, Nielsen LK, Kromer JO (2009) Microb Cell Fact 8:25

102. Wiechert W, Mollney M, Petersen S, de Graaf AA (2001) Metab Eng 3:265–283
103. Zamboni N, Fischer E, Sauer U (2005) BMC Bioinformatics 6:209
104. Goudar C, Biener R, Boisart C, Heidemann R, Piret J, de Graaf A, Konstantinov K (2010) Metab Eng 12:138–149
105. Metallo CM, Walther JL, Stephanopoulos G (2009) J Biotechnol 144:167–174
106. Eagle H (1959) Science 130:432–437
107. Seth G, Hossler P, Yee JC, Hu WS (2006) Adv Biochem Eng Biotechnol 101:119–164
108. Pavlou AK, Reichert JM (2004) Nat Biotechnol 22:1513–1519
109. Chu L, Robinson DK (2001) Curr Opin Biotechnol 12:180–187
110. Pau MG, Ophorst C, Koldijk MH, Schouten G, Mehtali M, Uytdehaag F (2001) Vaccine 19:2716–2721
111. Lim Y, Wong NS, Lee YY, Ku SC, Wong DC, Yap MG (2010) Biotechnol Appl Biochem 55:175–189
112. Godia F, Cairo JJ (2002) Bioprocess Biosyst Eng 24:289–298
113. Xie L, Wang DI (1996) Biotechnol Bioeng 52:579–590
114. Xie L, Wang DI (1997) Trends Biotechnol 15:109–113
115. Follstad BD, Balcarcel RR, Stephanopoulos G, Wang DI (1999) Biotechnol Bioeng 63:675–683
116. Europa AF, Gambhir A, Fu PC, Hu WS (2000) Biotechnol Bioeng 67:25–34
117. Gambhir A, Korke R, Lee J, Fu PC, Europa A, Hu WS (2003) J Biosci Bioeng 95:317–327
118. Bonarius HP, Ozemre A, Timmerarends B, Skrabal P, Tramper J, Schmid G, Heinzle E (2001) Biotechnol Bioeng 74:528–538
119. Bonarius HP, Houtman JH, Schmid G, de Gooijer CD, Tramper J (2000) Cytotechnology 32:97–107
120. Omasa T, Furuichi K, Iemura T, Katakura Y, Kishimoto M, Suga K (2010) Bioprocess Biosyst Eng 33:117–125
121. Dorka P, Fischer C, Budman H, Scharer JM (2009) Bioprocess Biosyst Eng 32:183–196
122. Selvarasu S, Wong VV, Karimi IA, Lee DY (2009) Biotechnol Bioeng 102:1494–1504
123. Puck TT, Cieciura SJ, Robinson A (1958) J Exp Med 108:945–956
124. Nyberg GB, Balcarcel RR, Follstad BD, Stephanopoulos G, Wang DI (1999) Biotechnol Bioeng 62:324–335
125. Nyberg GB, Balcarcel RR, Follstad BD, Stephanopoulos G, Wang DI (1999) Biotechnol Bioeng 62:336–347
126. Altamirano C, Illanes A, Becerra S, Cairo JJ, Godia F (2006) J Biotechnol 125:547–556
127. Altamirano C, Illanes A, Casablancas A, Gamez X, Cairo JJ, Godia C (2001) Biotechnol Prog 17:1032–1041
128. Altamirano C, Paredes C, Illanes A, Cairo JJ, Godia F (2004) J Biotechnol 110:171–179
129. Goudar C, Biener R, Zhang C, Michaels J, Piret J, Konstantinov K (2006) Adv Biochem Eng Biotechnol 101:99–118
130. Sengupta N, Rose ST, Morgan JA (2010) Biotechnol Bioeng 108:82–92
131. Mancuso A, Sharfstein ST, Tucker SN, Clark DS, Blanch HW (1994) Biotechnol Bioeng 44:563–585
132. Audsley JM, Tannock GA (2008) Drugs 68:1483–1491
133. Genzel Y, Dietzsch C, Rapp E, Schwarzer J, Reichl U (2010) Appl Microbiol Biotechnol 88:461–475
134. Kessler N, Thomas-Roche G, Gerentes L, Aymard M (1999) Dev Biol Stand 98:13–21 (discussion 73–74)
135. Le Ru A, Jacob D, Transfiguracion J, Ansorge S, Henry O, Kamen AA (2010) Vaccine 28:3661–3671
136. Jordan I, Vos A, Beilfuss S, Neubert A, Breul S, Sandig V (2009) Vaccine 27:748–756
137. Ritter JB, Wahl AS, Freund S, Genzel Y, Reichl U (2010) BMC Syst Biol 4:61
138. Henry O, Perrier M, Kamen A (2005) Metab Eng 7:467–476
139. Martinez V, Gerdtzen ZP, Andrews BA, Asenjo JA (2010) Metab Eng 12:129–137

140. Ramakrishna R, Edwards JS, McCulloch A, Palsson BO (2001) Am J Physiol Regul Integr Comp Physiol 280:R695–R704
141. Lee K, Berthiaume F, Stephanopoulos GN, Yarmush ML (1999) Tissue Eng 5:347–368
142. Forbes NS, Clark DS, Blanch HW (2001) Biotechnol Bioeng 74:196–211
143. Zwingmann C, Leibfritz D (2003) NMR Biomed 16:370–399
144. Cakir T, Alsan S, Saybasili H, Akin A, Ulgen KO (2007) Theor Biol Med Model 4:48
145. Teixeira AP, Santos SS, Carinhas N, Oliveira R, Alves PM (2008) Neurochem Int 52:478–486
146. Amaral AI, Teixeira AP, Martens S, Bernal V, Sousa MF, Alves PM (2010) J Neurochem 113:735–748
147. Kramer JA, Sagartz JE, Morris DL (2007) Nat Rev Drug Discov 6:636–649
148. Kola I, Landis J (2004) Nat Rev Drug Discov 3:711–715
149. Nicholson JK, Connelly J, Lindon JC, Holmes E (2002) Nat Rev Drug Discov 1:153–161
150. O'Connell TM, Watkins PB (2010) Clin Pharmacol Ther 88:394–399
151. Winnike JH, Li Z, Wright FA, Macdonald JM, O'Connell TM, Watkins PB (2010) Clin Pharmacol Ther 88:45–51
152. Hollemeyer K, Velagapudi VR, Wittmann C, Heinzle E (2007) Rapid Commun Mass Spectrom 21:336–342
153. Wittmann C, Kim HM, Heinzle E (2004) Biotechnol Bioeng 87:1–6
154. Balcarcel RR, Clark LM (2003) Biotechnol Prog 19:98–108

Adv Biochem Engin/Biotechnol (2012) 127: 133–163
DOI: 10.1007/10_2010_98

Published Online: 3 February 2011

Advancing Biopharmaceutical Process Development by System-Level Data Analysis and Integration of Omics Data

Jochen Schaub, Christoph Clemens, Hitto Kaufmann and Torsten W. Schulz

Abstract Development of efficient bioprocesses is essential for cost-effective manufacturing of recombinant therapeutic proteins. To achieve further process improvement and process rationalization comprehensive data analysis of both *process* data and *phenotypic cell-level* data is essential.

Here, we present a framework for advanced bioprocess data analysis consisting of multivariate data analysis (MVDA), metabolic flux analysis (MFA), and pathway analysis for mapping of large-scale gene expression data sets. This data analysis platform was applied in a process development project with an IgG-producing Chinese hamster ovary (CHO) cell line in which the maximal product titer could be increased from about 5 to 8 g/L.

Principal component analysis (PCA), *k*-means clustering, and partial least-squares (PLS) models were applied to analyze the *macroscopic* bioprocess data. MFA and gene expression analysis revealed *intracellular* information on the characteristics of high-performance cell cultivations. By MVDA, for example, correlations between several essential amino acids and the product concentration were observed. Also, a grouping into rather *cell specific productivity*-driven and *process control*-driven processes could be unraveled. By MFA, phenotypic characteristics in glycolysis, glutaminolysis, pentose phosphate pathway, citrate cycle, coupling of amino acid metabolism to citrate cycle, and in the energy yield could be identified. By gene expression analysis 247 deregulated metabolic genes were identified which are involved, inter alia, in amino acid metabolism, transport, and protein synthesis.

Keywords CHO · Mammalian cell culture · Biopharmaceuticals · Bioprocess development · Advanced data analysis · Omics technologies

J. Schaub (✉) · C. Clemens · H. Kaufmann · T. W. Schulz
Department of Biopharmaceutical Process Science, Boehringer Ingelheim Pharma GmbH & Co. KG, 88397 Biberach an der Riss, Germany
e-mail: jochen.schaub@boehringer-ingelheim.com

Contents

1 Introduction

Today, important biopharmaceuticals such as monoclonal antibodies can be produced in mammalian cell culture with final product concentrations above 5 g/L. Development of efficient bioprocesses requires high-producer cell lines, powerful cell culture media, and adequate process control strategies [1–3]. Besides product quality attributes (e.g., glycosylation), the achieved product titers, respectively productivities are crucial due to their direct impact on the overall process economics in large-scale manufacturing.

For further process improvement and process rationalization, the generation, analysis, and interpretation of bioprocess data play a key role. An integrated analysis comprising both *process* data and *phenotypic cell-level* data is essential. The increasing availability of omics technologies such as transcriptomics, proteomics, metabolomics, and fluxomics can provide *intracellular* data on the state of a biological system in a cell cultivation environment and promise to contribute towards the goal of a more rational cell culture process development using high-yielding mammalian cell lines [4, 5].

Currently, bioprocess development mainly relies on *macroscopic* data. In the development of mammalian cell culture processes, typically, time course data of viable/total cell concentration, viability, recombinant product concentration, metabolite concentrations of glucose, lactate, ammonium, and amino acids as well as osmolality and pCO_2 are determined and used for the design of improved processes. On-line measurements usually are available for temperature, dissolved oxygen, and pH. Conversion of time course *concentration* data into uptake and excretion *rates* provides additional information. The computation of characteristic stoichiometric ratios, such as lactate/glucose, is also useful for analysis of cellular metabolism, for example, in the investigation of metabolic shift phenomena [6, 7]. Acquisition and analysis of such *macroscopic* data sets is the first step in bioprocess development.

With an increasing number of fermentation experiments to be performed, bioprocess data analysis is becoming more elaborate. However, comprehensive data analysis is necessary to extract as much information as possible both from a single fermentation experiment and from comparison with other fermentations. For the purpose of bioprocess design, the *data* need to be appropriately analyzed and, with recourse to bioprocess knowledge, transformed into *information*. By means of comprehensive data analysis, in the end, also the number of fermentation experiments can be reduced.

The generation and analysis of bioprocess data is also brought forward in the process analytical technology (PAT) guidance launched by the Food and Drug Administration (FDA) in 2004. The PAT initiative promotes the implementation of modern process monitoring concepts using on-line and/or at-line analysis of key process variables in biomanufacturing in order to achieve a predefined product quality [8–12]. By detailed analysis of the process data, critical quality attributes (CQAs) with respect to the product properties can be identified and, finally, a quality by design (QbD) approach can be implemented [13]. For on-line bioprocess monitoring mainly fluorescence and spectroscopic techniques are discussed in the PAT context. However, application in the biopharmaceutical industry is currently limited [11], mainly due to the complexity of the signals and the signal data processing, and due to changes in the cultivation environment in a bioreactor over process time which can hamper robust on-line monitoring. Also, the physiological relevance of the signals as to mechanistic information is considered to be rather low [8].

Here, we report on a bioprocess development project with an IgG-producing Chinese hamster ovary (CHO) cell line in which the maximal product titer could be increased from about 5 to 8 g/L. Starting from a total of 45 fermentations, a reduced set of high-performance cultivations was identified and selected for in-depth bioprocess analysis. In addition to standard bioprocess data analysis (e.g., viable cell concentration, viability, product titer, glucose/lactate, glutamine/glutamate) advanced data analysis methods were developed and applied. The established data analysis platform currently comprises multivariate data analysis (MVDA), the mechanistic framework of metabolic flux analysis (MFA), and pathway analysis for mapping of large-scale gene expression data sets to metabolic and signaling pathways. The MVDA tools principal component analysis (PCA) and *k*-means clustering were used to reduce the number of 45 fermentations to nine high-performance processes that were further grouped into three classes. Partial least-squares (PLS) models were used to analyze the *macroscopic* data of these high-performance processes in more detail. Distribution of *intracellular* metabolic fluxes was computed by MFA for representative fermentation runs. Finally, gene expression analysis was performed for one of the high-titer runs. The results of the application of our data analysis workflow will be described and the use of comprehensive data analysis tools for the analysis and the design of high-titer industrial bioprocesses will be discussed.

2 Overview of System-Level Data Analysis and Omics Data

2.1 Multivariate Data Analysis

Besides data-mining techniques employing pattern recognition, clustering, artificial neural networks (ANN), decision trees (DT), and support vector machines (SVM) [14–17], multivariate statistical process monitoring [18] and control [19], the framework of multivariate data analysis (MVDA), has become an important tool in bioprocess data analysis [20, 21]. MVDA is also applied in the context of design-of-experiments approaches (DoE) [22].

2.1.1 Principal Component Analysis (PCA)

A widely applied MVDA method is PCA. Besides reducing the dimensionality of a data set (i.e., removal of redundant information), one aim of performing PCA in bioprocess analysis is the assignment of certain classes or groups in screening of process data and to uncover the discriminable structure in complex data sets. Other applications of PCA refer to the establishment of multivariate calibration models for on-line monitoring of bioprocesses by different spectroscopic techniques such as near infrared (NIR), mid infrared (MIR), 2D fluorescence, and dielectric capacitance spectroscopy, and to statistical process control [11, 19, 23].

2.1.2 Clustering

Use of clustering algorithms is well established in large-scale gene expression analysis, with hierarchical clustering, *k*-means clustering, and self-organizing maps (SOM), being the most widely used clustering methods [24]. Besides application in other omics areas such as metabolomics [25], clustering methods are also used for mining of bioprocess data [16] and data-based bioprocess modeling [17].

2.1.3 Partial Least-Squares (PLS) Models

Partial least-squares models are based upon multiple regression methods that can cope with several collinear input and output variables and can consider the covariances in the data [26]. In bioprocessing, PLS is used to establish chemometric models (multivariate calibration due to collinear variables) for bioprocess monitoring using on-line fluorescence and infrared spectroscopy [8, 11, 27–29] and for design of experiments [22].

Given a large number of bioprocesses to be compared and large amounts of bioprocess data to be analyzed, MVDA should be the second step in advanced bioprocess analysis after completion of basic *macroscopic* data collection and analysis. However, the limitation of MVDA is that it is solely based upon statistical calculations. Since MVDA does not contain any process, bio(techno)logical, or mechanistic knowledge, interpretation is not straightforward and detailed process understanding is required to draw appropriate conclusions.

Nowadays, large-scale omics data sets are generated and exploited to analyze and optimize mammalian cell culture processes both with respect to strain and to process improvement [4].

In the following section the acquisition of cell-level data is described. These data are analyzed by using mechanistic models of cell metabolism and knowledge of biological pathways and functions.

2.2 Omics Approaches

2.2.1 Metabolic Flux Analysis (MFA)

In contrast to MVDA methods, the MFA methodology enables mechanistic analysis of cellular metabolism, i.e., the quantification of intracellular metabolic fluxes. The framework of MFA is well established [30] and widely accepted in the analysis and the optimization of microbial production strains (metabolic engineering) and production processes [31–35]. Different modeling and simulation tools such as stoichiometric MFA (here defined as metabolite balancing without use of additional ^{13}C labeling information), ^{13}C MFA (isotopic stationary versus isotopic instationary), and flux balance analysis are available [36, 37]. In short, the application of these methods depends on the available analytical data (e.g., ^{13}C labeling patterns), the *structure* of the metabolic network model (e.g., over- versus underdetermined model), and the *size* of the metabolic network to be analyzed (e.g., central carbon metabolism versus genome-scale models). In mammalian cell culture, mainly stoichiometric MFA is used for quantitative analysis of metabolism. In hybridoma, for example, stoichiometric MFA was applied in metabolic screening [38], to analyze the effects of different cell culture media [39] and oxidative/reductive stress [40], to investigate the phenomenon of steady state multiplicity [41], to analyze distinct physiological states [6, 7], and to optimize perfusion processes [42]. MFA in hybridoma was also achieved by ^{13}C MFA [43]. In CHO, for example, the metabolic redistribution upon glutamate feeding [44] and the effect of galactose supplementation on the lactate metabolism [45] were examined by stoichiometric MFA. Also for CHO, Goudar [46] presented a stoichiometric MFA model suitable to perform quasi real-time MFA in an industrial cell culture process.

Clearly, application of MFA in mammalian cell culture is more challenging than in microbial systems. Essentially, this is due to cell compartmentation, the use of complex cell culture media (and associated analytical challenges), as well as changes in cell size and in viable cell fraction over cultivation time. Relevant aspects of the application of MFA in mammalian cell culture were recently reviewed by Niklas [47] and Quek [48], and the importance of MFA in pharmaceutical production was reviewed by Boghigian [49]. Application of ^{13}C MFA provides a more detailed and accurate picture of metabolism than stoichiometric MFA and can, for example, resolve parallel and circular pathways, but it comes currently at the cost of elaborate (and expensive) additional analyses and a limitation to central carbon metabolism. At present, a suitable approach for use of MFA in mammalian bioprocess development is the application of stoichiometric MFA using a metabolic network model that was validated by ^{13}C MFA [48, 50]. By this means, costs for laborious analyses of ^{13}C labeling patterns in metabolites by gas chromatography–mass spectrometry (GC–MS), liquid chromatography–mass spectrometry (LC–MS), or 2D correlation spectroscopy (COSY) nuclear magnetic resonance (NMR) can be reduced, and the intracellular flux distribution can be computed only on the basis of measured metabolite uptake and production rates.

2.2.2 Gene Expression Analysis

Although the industrially important CHO cell still lacks genome sequence information, the technically most mature omics approach in mammalian cell culture, currently, is gene expression profiling using either customized DNA chips or next generation sequencing [51–53]. CHO genome sequencing projects are already foreseeable for the near future and can be expected to have a major impact on CHO cell genomics.

Gene expression analysis has been widely applied to investigate different cell culture technology aspects such as metabolic process conditions [54], cell culture media [55–57], clone selection [58], glycosylation [59], apoptosis [60, 61], specific productivity related topics such as osmotic stress [62, 63], temperature shift [64, 65], or sodium butyrate treatment [66], and even process characterization [67].

2.2.3 Proteomics

Proteomic technologies aim to comprehensively characterize protein products. Properties that are not accessible by genomic methods such as post-translational modifications need to be examined on the protein level. Yet, proteomic studies in CHO presently rely on protein sequence information of related organisms. Compared with gene expression analysis, proteomics approaches are currently more limited in cell culture applications [4].

Proteomics approaches have been used to investigate proteome changes during the time course in fed-batch processes for recombinant IgG production [68], to investigate the phenomenon of metabolic shift in recombinant antibody-producing CHO cell lines [69, 70], to study differential protein expression in CHO upon a temperature shift [71], to investigate the correlation between the abundance of intracellular proteins and specific mAb productivity in engineered mammalian cell lines [72], to analyze the protein secretion pathways in high-producing mammalian cell lines [73], to study the effects of the insertion of the apoptosis inhibitor Bcl-x_L [74, 75], and to profile host cell proteins for analysis of potential implications to downstream protein purification [76].

2.2.4 Metabolomics

Metabolomics is the most recent omics technology. It allows the analysis of the observed phenotype which is the ultimate result of all hierarchical cellular regulation processes due to genetic or environmental changes.

Because of the diverse chemical structures of metabolic intermediates GC–MS and LC–MS methods are typically applied. For microbial systems, analytical platforms have been developed that enable the analysis of a considerable number (about 380) of metabolites [77]. Focus of the application of metabolomics in mammalian cell culture, so far, is the analysis of extracellular metabolites for the purpose of medium optimization [78–80]. Besides spent medium analysis, intracellular metabolite analysis is increasingly used in mammalian cell culture to identify possible metabolic bottlenecks, for example, in central carbon metabolism in recombinant protein-producing CHO cell lines [81]. Because of the relatively instable cell membrane and the associated risk of a leakage of intracellular metabolites into the supernatant, special attention has to be directed to the development of suitable quenching and extraction protocols for intracellular metabolites. Currently, such sampling protocols are developed for application in mammalian cell culture [82, 83].

2.2.5 Integrated Omics Approaches

Combined approaches consisting of gene expression profiling and proteomics are applied, for example, to examine the effects of temperature shift, butyrate treatment, and different specific growth rates on antibody-producing CHO cell lines [84–87].

Gene expression profiling has the advantage that experimental and analytical platforms are well established, whereas more recent proteomics technologies currently are much more laborious and intricate to perform. However, potential biases in gene expression analysis, for example, due to post-translational modifications, can be reduced when proteome data are also available. In fact, transcript and proteome data provide complementary information. But the number of proteins that can be identified at present is much smaller than for gene expression analysis. For example, Carlage [74] identified roughly 400 proteins in a high-producer CHO cell line. In contrast, application of next generation sequencing technology enables the identification of more than 13,000 genes in CHO [51].

3 A Framework for Advanced Data Analysis and Its Application

3.1 General Description and Components of the Framework

3.1.1 Overview

The established data analysis platform currently consists of the MVDA methods PCA and *k*-means clustering. These are applied to reduce data complexity and to group data of large numbers of fermentation runs. PLS models are used to analyze the *macroscopic* data of identified high-performance processes in more detail.

Mechanistic insights provide the omics tools MFA (intracellular flux distributions) as well as gene expression profiling and pathway analysis (mapping of large-scale gene expression data sets to metabolic/signaling pathways and biological functions). The details of the developed framework are described in the following sections.

3.1.2 Principal Component Analysis (PCA)

The aim of PCA [88] is the reduction of the dimensionality in a given data set without a significant loss of information contained in the data. PCA was performed by using the software Spotfire (Decision Site 9.1.1, TIBCO Spotfire, Somerville, MA, USA). In short, by PCA a high-dimensional set of correlated variables is transformed into a lower-dimensional space of orthogonal, uncorrelated variables called the principal components (PCs). The PCs are linear combinations of the original data set and maintain most of the variance in the data. The first PC contains the maximum variance of the original data set and as much of the remaining variance in the data as possible is collected in the subsequent PCs. As a linear transformation method, use of PCA might not be adequate in the case of highly non-linear process variables and non-linear statistical techniques should be applied [89].

Here, for PCA about 3,400 time course process data (e.g., cell count and cell viability, metabolite concentration data, specific uptake, and excretion rates), on-line data not included, were used. In a second PCA only time course product titer data were considered.

3.1.3 Clustering

Clustering analyses were performed by using the software package Spotfire (Decision Site 9.1.1, TIBCO Spotfire, Somerville, MA, USA). Non-hierarchical *k*-means clustering was applied since some studies indicate that it outperforms popular hierarchical clustering [24]. The former partitioning method groups data points into a predetermined number of clusters on the basis of similarity in an iterative process until no more data points change membership to a certain cluster in a further round of iteration. It has to be noted that both the cluster algorithm and the chosen similarity measure significantly affect the clustering results. Therefore, clustering results should be replicated by different methods. Here, *k*-means clustering was used for time course product titer measurements. Two cluster initialization methods (data centroid-based search, evenly spaced profiles) and four similarity measures (euclidean distance, correlation, cosine correlation, city block distance) were used. Also the maximal number of clusters was varied.

3.1.4 Partial Least-Squares (PLS) Models

Partial least-squares is a method to relate two matrices X (containing the input variables) and Y (containing the output variables) to each other by a multivariate linear PLS model. The method can be regarded as a multiple regression approach that can deal with a complex data structure consisting of several collinear input and output variables, and that takes the covariances into account (for details on the methodology the reader is referred to textbooks on MVDA, e.g., [26]). Like PCA, PLS models are based upon data projection methods but, by correlating the input and output matrices X and Y, PLS exceeds PCA.

Here, the commercially available software package SIMCA-P+ 11 version 11.0.0.0 (Umetrics AB, Malmö, Sweden) was used for MVDA and PLS modeling. Briefly, using the Simca terminology, for the X and Y matrices the corresponding spaces are constructed where each X and Y variable constitutes a (scaled) coordinate axis. Every observation in a data set is represented in both the X and Y space. The data are mean-centered and the projection coordinates (or *scores*) $t1$ (for X, t are linear combinations of X) and $u1$ (for Y, u are linear combinations of Y) are then computed such that they (i) give a good approximation of the shape of X and Y data, and (ii) maximize the correlation between X and Y. A second PLS component is calculated in the same way (associated with a second set of score vectors $t2$ and $u2$). In doing so the projection coordinates are replaced by model planes (the second projection coordinate $t2$ in the X space being orthogonal to the first $t1$). By this means, both the approximation and correlation can be improved. Usually, the strongest correlation structure is kept in the first score vector pair (i.e., $t1$, $u1$). The *weights* w (for X) and c (for Y) are important variable-related parameters and a result of a PLS analysis. The PLS weights give information about how the X variables combine to form the scores t. Simplified, the weights w for the X variables indicate how much they (relatively) contribute to the modeling of Y. In a *loading* ($w*c$) plot this relationship between X and Y can be visualized in order to evaluate the relative importance and influence (positive or negative) of the X variables on Y. Regression coefficient plots show the influence (extent, sign) of each variable for a chosen response. For analysis of a PLS model with a large number of components and responses, the computation of the so-called *variable influence on projection* (*VIP*) vector is useful. The VIP values represent the importance of the terms in the model with respect to both Y (correlation to all the responses) and *X* (the projection). Thus, by pooling over all components and all responses, VIP gives a more condensed view for PLS model analysis and interpretation than coefficient plots (pooling over all components).

3.1.5 Metabolic Flux Analysis (MFA)

A comprehensive metabolic network model of CHO metabolism was reconstructed by using a commercial software package for in silico modeling and simulation of cellular metabolic networks. The compartmented model (compartments for cytosol, mitochondria, and endoplasmic reticulum; stoichiometry/mechanism of transport steps between compartments is considered; metabolites involved in inter-compartmental transport are balanced in each compartment separately) includes 44 metabolic pathways (e.g., carbohydrate metabolism with glycolysis, pentose phosphate pathway (PPP), citrate cycle, amino acid metabolism, energy metabolism, lipid metabolism, nucleotide metabolism, and synthesis of macromolecules such as DNA, RNA, biomass, cell protein, and recombinant product protein) and comprises 338 balanced compounds. The total number of outer degrees of freedom was 31, the number of measured specific uptake and excretion rates was 33. The model was overdetermined and observable. Intracellular metabolic fluxes were calculated by metabolite balancing. For short time intervals of one cultivation day steady state conditions were assumed. Validity of the metabolite balancing approach was verified by isotopic stationary ^{13}C MFA (IS-^{13}C MFA) [90] using ^{13}C mass isotopomer labeling patterns in metabolic intermediates from central carbon metabolic pathways glycolysis, PPP, and citrate cycle (data not shown). The framework of MFA is well established and described in detail, for example, by Stephanopoulos and Nielsen [30].

Specific rates of substrate uptake and (by-)product excretion were calculated on the basis of two consecutive off-line measurements and averaged cell counts. Additionally, specific rates estimates were approximated by polynomial fitting (3rd or 4th order regression) to the cumulative consumption/production curves of cells, metabolites, and nutrients over process time. Smoothed specific rates for each time point could then be computed from the first derivative of the polynomial equation and division by the cell count.

3.1.6 Gene Expression and Pathway Analysis

Sampling, extraction of RNA, microarray hybridization, and data processing were performed as described previously [55]. Time-series gene expression was profiled over different process phases (cultivation days (d) 0, 4, 6, 8, 11). Gene expression analysis was done for process A only. In short, the Qiagen RNeasy Kit (Qiagen GmbH, Hilden, Germany) was used according to the manufacturer's protocol, the extracted cRNA was hybridized on a custom-made CHO-specific Affymetrix microarray containing 22,827 probe sets, the samples were scanned on an Affymetrix GeneChip Scanner 3000 system (Affymetrix, Inc., Santa Clara, CA), and the MAS 5 algorithm was applied for raw data normalization. Fold changes (FCs) were calculated as $\log_2$-transformed ratios relative to the control (d0). Signal intensities below 80 were discarded as noise. Genes were considered to be differentially expressed over process time if FC $\geq |2|$ for at least one sample (d4/d0, d6/d0, d8/d0, or d11/d0) and if $p \leq 0.05$ (d6/d0).

Pathway analysis was performed by using Ingenuity Pathway Analysis (Ingenuity Systems, Inc., Redwood City, CA) and the curated sequence database RefSeq for gene annotation [91].

3.2 *Cell Cultivation and Data Generation*

3.2.1 Cell Line and Cell Cultivation

The same recombinant IgG-producing CHO cell line was used in all fermentations. The viable inoculation cell concentration was 3.0×10^5 cells/mL. Cultivations were performed in fed-batch mode in controlled bioreactors with 5.5-L start volume. Temperature was controlled at 37 °C, pH at physiological optimum, and dissolved oxygen concentration at 60% air saturation by adaptation of stirrer speed and oxygen fraction in the nitrogen/oxygen gas mixture. Basal and fed-batch media (proprietary chemically defined, serum-free media) as well as process control schemes were adjusted in order to optimize bioprocess performance.

3.2.2 Fermentation Experiments

Over the course of a bioprocess development project a total of 45 fermentation runs were performed. Using the same BI HEX® CHO cell line, we increased the maximal product titer from 4.7 g/L (data not shown) to 7.9 g/L. By application of our data analysis workflow the number of fermentation runs to be analyzed in more detail was successively reduced on the basis of bioprocess performance and the advanced data analysis methods as described. By this means, nine high-performing fermentation experiments (three process generations A, B, and C) were selected for in-depth analysis.

3.2.3 Analytical Methods

Metabolite concentrations in the fermentation supernatant and process state variables were measured off-line on a daily basis. Cell concentration and cell viability were determined by the trypan blue exclusion method using a CEDEX automated cell analyzer (Roche Innovatis AG, Bielefeld, Germany). Recombinant IgG antibody concentration was quantified by surface plasmon resonance detection of an antibody–antigen complex (Biacore C instrument, GE Healthcare Europe GmbH, Germany). An YSI 2700 analyzer (YSI Incorp., Yellow Springs, USA) was used for quantification of glucose, lactate, glutamine, and glutamate in the fermentation supernatant. Ammonium was enzymatically determined according to assay instructions (ammonia test kit, 11112732035, Roche Diagnostics GmbH, Mannheim, Germany) on a Konelab 20i (Thermo Fisher Scientific, Waltham, USA). Amino acid concentrations were quantified by gas chromatography (6890 N GC, Agilent Technologies, Waldbronn, Germany) after derivatization (EZ:faast, Phenomonex, Torrance, CA, USA). Off-line pH, pCO_2, and pO_2 were determined on a blood gas analyzer (RAPIDLab 348 System, Siemens Healthcare Diagnostics GmbH, Eschborn, Germany). Osmolality of the fermentation supernatant was measured by the freezing point depression method (Osmomat auto, Gonotec GmbH, Berlin, Germany). Oxygen and carbon dioxide fractions in the off-gas were measured on-line by a process mass spectrometer (ProMaxion, Ametek GmbH, Meerbusch, Germany).

4 Results and Discussion

4.1 Principal Component Analysis

Out of 45 fermentation runs nine high-performing fermentation experiments were selected for further analysis. For these nine fermentation runs the PCs as shown in Fig. 1a were computed to reduce the dimensionality of the data set (about 3,400 data). Whereas all fermentations followed the same trajectory until cultivation day 6, a separation of generation B processes (B1, B3) and generation C processes (C1, C2) could be observed for subsequent cultivation days. Generation A processes (A2, A3) diverged during fermentation (days 7–13) from each other and then converged again (days 14–17). Process A1 was similar to process A2 for days 10–13 but different for the last 2 days of cultivation (days 14 and 15) compared with A2 and A3. Process B2 was comparable to B1 and B3 until day 10, and then more similar to process C3, however, at a different timescale as the achieved product titers show (Fig. 2a). The PCA time course of process C3 was different from processes C1 and C2 for cultivation days 7–12 but comparable for cultivation days 13, 14, and 17. PCs of process C3 at cultivation days 18 and 19 were similar to those of processes B1 and B3 at cultivation days 17 and 18.

When time course product concentration data were used in the PCA (the product titer eigenvector was characterized by the largest contribution to the PCs; normalized data), a separation of the process generations A–C into three different quadrants could be observed (Fig. 1b). Yet, it has to be noted that PCA results depend on the amount and structure of the input data and should be repeated with different settings to strengthen the significance of the obtained results.

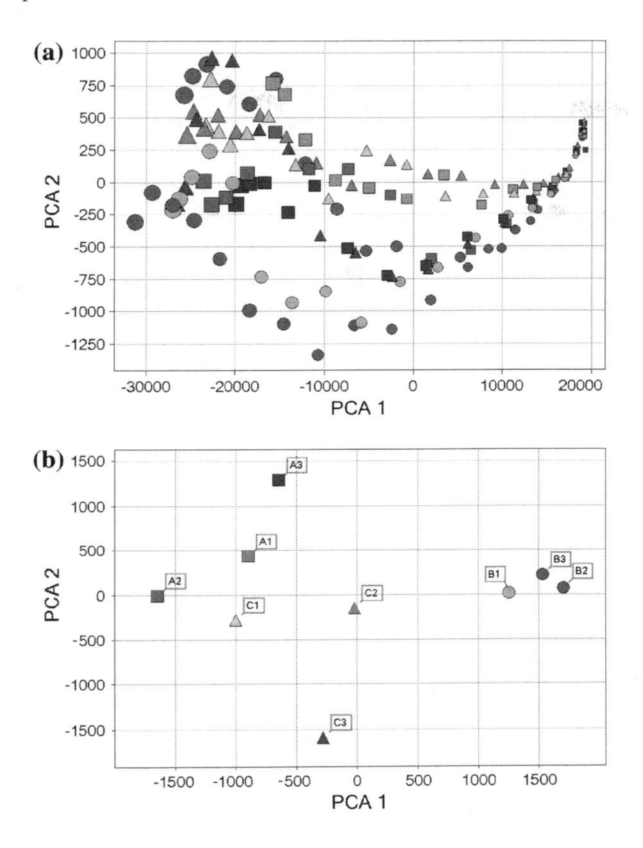

Fig. 1 Principal component analysis for process (**a**) and for product titer (**b**) data. **a** In the projection of the bioprocess data the time course was maintained (note that the marker size increases with run time). The first two principal components contained about 67% of the variability in the data set (normalized data); the process variables associated with growth and product formation contributed the most to the eigenvectors of PC1 and PC2. **b** The time course product concentration data were projected into one value per fermentation run. About 91% of the data variability was preserved in PC1 and PC2 (the largest loadings were observed from day 9 on). The legend is depicted in **b**; different processes use different symbols (*squares* process A, runs A1–A3; *circles* process B, runs B1–B3; *triangles* process C, runs C1–C3); the same color/shading is used in both graphs

4.2 Clustering

k-means clustering was performed for product titer time course data in order to reproduce the assignment of the fermentation experiments (that reflect the achieved progress in the course of process development) as identified by PCA. Three out of eight *k*-means clusterings resulted in the same grouping as in the PCA. The results are shown in Fig. 2. The application of two clustering

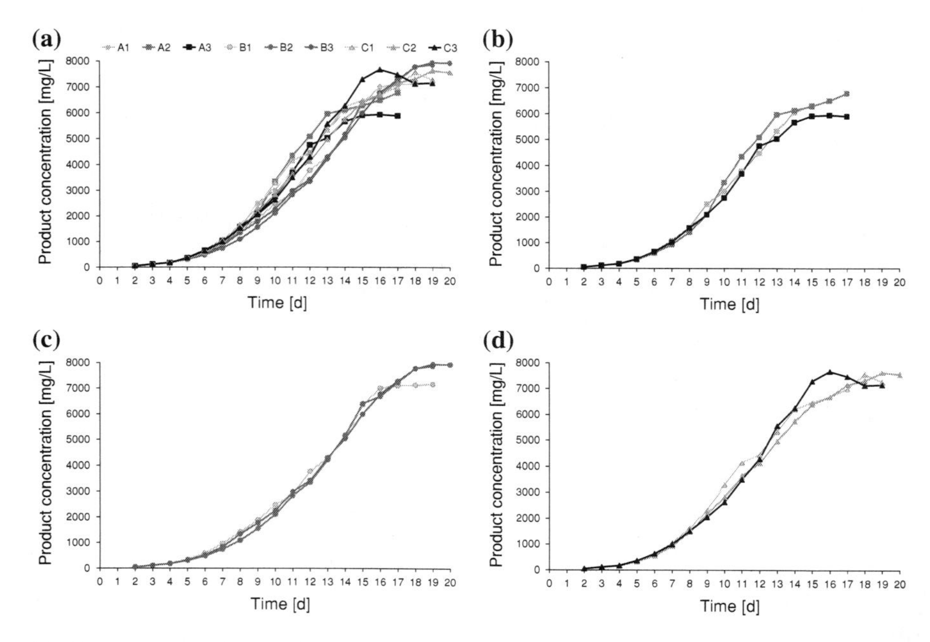

Fig. 2 *k*-means clustering of product titer time course data for the nine high-performing fermentation runs (**a**). The clustering into processes A (A1–A3), B (B1–B3), and C (C1–C3) is shown in figures **b**–**d**. The legend is shown in **a**; the same coloring/shading and symbols are used as in Fig. 1

approaches (evenly spaced profiles/correlation and evenly spaced profiles/cosine correlation) gave only two clusters (versus three as in the other methods). In three clusterings one A process was clustered to the C processes and vice versa. The results indicate that replication of cluster analysis is necessary for process classification even when a reduced data set (here time course product titer data) is used.

For this reason, we followed a dual approach in the pursuit to (i) reduce dimensionality in a data set with minimal loss of information and (ii) identify patterns/groupings in the data. First, different MVDA methods (such as PCA and *k*-means clustering) were applied for verification of clustering results. Second, the same data set was analyzed on different complexity levels (from large-scale bioprocess data with about 3,400 entries to a reduced set of key process data such as product titer with about 150 values).

4.3 Partial Least-Squares Models

With the obtained data set adequate PLS models in terms of both goodness of fit (R^2) and goodness of predictive ability (Q^2) could be computed and two PLS

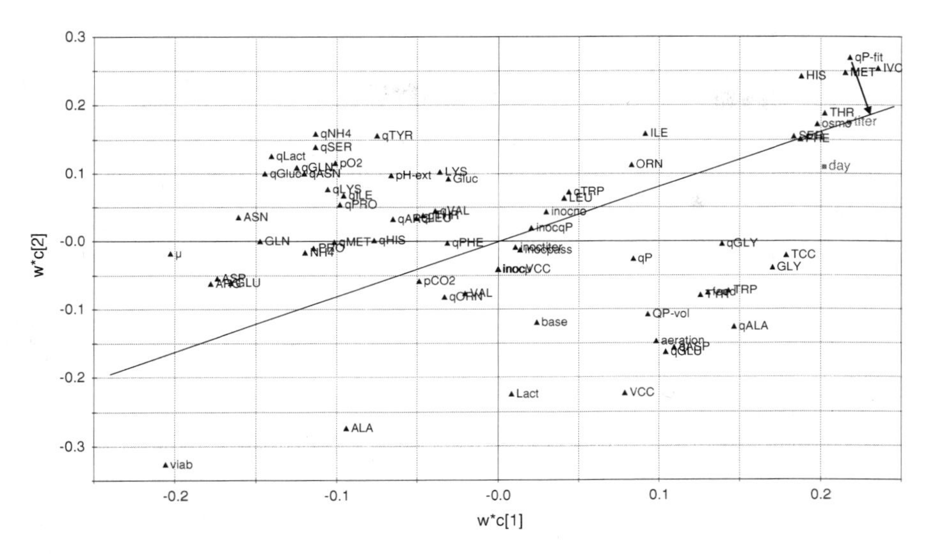

Fig. 3 Loading plot for processes A–C (total of nine fermentation runs) showing how the process variables relate to the product titer. All process variables that are plotted on the same side as the product titer are positively related to the product concentration. The distance between the orthogonal projection of an *X* variable on the straight line connecting (0,0) with the product titer and the origin determines how important this correlation is. Obviously, the specific productivity q_p is significantly positively related to the product titer as exemplarily shown

components were sufficient to obtain a R^2Y(cum) > 0.95 and a Q^2(cum) > 0.95. The loading plot shown in Fig. 3 gives a summarized overview of the relative importance of the main process variables (*X*), e.g., concentrations and rates, on the product titer (*Y*). The same data are visualized in Fig. 4 in a coefficient plot and in a VIP plot. While the coefficient plot gives information on the extent and the sign of a correlation, the VIP plot gives a more concentrated view of the data without information about the sign of a relationship but provides an additional measure of statistical significance for data ranking. The analyses demonstrate the correlation between several amino acids and the product titer. As shown in the VIP plot (Fig. 4b) the essential amino acids methionine, threonine, histidine, phenylalanine, and arginine had the largest influence on the product concentration; also the non-essential amino acids serine, aspartate, glycine, glutamate, and asparagine showed a significant relationship. This correlation structure could be consistently observed and was independent from the applied analysis method (loading plot, coefficient plot, or VIP plot). However, the influence of glycine and glutamate appeared questionable when based on the coefficients considering the 95% confidence intervals for these two amino acids. Similarly, tryptophan had a VIP value of about 1.15 and a moderate correlation in the loading plot but the coefficient for tryptophan was not significant. Conversely, a negative correlation of the lactate concentration on the product titer can be assumed when the analysis is based upon the coefficient plot; however, the VIP value is only about 0.7.

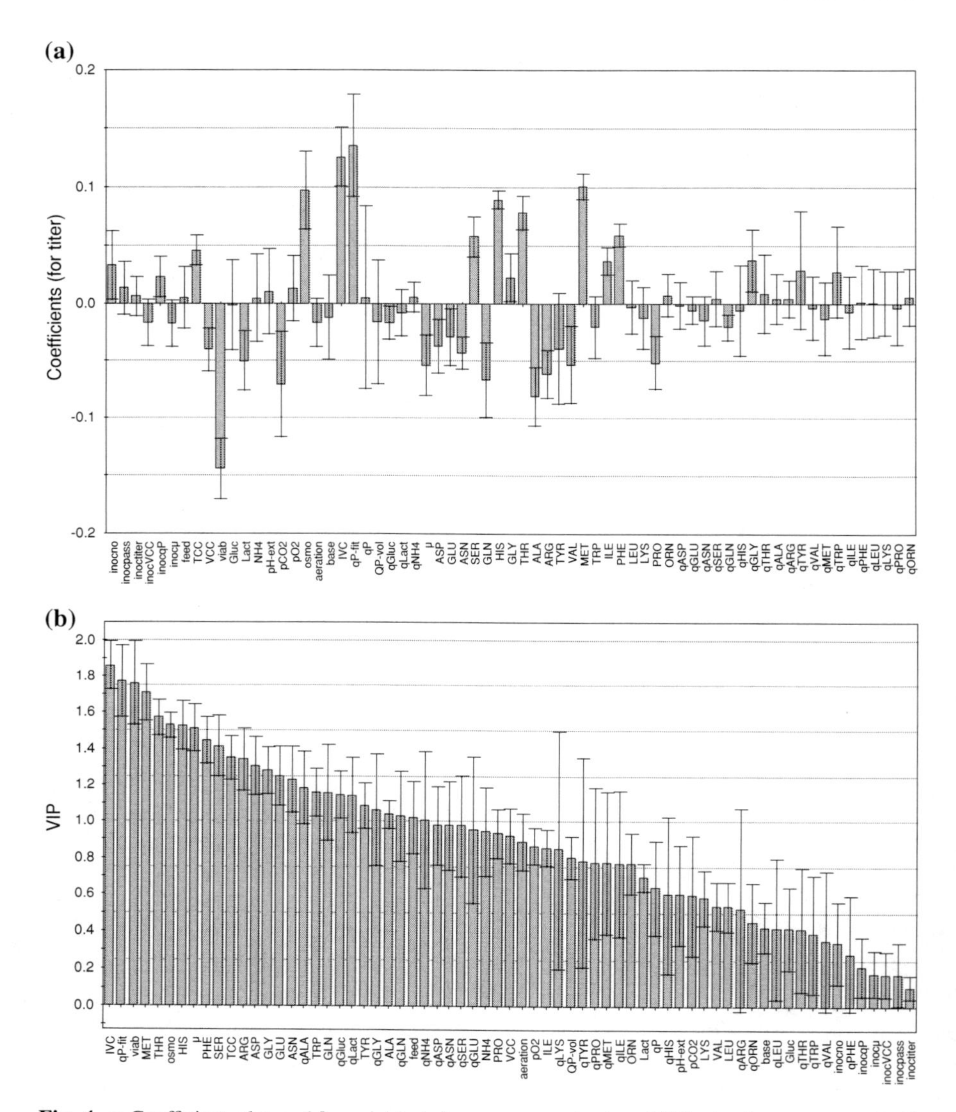

Fig. 4 **a** Coefficient plot and **b** variable influence on projection (VIP) plot for processes A–C. Data were centered and scaled; statistical significance is indicated by the 95% confidence interval. VIP values larger than 1 indicate statistical significance

To put it more generally, most of the process variables that were identified to be significant in the VIP plot were also found to be important in the coefficient plot and in the loading plot; however, the contrary was not always the case as shown. This points out that not only the setup of an adequate PLS model is essential, but it is also advisable to use different analyses methods in MVDA to evaluate the complex correlation structure contained in bioprocess data sets. Furthermore, mainly the concentration data were found to be significantly related to the product titer. Obviously metabolic *rate* data are less correlated to the product *concentration*, but one needs to keep in mind that metabolite uptake and excretion rates contain important information (e.g., different metabolite concentrations in the fermenter often result in the same rate). Ideally, multivariate data analyses should also consider the cell specific productivity and the product quality attributes.

The processes A–C were analyzed in more detail considering both the product titers and the cell specific productivities. The coefficients for the product concentration (c_{P}) and the cell specific productivity (q_{P}) are shown in Fig. 5. Consistent coefficients were observed for both c_{P} and q_{P} in processes A–C. Noticeable is the positive coefficient of q_{P} on c_{P} in process A, which is less distinct in process B and not significant in process C (Fig. 5a). In contrast, the significance of the titer coefficients for important metabolite concentrations such as glucose and lactate decreased from processes B and C, to process A. Similarly, more amino

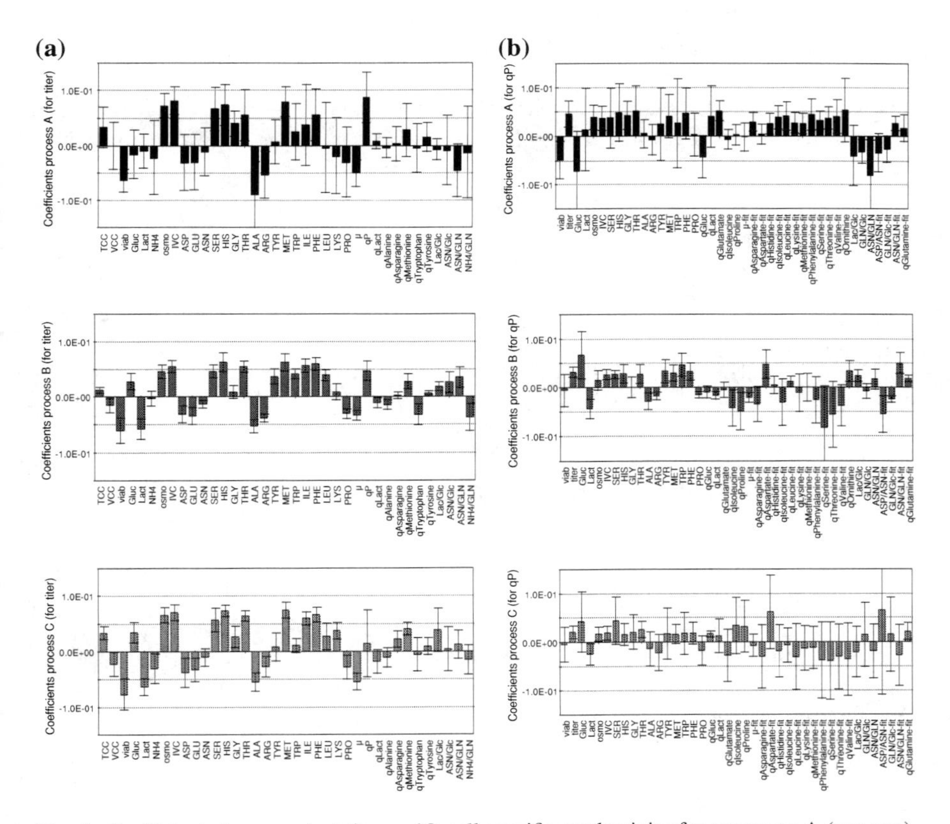

Fig. 5 Coefficients for **a** product titer and **b** cell specific productivity for processes A (*top row*), B (*middle row*), and C (*bottom row*). Note that the coefficients are shown only for those variables that were significantly different from zero in at least one of the three processes based on 95% confidence levels. Scaled and centered data were used for calculation of the coefficients

acids coefficients were significant in processes B and C than in process A. Analysis of the coefficients for q_p (Fig. 5b) of process A revealed significant positive correlations mainly for the amino acids rates (which are associated with cell protein and recombinant protein synthesis). Coefficients of amino acids concentrations and metabolite concentrations and rates were not significant in process A. In process B the relevant coefficients shifted from the amino acids rates to the amino acids concentrations, whereas no significant coefficients were identified in process C. The coefficients of the stoichiometric metabolite ratios such as lactate/glucose for c_p and q_p were small except for process B.

Taken together the coefficients might be an indication that process A is rather q_p *driven*, whereas process C is rather *process driven*, i.e., in process C the process performance is mainly governed by the process control and feeding scheme ensuring an optimal metabolic and nutritional environment for product formation (q_p is the same in processes A and C, data not shown).

It has to be noted that all MVDA results are solely based upon statistical computations without incorporating any mechanistic, cell biological, biochemical, or bioprocess knowledge. Though undoubtedly useful, these results have to be handled with care in terms of interpretation. Generally, to avoid drawing misleading conclusions, results from MVDA always need to be interpreted by taking recourse to process knowledge. To give a very basic example, a low viability does not result in increased product titers, instead the negative coefficient of the viability for c_p (Fig. 5a) simply refers to the fact that the product concentration increases over process time while the cell viability typically declines.

Furthermore, so far only *extracellular* macroscopic data were considered in the analyses. By application of mechanistic and omics approaches and integration of *intracellular* information additional insights can be obtained and a more cell-centered bioprocess analysis and design can be achieved as described in detail in the following section.

4.4 Metabolic Flux Analysis

In addition to the analysis of the metabolic rates of nutrient uptake and (by-) product excretion, the computation of intracellular flux distributions gives further insight into the metabolism of the employed production cell line for different process setups and over the time course of a fermentation run. Since the required input data (i.e., the metabolic rates) are typically determined within a bioprocess development project, no further experimental analyses need to be performed. Thus, MFA can be considered as a means of advanced bioprocess data analysis using already available data sets. Prior to the computation of intracellular fluxes by MFA, a statistical chi-square test was performed to detect potential gross measurement errors [92, 93], and to validate the suitability of the CHO metabolic network model. For process A, the average (days 4–11) chi-square value was 68%, for process B 84%, and for process C 73%, indicating the validity of the model.

4.4.1 Upper Glycolysis and Pentose Phosphate Pathway

The phenomenon of overflow metabolism from glucose to lactate is well known in mammalian cell cultivation [94]. Since accumulation of the by-product lactate in the fermentation broth to high concentrations has adverse effects on the process performance, both metabolic engineering approaches were applied and fed-batch process control and nutrient feeding strategies were designed to achieve low lactate concentrations [2, 95]. By MFA it could be shown that even in a process phase associated with a strong increase in viable cell concentration (about cultivation days 4–8), the glycolytic flux exceeds the flux entering the PPP by far (Fig. 6a, b). This means, that the metabolic demand for PPP intermediates (e.g., D-ribose 5-phosphate) that are required for nucleotide synthesis in biomass formation should be easily met by the glucose metabolism. Even when the cells are shifted by the process control scheme to a more efficient nutrient use (about days 5–7), which is characterized by a decrease in the glycolytic flux (Fig. 6a), the PPP flux is initially maintained and then decreases with decline in specific growth rate μ. Accordingly, the split ratios between glycolysis and PPP changed from about 90%:8% (day 4) to 97%:2% (day 11). These split ratios were confirmed by IS-^{13}C MFA (data not shown). Interestingly, the glycolytic flux profile of the generation C process showed a relatively high glycolytic flux profile in the early process phase (similar to the generation B process) but then switched to a lower glycolytic flux state as in the generation A process.

Since intracellular flux distributions strongly depend on the specific cell line, the recombinant product, the fermentation system (e.g., shake flask versus controlled bioreactor, cultivation scale), the cultivation mode (batch, fed-batch, continuous, or perfusion), the process control scheme, and the applied cell culture media, comparisons with literature data are usually difficult. Applying metabolite balancing Altamirano found in recombinant CHO cell lines that over 90% of the flux entered glycolysis, whereas less than about 10% entered the PPP [44, 45]. In perfusion CHO cultivation processes Goudar [46] observed a partitioning of about 70% glycolysis and about 20% PPP by applying metabolite balancing; in another CHO study fractions of 55% glycolysis and 41% PPP were determined by use of metabolite balancing and 2D COSY NMR spectroscopy [50].

4.4.2 Lower Glycolysis, Lactate Formation, and Citrate Cycle

Glycolysis and PPP pathways merge again at the D-glyceraldehyde 3-phosphate branch point. The flux profile for the lower glycolytic part is shown in Fig. 6c for the conversion of phosphoenolpyruvate to pyruvate by the enzyme pyruvate kinase. The small PPP fractions of the carbon fluxes resulted in a flux pattern quite similar to the upper part of glycolysis. Note that the approximate doubling of the flux values is due to the fact that C6 molecules (e.g., glucose) are split into two C3 molecules (e.g., pyruvate). The pyruvate node is an important branch point of the central carbon metabolism in mammalian cells since, on the one hand, pyruvate

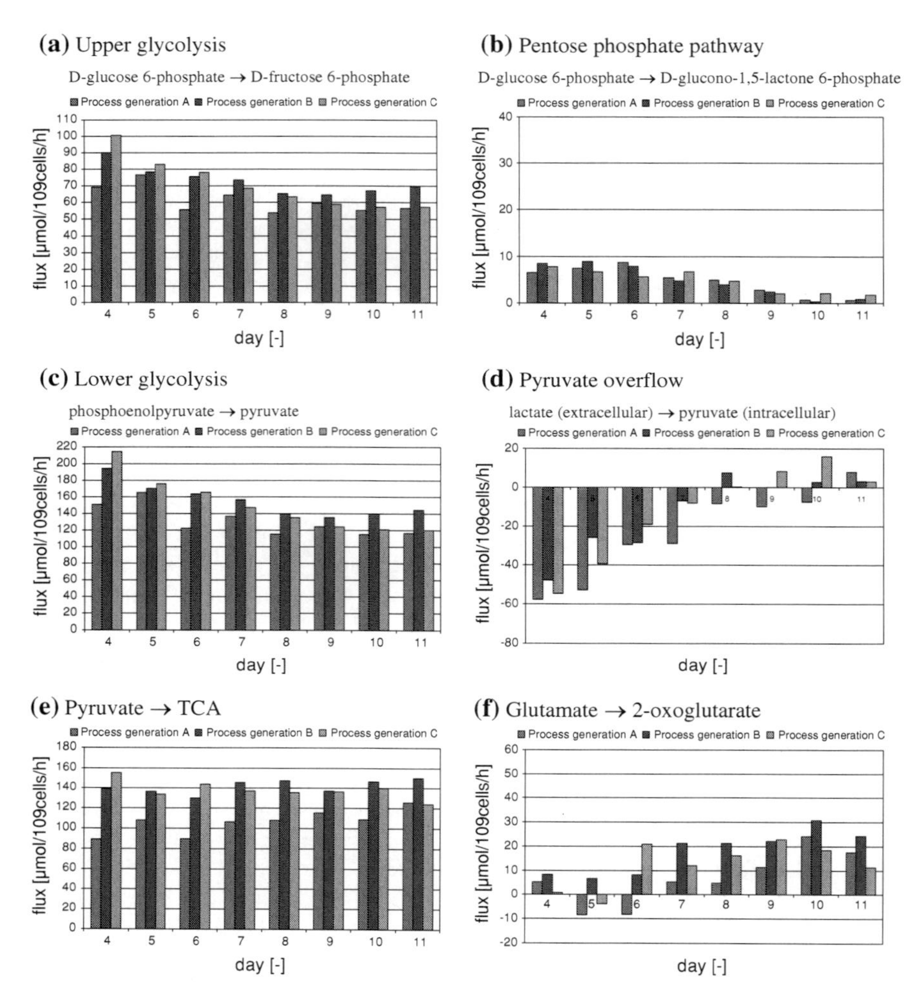

Fig. 6 Rates of intracellular metabolic fluxes as computed by MFA. Flux values for important metabolic branch points and enzymatic reactions of the central carbon metabolism are shown. Positive values denote a flux in the direction as indicated in the figure caption (e.g., a negative rate in **d** indicates that intracellular pyruvate is converted to lactate by the enzyme L-lactate dehydrogenase and excreted into the fermentation supernatant)

marks the entry point into the citrate cycle (TCA) and, on the other hand, pyruvate can be converted to lactate in one step by the enzyme L-lactate dehydrogenase and is then excreted into the fermentation broth. Although the glycolytic flux in the generation C process was relatively high in the early process phase, lactate formation via L-lactate dehydrogenase (Fig. 6d) was comparable to the generation A process at day 4 but significantly lower in the following days (about days 5–7), almost zero at day 8, and lactate formation even turned into lactate uptake from day 9 on in the generation C process. Similar to process B (and in contrast to process A), the generation C process was characterized by a large fraction of the lower glycolytic flux entering the TCA cycle from the pyruvate node (Fig. 6e). This fraction increased from about 65% (day 4) to more than about 80% (day 7).

Reported values for the flux fraction entering the TCA in CHO range from 17 to 49% [46, 50], and from 39 to 79% [44]. The dependency of flux results on cell line, process, and media was already emphasized in the previous paragraph.

4.4.3 Glutaminolysis

Besides glycolysis the TCA cycle is fueled by glutaminolysis, and the complex interplay between these two major pathways [94] plays an important role in optimizing mammalian cell culture, mainly because of undesired accumulation of ammonium [96]. Essentially, in glutaminolysis one molecule of ammonium is formed in the conversion of glutamine to glutamate by the enzyme glutaminase; the second ammonium molecule is released when glutamate is enzymatically converted into 2-oxoglutarate by glutamate dehydrogenase. Alternatively, the formation of 2-oxoglutarate from glutamate can proceed via aspartate transaminase where the amino group is transferred from glutamate to oxaloacetate to yield aspartate. Also, it has to be noted that glutamate is not only converted to 2-oxoglutarate but is involved in several other reactions, for example, in alanine transaminase or asparagine synthetase reactions, indicating that detailed metabolic models are needed for quantitative analysis. Here, the ratio between glycolysis and glutaminolysis was about 80%:20% and larger (based on the glycolytic flux entering the TCA cycle versus the summarized glutamate flux that is converted to 2-oxoglutarate). By reducing the glutaminolysis to a proportion below about 20%, ammonium concentrations of less than 100 mg/L could be obtained in the bioreactor. Goudar observed in CHO a glutaminolysis fraction larger than about 30% [46], whereas Altamirano reported values between 15 and 25% [44], and of about 30% [45]. Again, the experimental conditions were not comparable.

4.4.4 Coupling of Amino Acid Metabolism and Citrate Cycle

Amino acids are important, with respect to the essential amino acids necessary, cell culture media components. As amino acids provide the building blocks for the synthesis of the recombinant proteins to be manufactured, they are of major importance in bioprocess development and both their concentration levels and concentration ratios need to be stoichiometrically adjusted for optimal product yields. In this regard, a quantitative knowledge of the amino acid demands required for the synthesis of cell protein, of recombinant protein, and of the fluxes between TCA and amino acid metabolism supports such rational media design strategies.

By MFA this interconnection between TCA and amino acid metabolism could be analyzed in detail. Acetyl-CoA and the TCA intermediates 2-oxoglutarate, succinyl-CoA, fumarate, and oxaloacetate are connected with the amino acid metabolism (anabolism and catabolism). The highest fluxes were determined between 2-oxoglutarate and glutamate (glutamate is further interconnected with the amino acids arginine, glutamine, histidine, and proline). In processes A and C the TCA intermediate 2-oxoglutarate was used in anabolic reactions of amino acids at days 5 and 6 (process A), or day 5 (process C), whereas process B was characterized by amino acid catabolism at a relatively high level throughout the fermentation run (Fig. 6f). Fluxes between amino acid metabolism and the corresponding TCA branch point metabolites acetyl-CoA, succinyl-CoA, and oxaloacetate were below 10 μmol/10^9 cells/h, and between amino acid metabolism and fumarate below 5 μmol/10^9 cells/h. For glucose uptake rates of about 60 and 71 μmol/10^9 cells/h, Altamirano [44] reported metabolic rates in the range of about 1 μmol/10^9 cells/h between amino acid metabolism and corresponding TCA metabolites.

Citrate is not only catabolized in the TCA but also transported from the mitochondria into the cytosol and converted to acetyl-CoA by ATP citrate (pro-S)-lyase for biosynthesis of fatty acids and cholesterol. The coproduct oxaloacetate is, inter alia, further converted to malate by cytosolic malate dehydrogenase and, in turn, is either transformed to pyruvate (cytosolic malic enzyme) or shuttled into the TCA. Here, only a relatively small citrate metabolic flux from the mitochondria to the cytosol of about 5 μmol/10^9 cells/h was determined by MFA. Altamirano [44] reported values between about 1 and 2 μmol/10^9 cells/h.

Summarizing, a relatively high incorporation of acetyl-CoA into the TCA, i.e., a high glycolytic flux entering the TCA cycle from the pyruvate node, was observed whereas TCA depletion by citrate outflow was rather low. Accordingly, TCA replenishment by glutaminolysis likewise was rather moderate. Thus, a truncation of the citrate cycle was observed only to a minor extent. However, in CHO cell cultivation a partially truncated TCA cycle was suggested on the basis of intracellular and extracellular measurements of the citrate concentration, the latter being up to tenfold higher [81].

4.4.5 Energy Yield

The ATP energy yield is a quite informative, comprehensive value to assess the overall metabolic state in a cell culture process. Protein synthesis is an energy-intensive process for the cell and, thus, renders the ATP energy yield to an important value to evaluate the process performance in recombinant protein production. As in Stouthamer [97], it is assumed here that four high-energy phosphate bonds are required to incorporate one amino acid into cell or recombinant protein (one ATP is converted into one AMP in amino acid activation and two GTP are converted into two GDP in ribosomal peptide bond formation). However, as ATP is involved in many reactions throughout cell metabolism, this

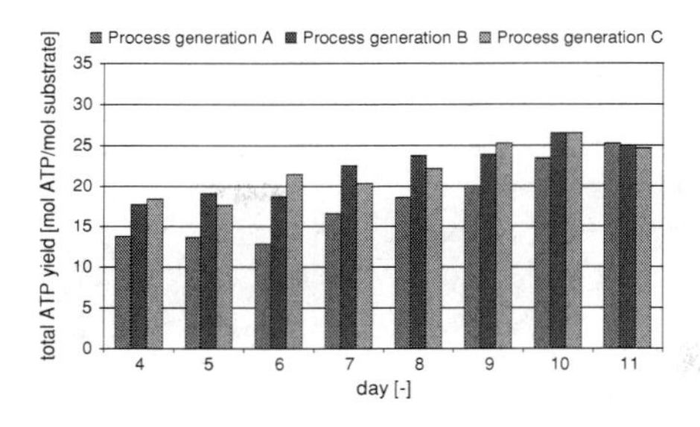

Fig. 7 ATP energy yields in processes A–C as calculated by MFA. The ATP energy yield was calculated as the ratio of the ATP net production rate and the sum of the measured extracellular glucose and amino acid uptake/excretion rates by using the established CHO metabolic model as described. In the model about 50 biochemical reactions (enzymatic reactions, polymerization of macromolecules, transport steps) involve ATP

value is difficult to determine, requiring a detailed analysis of the metabolic reaction network. Even for the relatively simple case of ATP generation from glutamine entering the TCA, the use of alternative pathways results in quite different ATP energy yields [94].

MFA was applied to compute the ATP energy yields based on the established CHO metabolic model. As shown in Fig. 7, in processes B and C higher ATP energy yields, indicating an energetically more efficient metabolic state, were achieved compared with process A. For example, in process C an energy yield above 20 mol ATP/mol substrate was obtained from day 6 on, in process B from day 7 on, and in process A from about day 9 on. Maximal ATP production rate in process A was about 2.5 mmol/10^9 cells/h, whereas maximal values of about 3.1 mmol/10^9 cells/h were attained in processes B and C. In CHO an ATP production rate of about 0.39 mmol/10^9 cells/h was determined by Goudar [50]. For an IgG-producing hybridoma cell line grown in continuous culture Gambhir [7] calculated ATP production rates between about 1.1 and 2.4 mmol/10^9 cells/h ($\mu = 0.031$–$0.033\ h^{-1}$) depending on the achieved metabolic states. It has to be noted that calculation of ATP rates depends on the level of detail of the applied metabolic model but also, for example, on the assumptions for the P/O ratio in the conversion of NADH and FADH to ATP.

Summarizing, MFA can translate measured metabolic exchange rates between the cell and the fermentation environment into intracellular metabolic flux distributions. By this means, bioprocess analysis can be performed down to the level of cell metabolism. This approach enables the design of stoichiometrically balanced cell culture media, optimized feeding strategies, and advanced process control schemes. It also supports the identification of improved process conditions for a desired phenotype, e.g., minimal accumulation of unwanted by-products such as lactate. To circumvent the limitations of the metabolite balancing approach, currently a suitable approach for application of MFA in bioprocess development is the use of the metabolite balancing concept and a metabolic network model that was validated in its core structure by IS-^{13}C MFA as was done here.

4.5 Gene Expression and Pathway Analysis

In gene expression analysis of process A out of a total of 22,827 sequences 8,593 sequences had signal intensities above 80 and displayed a significant change over the time course of the fermentation run ($p \leq 0.05$ for d6/d0). Out of these 8,593 sequences 2,052 sequences had FC $\geq |2|$ for at least one sampling point. The potential of gene expression profiling for cell culture process analysis and design is shown in Fig. 8. In metabolism, 247 genes were found to be deregulated (note that single gene products can be involved in several metabolic pathways). Interestingly, a large number of genes were associated with amino acid pathways (synthesis, metabolism, and degradation) and transport (Fig. 8a). Since gene expression analysis in metabolic pathways is *close* to bioprocess design issues, it provides useful information for process optimization, e.g., for rational media improvement [55]. Mapping of deregulated genes to biological functions gives a more comprehensive overview on intracellular processes (Fig. 8b). The biological functions cell cycle/ cell cycle control, cell death, and cellular growth/proliferation were characterized by a large number of deregulated genes although the cell viability decreased only moderate to 89% at cultivation day 11 (data not shown). Gene expression and post-translational modification were further biological functions with many deregulated genes. Though smaller in number, the functions protein synthesis, protein trafficking, and protein degradation obviously are relevant with respect to recombinant protein production.

Mapping of gene expression data to metabolic pathways (Fig. 8c) facilitates the biological analysis of the data as shown for the example of the citrate cycle (TCA) of process A. The TCA genes encoding the enzymes isocitrate dehydrogenase 1, $NADP^+$, soluble (gene IDH1) and succinate dehydrogenase complex, subunit A (gene SDHA) were found to be significantly upregulated over the process time. FCs for IDH1 increased considerably from 1.7 (d4/d0) to 2.8 (d11/d0) and for SDHA slightly from 1.8 (d4/d0) to 2.1 (d11/d0). If a less conservative cutoff of FC $\geq |1.4|$ [54, 64, 66] is applied another five TCA genes could be analyzed. These genes encode ATP citrate lyase, citrate synthase, dihydrolipoamide dehydrogenase, dihydrolipoamide S-succinyltransferase, and fumarate hydratase (data not shown). The gene OGDH encoding 2-oxoglutarate dehydrogenase/lipoamide (EC 1.2.4.2) is also spotted on the customized CHO chip but had signal intensity values below the noise threshold in this specific experiment. As for MFA, comparisons with literature data are hardly possible. For example, in a recombinant IgG-producing recombinant CHO cell line the IDH1 gene (encoding isocitrate dehydrogenase 1) was not found to be deregulated upon a temperature shift from 37 to 33 °C [64]. When four mAb-producing CHO cell lines (two cell lines, each fast versus slow growing) were cultivated, a downregulation (slow versus fast growing) of IDH3A was observed in both cases.

For further analysis, omics data such as gene expression data should be analyzed in the context with other data sets to identify targets for bioprocess optimization. In this specific example, it was found that the metabolic flux through the

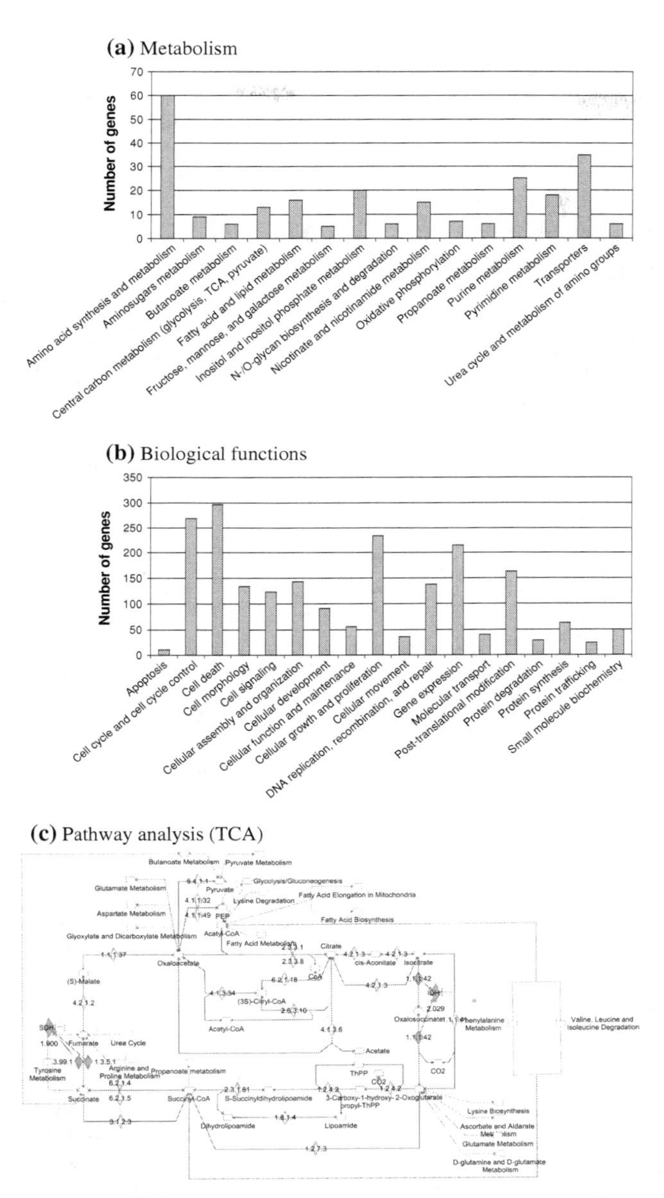

Fig. 8 Number of differentially expressed genes in process A over the time course of the fermentation run that are associated with **a** metabolism and **b** biological functions. Note that the mapping of single genes is not unique, i.e., a gene product can play a role in different biological functions (e.g., cell death and apoptosis). **c** Mapping of gene expression data to metabolic pathways. Citrate cycle genes IDH1 (encoding isocitrate dehydrogenase 1) and SDHA (encoding succinate dehydrogenase complex, subunit A) were found to be significantly upregulated in process A over fermentation time. Note that only gene expression data with FC $\geq |2|$ and $p \leq 0.05$ are shown

TCA increased with process time (compare Fig. 6e). When compared with gene expression data, a consistent upregulation of the TCA genes IDH1 and SDHA could be observed which results in a more detailed picture of the metabolism.

Ultimately, such integrated, data-driven analyses will support the identification of metabolic bottlenecks.

5 Conclusion and Outlook

Bioprocess development is increasingly science and data driven. Currently, two trends can be observed that generate large-scale bioprocess data sets that both bear the potential to provide novel information for rational process development.

On the one hand, at-line instruments and on-line sensors (e.g., infrared spectroscopy, 2D fluorescence, dielectric spectroscopy) are becoming more established in bioprocess analysis, development, and monitoring [11] in addition to standard bioprocess analytics. By application of the favored in situ monitoring tools mainly *process* data are obtained. However, elaborate chemometric models are required to analyze the complex data sets and the mechanistic information contained in the data is often limited.

The second trend refers to genome-scale technologies [4] such as transcriptomics, proteomics, metabolomics, and fluxomics which are increasingly used in both academia and industry. These tools provide large-scale *intracellular* data sets that enable *cell-level* bioprocess analysis and contribute to the development of high-producer mammalian cell lines. Though expected soon, the genome sequence for these technologies currently do not provide complete information with respect to mammalian cell culture applications (e.g., the genome sequence for the industrially relevant CHO cell is not published up to now), the application of these tools has already given profound insight into cell culture processes and can be expected to advance rapidly, for example, with respect to CHO genome sequencing or recent progress in metabolomics approaches in mammalian cell culture [78, 81]. Finally, the integration of these omics tools will support data-driven, rational biopharmaceutical process development ranging from cell line development to bioprocess monitoring.

With respect to both trends, advanced data analysis is key to fully exploiting the information contained in measured bioprocess data and to generating novel process knowledge. Efficient and comprehensive data analysis becomes even more important if large numbers of processes have to be analyzed in order to identify, for example, correlations associated with specific cell lines, products, process modes, cell culture media, process performance, or product quality attributes.

In this contribution, a framework consisting of MVDA and systems biotechnology tools was established and the potential of these technologies was presented. It was exemplarily shown how these methods can be used to address important aspects of biopharmaceutical process development in the pursuit to develop high-performance processes. It will be interesting to see how these novel technologies will change the way bioprocess development is performed in the near future.

References

1. Seth G, Hossler P, Yee JC et al (2006) Engineering cells for cell culture bioprocessing—physiological fundamentals. Adv Biochem Eng Biotechnol 101:119–164
2. Wlaschin KF, Hu WS (2006) Fedbatch culture and dynamic nutrient feeding. Adv Biochem Eng Biotechnol 101:43–74
3. Wurm FM (2004) Production of recombinant protein therapeutics in cultivated mammalian cells. Nat Biotechnol 22(11):1393–1398
4. Griffin TJ, Seth G, Xie H et al (2007) Advancing mammalian cell culture engineering using genome-scale technologies. Trends Biotechnol 25(9):401–408
5. O'Callaghan PM, James DC (2008) Systems biotechnology of mammalian cell factories. Brief Funct Genomic Proteomic 7(2):95–110
6. Europa AF, Gambhir A, Fu PC et al (2000) Multiple steady states with distinct cellular metabolism in continuous culture of mammalian cells. Biotechnol Bioeng 67(1):25–34
7. Gambhir A, Korke R, Lee J et al (2003) Analysis of cellular metabolism of hybridoma cells at distinct physiological states. J Biosci Bioeng 95(4):317–327
8. Clementschitsch F, Bayer K (2006) Improvement of bioprocess monitoring: development of novel concepts. Microb Cell Fact 5:19
9. Read EK, Park JT, Shah RB et al (2010) Process analytical technology (PAT) for biopharmaceutical products: Part I. concepts and applications. Biotechnol Bioeng 105(2):276–284
10. Read EK, Shah RB, Riley BS et al (2010) Process analytical technology (PAT) for biopharmaceutical products: Part II. Concepts and applications. Biotechnol Bioeng 105(2):285–295
11. Teixeira AP, Oliveira R, Alves PM et al (2009) Advances in on-line monitoring and control of mammalian cell cultures: supporting the PAT initiative. Biotechnol Adv 27(6):726–732
12. Gnoth S, Jenzsch M, Simutis R et al (2007) Process analytical technology (PAT): batch-to-batch reproducibility of fermentation processes by robust process operational design and control. J Biotechnol 132(2):180–186
13. Rathore AS (2009) Roadmap for implementation of quality by design (QbD) for biotechnology products. Trends Biotechnol 27(9):546–553
14. Stephanopoulos G, Locher G, Duff MJ et al (1997) Fermentation database mining by pattern recognition. Biotechnol Bioeng 53(5):443–452
15. Charaniya S, Hu WS, Karypis G (2008) Mining bioprocess data: opportunities and challenges. Trends Biotechnol 26(12):690–699
16. Kamimura RT, Bicciato S, Shimizu H et al (2000) Mining of biological data II: assessing data structure and class homogeneity by cluster analysis. Metab Eng 2(3):228–238
17. Karim MN, Hodge D, Simon L (2003) Data-based modeling and analysis of bioprocesses: some real experiences. Biotechnol Prog 19(5):1591–1605
18. Albert S, Kinley RD (2001) Multivariate statistical monitoring of batch processes: an industrial case study of fermentation supervision. Trends Biotechnol 19(2):53–62
19. Lennox B, Montague GA, Hiden HG et al (2001) Process monitoring of an industrial fed-batch fermentation. Biotechnol Bioeng 74(2):125–135
20. Kirdar AO, Conner JS, Baclaski J et al (2007) Application of multivariate analysis toward biotech processes: case study of a cell-culture unit operation. Biotechnol Prog 23(1):61–67
21. Kirdar AO, Green KD, Rathore AS (2008) Application of multivariate data analysis for identification and successful resolution of a root cause for a bioprocessing application. Biotechnol Prog 24(3):720–726
22. Mandenius CF, Brundin A (2008) Bioprocess optimization using design-of-experiments methodology. Biotechnol Prog 24(6):1191–1203
23. Gnoth S, Jenzsch M, Simutis R et al (2008) Control of cultivation processes for recombinant protein production: a review. Bioprocess Biosyst Eng 31(1):21–39

24. D'haeseleer P (2005) How does gene expression clustering work? Nat Biotechnol 23(12):1499–1501
25. Steuer R, Morgenthal K, Weckwerth W et al (2007) A gentle guide to the analysis of metabolomic data. Methods Mol Biol 358:105–126
26. Izenman AJ (2010) Modern multivariate statistical techniques: regression, classification, and manifold learning. Springer, New York
27. Shaw AD, Winson MK, Woodward AM et al (2000) Rapid analysis of high-dimensional bioprocesses using multivariate spectroscopies and advanced chemometrics. Adv Biochem Eng Biotechnol 66:83–113
28. Skibsted E, Lindemann C, Roca C et al (2001) On-line bioprocess monitoring with a multi-wavelength fluorescence sensor using multivariate calibration. J Biotechnol 88(1):47–57
29. Stark E, Hitzmann B, Schugerl K et al (2002) In situ-fluorescence-probes: a useful tool for non-invasive bioprocess monitoring. Adv Biochem Eng Biotechnol 74:21–38
30. Stephanopoulos G, Nielsen J (1998) Metabolic engineering: principles and methodologies. Academic, San Diego
31. Christensen B, Nielsen J (2000) Metabolic network analysis. A powerful tool in metabolic engineering. Adv Biochem Eng Biotechnol 66:209–231
32. Iwatani S, Yamada Y, Usuda Y (2008) Metabolic flux analysis in biotechnology processes. Biotechnol Lett 30(5):791–799
33. Nielsen J (2001) Metabolic engineering. Appl Microbiol Biotechnol 55(3):263–283
34. Koffas M, Stephanopoulos G (2005) Strain improvement by metabolic engineering: lysine production as a case study for systems biology. Curr Opin Biotechnol 16(3):361–366
35. de Graaf AA, Eggeling L, Sahm H (2001) Metabolic engineering for L-lysine production by *Corynebacterium glutamicum*. Adv Biochem Eng Biotechnol 73:9–29
36. Wiechert W (2002) Modeling and simulation: tools for metabolic engineering. J Biotechnol 94(1):37–63
37. Reed JL, Palsson BO (2003) Thirteen years of building constraint-based in silico models of *Escherichia coli*. J Bacteriol 185(9):2692–2699
38. Balcarcel RR, Clark LM (2003) Metabolic screening of mammalian cell cultures using well-plates. Biotechnol Prog 19(1):98–108
39. Bonarius HP, Hatzimanikatis V, Meesters KP et al (1996) Metabolic flux analysis of hybridoma cells in different culture media using mass balances. Biotechnol Bioeng 50(3):299–318
40. Bonarius HP, Houtman JH, Schmid G et al (2000) Metabolic-flux analysis of hybridoma cells under oxidative and reductive stress using mass balances. Cytotechnology 32(2):97–107
41. Follstad BD, Balcarcel RR, Stephanopoulos G et al (1999) Metabolic flux analysis of hybridoma continuous culture steady state multiplicity. Biotechnol Bioeng 63(6):675–683
42. Dalm MC, Lamers PP, Cuijten SM et al (2007) Effect of feed and bleed rate on hybridoma cells in an acoustic perfusion bioreactor: metabolic analysis. Biotechnol Prog 23(3):560–569
43. Bonarius HP, Ozemre A, Timmerarends B et al (2001) Metabolic-flux analysis of continuously cultured hybridoma cells using (13)CO(2) mass spectrometry in combination with (13)C-lactate nuclear magnetic resonance spectroscopy and metabolite balancing. Biotechnol Bioeng 74(6):528–538
44. Altamirano C, Illanes A, Casablancas A et al (2001) Analysis of CHO cells metabolic redistribution in a glutamate-based defined medium in continuous culture. Biotechnol Prog 17(6):1032–1041
45. Altamirano C, Illanes A, Becerra S et al (2006) Considerations on the lactate consumption by CHO cells in the presence of galactose. J Biotechnol 125(4):547–556
46. Goudar C, Biener R, Zhang C et al (2006) Towards industrial application of quasi real-time metabolic flux analysis for mammalian cell culture. Adv Biochem Eng Biotechnol 101:99–118
47. Niklas J, Schneider K, Heinzle E (2010) Metabolic flux analysis in eukaryotes. Curr Opin Biotechnol 21:63–69

48. Quek LE, Dietmair S, Kromer JO et al (2010) Metabolic flux analysis in mammalian cell culture. Metab Eng 12(2):161–171
49. Boghigian BA, Seth G, Kiss R et al (2010) Metabolic flux analysis and pharmaceutical production. Metab Eng 12(2):81–95
50. Goudar C, Biener R, Boisart C et al (2010) Metabolic flux analysis of CHO cells in perfusion culture by metabolite balancing and 2D [13C, 1H] COSY NMR spectroscopy. Metab Eng 12(2):138–149
51. Birzele F, Schaub J, Rust W et al (2010) Into the unknown: expression profiling without genome sequence information in CHO by next generation sequencing. Nucleic Acids Res 38(12):3999–4010
52. Jacob NM, Kantardjieff A, Yusufi FN et al (2009) Reaching the depth of the Chinese hamster ovary cell transcriptome. Biotechnol Bioeng 105(5):1002–1009
53. Kantardjieff A, Nissom PM, Chuah SH et al (2009) Developing genomic platforms for Chinese hamster ovary cells. Biotechnol Adv 27(6):1028–1035
54. Korke R, Gatti ML, Lau AL et al (2004) Large scale gene expression profiling of metabolic shift of mammalian cells in culture. J Biotechnol 107(1):1–17
55. Schaub J, Clemens C, Schorn P et al (2010) CHO gene expression profiling in biopharmaceutical process analysis and design. Biotechnol Bioeng 105(2):431–438
56. Wong VV, Nissom PM, Sim SL et al (2006) Zinc as an insulin replacement in hybridoma cultures. Biotechnol Bioeng 93(3):553–563
57. Spens E, Haggstrom L (2009) Proliferation of NS0 cells in protein-free medium: the role of cell-derived proteins, known growth factors and cellular receptors. J Biotechnol 141(3–4):123–129
58. Trummer E, Ernst W, Hesse F et al (2008) Transcriptional profiling of phenotypically different Epo-Fc expressing CHO clones by cross-species microarray analysis. Biotechnol J 3(7):924–937
59. Clark KJ, Griffiths J, Bailey KM et al (2005) Gene-expression profiles for five key glycosylation genes for galactose-fed CHO cells expressing recombinant IL-4/13 cytokine trap. Biotechnol Bioeng 90(5):568–577
60. Wong DC, Wong KT, Lee YY et al (2006) Transcriptional profiling of apoptotic pathways in batch and fed-batch CHO cell cultures. Biotechnol Bioeng 94(2):373–382
61. Wong DC, Wong KT, Nissom PM et al (2006) Targeting early apoptotic genes in batch and fed-batch CHO cell cultures. Biotechnol Bioeng 95(3):350–361
62. Shen D, Kiehl TR, Khattak SF et al (2010) Transcriptomic responses to sodium chloride-induced osmotic stress: a study of industrial fed-batch CHO cell cultures. Biotechnol Prog 26(4):1104–1115
63. Wu MH, Dimopoulos G, Mantalaris A et al (2004) The effect of hyperosmotic pressure on antibody production and gene expression in the GS-NS0 cell line. Biotechnol Appl Biochem 40(Pt 1):41–46
64. Yee JC, Gerdtzen ZP, Hu WS (2009) Comparative transcriptome analysis to unveil genes affecting recombinant protein productivity in mammalian cells. Biotechnol Bioeng 102(1):246–263
65. Al-Fageeh MB, Marchant RJ, Carden MJ et al (2006) The cold-shock response in cultured mammalian cells: harnessing the response for the improvement of recombinant protein production. Biotechnol Bioeng 93(5):829–835
66. De Leon GM, Wlaschin KF, Nissom PM et al (2007) Comparative transcriptional analysis of mouse hybridoma and recombinant Chinese hamster ovary cells undergoing butyrate treatment. J Biosci Bioeng 103(1):82–91
67. Wang M, Senger RS, Paredes C et al (2009) Microarray-based gene expression analysis as a process characterization tool to establish comparability of complex biological products: scale-up of a whole-cell immunotherapy product. Biotechnol Bioeng 104(4):796–808
68. Stansfield SH, Allen EE, Dinnis DM et al (2007) Dynamic analysis of GS-NS0 cells producing a recombinant monoclonal antibody during fed-batch culture. Biotechnol Bioeng 97(2):410–424

69. Pascoe DE, Arnott D, Papoutsakis ET et al (2007) Proteome analysis of antibody-producing CHO cell lines with different metabolic profiles. Biotechnol Bioeng 98(2):391–410
70. Seow TK, Korke R, Liang RC et al (2001) Proteomic investigation of metabolic shift in mammalian cell culture. Biotechnol Prog 17(6):1137–1144
71. Kumar N, Gammell P, Meleady P et al (2008) Differential protein expression following low temperature culture of suspension CHO-K1 cells. BMC Biotechnol 8:42
72. Smales CM, Dinnis DM, Stansfield SH et al (2004) Comparative proteomic analysis of GS-NS0 murine myeloma cell lines with varying recombinant monoclonal antibody production rate. Biotechnol Bioeng 88(4):474–488
73. Alete DE, Racher AJ, Birch JR et al (2005) Proteomic analysis of enriched microsomal fractions from GS-NS0 murine myeloma cells with varying secreted recombinant monoclonal antibody productivities. Proteomics 5(18):4689–4704
74. Carlage T, Hincapie M, Zang L et al (2009) Proteomic profiling of a high-producing Chinese hamster ovary cell culture. Anal Chem 81(17):7357–7362
75. Baik JY, Lee GM (2010) A DIGE approach for the assessment of differential expression of the CHO proteome under sodium butyrate addition: effect of Bcl-x(L) overexpression. Biotechnol Bioeng 105(2):358–367
76. Jin M, Szapiel N, Zhang J et al (2010) Profiling of host cell proteins by two-dimensional difference gel electrophoresis (2D-DIGE): implications for downstream process development. Biotechnol Bioeng 105(2):306–316
77. van der Werf MJ, Overkamp KM, Muilwijk B et al (2007) Microbial metabolomics: toward a platform with full metabolome coverage. Anal Biochem 370(1):17–25
78. Chong WP, Goh LT, Reddy SG et al (2009) Metabolomics profiling of extracellular metabolites in recombinant Chinese Hamster Ovary fed-batch culture. Rapid Commun Mass Spectrom 23(23):3763–3771
79. Oldiges M, Lutz S, Pflug S et al (2007) Metabolomics: current state and evolving methodologies and tools. Appl Microbiol Biotechnol 76(3):495–511
80. Bradley SA, Ouyang A, Purdie J et al (2010) Fermentanomics: monitoring mammalian cell cultures with NMR spectroscopy. J Am Chem Soc 132(28):9531–9533
81. Ma N, Ellet J, Okediadi C et al (2009) A single nutrient feed supports both chemically defined NS0 and CHO fed-batch processes: improved productivity and lactate metabolism. Biotechnol Prog 25(5):1353–1363
82. Dietmair S, Timmins NE, Gray PP et al (2010) Towards quantitative metabolomics of mammalian cells: development of a metabolite extraction protocol. Anal Biochem 404(2):155–164
83. Sellick CA, Hansen R, Maqsood AR et al (2009) Effective quenching processes for physiologically valid metabolite profiling of suspension cultured Mammalian cells. Anal Chem 81(1):174–183
84. Baik JY, Lee MS, An SR et al (2006) Initial transcriptome and proteome analyses of low culture temperature-induced expression in CHO cells producing erythropoietin. Biotechnol Bioeng 93(2):361–371
85. Kantardjieff A, Jacob NM, Yee JC et al (2010) Transcriptome and proteome analysis of Chinese hamster ovary cells under low temperature and butyrate treatment. J Biotechnol 145(2):143–159
86. Doolan P, Meleady P, Barron N et al (2010) Microarray and proteomics expression profiling identifies several candidates, including the valosin-containing protein (VCP), involved in regulating high cellular growth rate in production CHO cell lines. Biotechnol Bioeng 106(1):42–56
87. Nissom PM, Sanny A, Kok YJ et al (2006) Transcriptome and proteome profiling to understanding the biology of high productivity CHO cells. Mol Biotechnol 34(2):125–140
88. Jolliffe IT (1986) Principal component analysis. Springer, New York
89. Zhang J, Martin EB, Morris AJ (1997) Process monitoring using non-linear statistical techniques. Chem Eng J 67(3):181–189

90. Schaub J, Mauch K, Reuss M (2008) Metabolic flux analysis in *Escherichia coli* by integrating isotopic dynamic and isotopic stationary 13C labeling data. Biotechnol Bioeng 99(5):1170–1185
91. Pruitt KD, Tatusova T, Maglott DR (2007) NCBI reference sequences (RefSeq): a curated non-redundant sequence database of genomes, transcripts and proteins. Nucleic Acids Res 35(Database issue):D61–D65
92. van der Heijden RT, Heijnen JJ, Hellinga C et al (1994) Linear constraint relations in biochemical reaction systems: I. Classification of the calculability and the balanceability of conversion rates. Biotechnol Bioeng 43(1):3–10
93. van der Heijden RT, Romein B, Heijnen J et al (1994) Linear constrain relations in biochemical reaction systems III. Sequential application of data reconciliation for sensitive detection of systematic errors. Biotechnol Bioeng 44(7):781–791
94. Haggstrom L, Ljunggren J, Ohman L (1996) Metabolic engineering of animal cells. Ann N Y Acad Sci 782:40–52
95. Wlaschin KF, Hu WS (2007) Engineering cell metabolism for high-density cell culture via manipulation of sugar transport. J Biotechnol 131(2):168–176
96. Schneider M, Marison IW, von Stockar U (1996) The importance of ammonia in mammalian cell culture. J Biotechnol 46(3):161–185
97. Stouthamer AH (1973) A theoretical study on the amount of ATP required for synthesis of microbial cell material. Antonie Van Leeuwenhoek 39(3):545–565

Adv Biochem Engin/Biotechnol (2012) 127: 165–185
DOI: 10.1007/10_2011_101

Published Online: 6 October 2011

Protein Glycosylation and Its Impact on Biotechnology

Markus Berger, Matthias Kaup and Véronique Blanchard

Abstract Glycosylation is a post-translational modification that is of paramount importance in the production of recombinant pharmaceuticals as most recombinantly produced therapeutics are N- and/or O-glycosylated. Being a cell-system-dependent process, it also varies with expression systems and growth conditions, which result in glycan microheterogeneity and macroheterogeneity. Glycans have an effect on drug stability, serum half-life, and immunogenicity; it is therefore important to analyze and optimize the glycan decoration of pharmaceuticals. This review summarizes the aspects of protein glycosylation that are of interest to biotechnologists, namely, biosynthesis and biological relevance, as well as the tools to optimize and to analyze protein glycosylation.

Keywords Biopharmaceuticals · Glycan analysis · Glycodesign · Glycoengineering · Glycosylation

Abbreviations

ASGPR	Asialoglycoprotein receptor
CE	Capillary electrophoresis
CHO	Chinese hamster ovary
CMP	Cytidine monophosphate
Dol-P	Dolichol phosphate
EPO	Erythropoietin
Fuc	Fucose
Gal	Galactose
GalNAc	*N*-Acetylgalactosamine
GDP	Guanosine diphosphate
Glc	Glucose
GlcNAc	*N*-Acetylglucosamine
GNE	Uridine diphosphate *N*-acetylglucosamine 2-epimerase/*N*-acetylmannosamine kinase

M. Berger (✉) · M. Kaup · V. Blanchard
Glycodesign and Glycoanalytics, Central Institute of Laboratory Medicine and Pathobiochemistry, Charité Berlin, Charitéplatz 1, 10117, Berlin Germany
e-mail: markus.berger@charite.de

HPAEC-PAD	High-performance anion-exchange chromatography coupled with pulsed amperometric detection
HPLC	High-performance liquid chromatography
Man	Mannose
Neu5Ac	*N*-Acetylneuraminic acid
Neu5Gc	*N*-Glycolylneuraminic acid
UDP	Uridine diphosphate

Contents

1 Introduction

More than 200 protein pharmaceuticals have been approved by authorities for therapeutic use and many more are in the development phases of clinical trials [1]. In 2009, the biopharmaceutical market was estimated to be worth $99 billion worldwide. Antibody-based products, which are glycosylated, represent more than a third of the market and five of the top ten sellers are antibody-based biopharmaceuticals [1]. The global market for protein-based therapeutics is estimated to grow by about 15% annually in the coming years [2, 3], and glycosylation is associated with 40% of all approved biopharmaceuticals. In view of the fact that

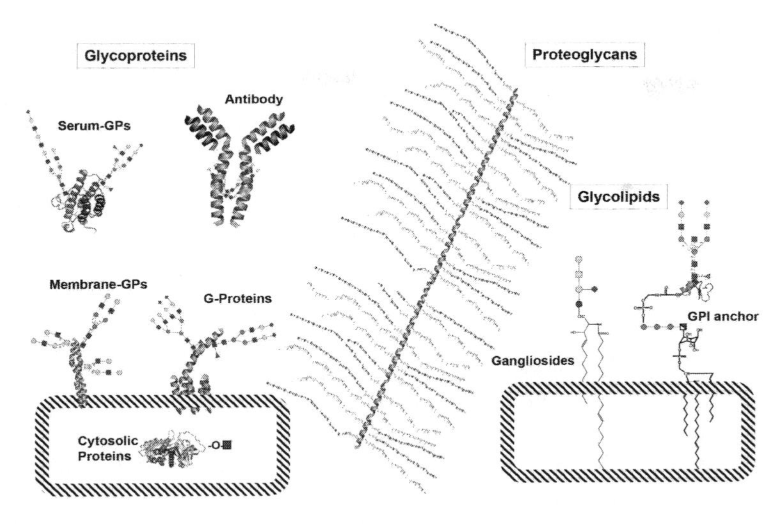

Fig. 1 Overview of the glycoconjugates present in eukaryotic systems: glycoproteins, e.g., serum glycoproteins, membrane-bound glycoproteins, cytosolic proteins, lipid-linked glycoproteins, and proteoglycans. *GPs* glycoproteins, *GPI* glycosylphosphatidylinositol, *green circles* mannose, *yellow circles* galactose, *blue squares* *N*-acetylglucosamine, *yellow squares* *N*-acetylgalactosamine, *red triangles* fucose, *purple diamonds* *N*-acetylneuraminic acid

glycosylation has a high impact on the activity and pharmacokinetics of therapeutics, academic and industrial research laboratories have been working on the improvement of therapeutic applications of glycosylation.

This review will first focus on the structure and biological significance of glycans. Then, the main strategies to optimize recombinant protein glycosylation will be examined. Finally, a brief overview of the techniques to analyze protein glycosylation will be given.

2 Structure and Biosynthesis

As constituents of glycoproteins and glycolipids, glycans play a central role in many essential biological processes (Fig. 1) [4, 5]. Glycoconjugates can be grouped into glycoproteins, e.g., serum glycoproteins (immune globulins), membrane-bound glycoproteins (cell adhesion molecules such as integrins or receptors), cytosolic proteins such as heat shock protein 70, lipid-linked glycoproteins (gangliosides, glycosylphosphatidylinositol-anchored proteins), and proteoglycans. They consist of a protein backbone which is heavily glycosylated with disaccharide repeating units (glycosaminoglycans), for instance, decorin, which forms one of the major components of the extracellular matrix. Over 50% of all proteins are glycoproteins and it is estimated that 1–2% of the genome encodes for glycan-related genes [6, 7].

2.1 Carbohydrate Diversity

Carbohydrates ($C_x(H_2O)_y$) can be defined as polyhydroxyaldehydes and polyhydroxyketones, the simplest ones found in nature being monosaccharides and disaccharides ("saccharide" is derived from *saccharon*, the Latin word for "sugar"). Glycans are composed of monosaccharides and are classified as oligosaccharides (two to 20 monosaccharides) or polysaccharides (more than 20 monosaccharides). The family of monosaccharides consists of 367 different members [8], which are named according to their number of carbon atoms ("triose" for three carbon atoms or "hexose" for six carbon atoms), their functional group ("aldose" for aldehydes and "ketose" for ketone), their ring size ("pyranose" for a six-membered ring and "furanose" for a five-membered ring), and their anomeric carbon atom (orientation of the hydroxyl group on the asymmetric center: D or L, α or β). After incorporation into glycoconjugates, oligosaccharides can be posttranslationally modified by phosphorylation, sulfation, or acetylation. The most abundant monosaccharide, glucose (Glc), is the repeating unit of the most widespread biopolymers. Glc polymers are the biggest resource of biomolecules. They mostly occur in nature in the form of cellulose (β1,4 linkage) and in the form of starch (α1,4 and α1,6 linkages). Their main function is to provide the host organism with energy. The most common monosaccharides found in *N*-glycans and *O*-glycans of higher animals are hexoses [galactose (Gal), mannose (Man)], deoxyhexoses [fucose (Fuc)], hexosamines [*N*-acetylglucosamine (GlcNAc) and *N*-acetylgalactosamine (GalNAc)], and sialic acids [*N*-acetylneuraminic acid (Neu5Ac) and *N*-glycolylneuraminic acid (Neu5Gc)]. N-acetylation at the C-2 position of Glc and Gal leads to GlcNAc and GalNAc. Deoxyhexoses lack a hydroxyl group at the C-6 position, and sialic acids have a backbone of nine carbon atoms and have a carboxyl group at C-1 (Fig. 2).

2.2 Glycoprotein Glycosylation

N-Glycans are covalently attached to the side chain of asparagine residues of glycoproteins via a GlcNAc. They share a common core structure, which consists of two GlcNAc followed by three Man residues. Further additions and trimming leads to three different *N*-glycan classes, namely high-Man, hybrid, and complex -*N*-glycans (Fig. 3). Protein glycosylation is initiated in the endoplasmic reticulum by a common consensus sequence motif, Asn-X-Ser/Thr, where X is any amino acid except Pro.

O-glycosylation of serine or threonine residues of glycoproteins occurs in the Golgi apparatus. Consensus sequences have not been reported yet, but some bioinformatics tools such as NetOGlyc allow O-glycosylation sites to be predicted [9]. NetOGlyc compares sequences with databases combining in vivo O-glycosylation of mammalian glycoproteins as well as the structure around the O-glycosylation sites.

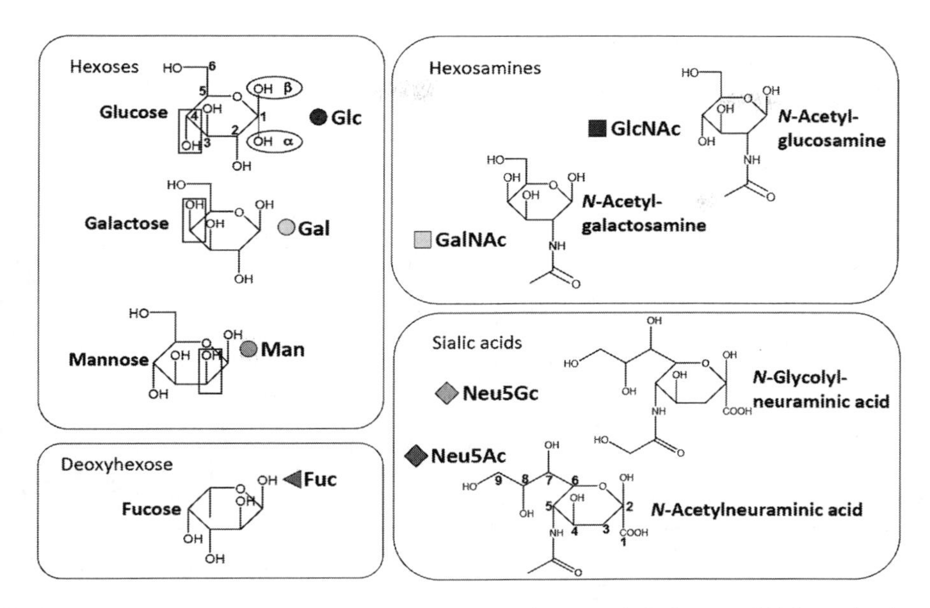

Fig. 2 Most common monosaccharides found in *N*-glycans and *O*-glycans of higher animals. The differences between hexoses are marked. Since 2005, most glycobiologists have adopted the symbol and color code proposed by EUROCarbDB to represent glycans (http://relax.organ.su.se:8123/eurocarb/home.action)

In contrast to *N*-glycans, O-linked glycans are classified by eight different core structures starting with a GalNAc residue (Fig. 3). Other types of O-glycosylation have been reported and occur as *O*-GlcNAc, *O*-Glc, *O*-Fuc, and *O*-Man at serine or threonine residues [10]. C-mannosylation [11] and phosphoserine glycosylation [12] are some of the newest types of protein glycosylation reported; phosphorylated serines are linked to GlcNAc, Man, Fuc, or xylose through the phosphodiester bond, and C-mannosylation occurs at tryptophan residues.

2.3 N-*Glycan and* O-*Glycan Biosynthesis*

The biosynthesis of *N*-glycans and *O*-glycans begins in the cytosol of vertebrates with the formation of activated monosaccharides as dolichol phosphate (Dol-P) or nucleotide derivates. The activated monosaccharides [Dol-P-Man, uridine diphosphate (UDP)–Gal, UDP-GlcNAc, UDP-GalNAc, guanosine diphosphate (GDP)–Man, GDP-Fuc, cytidine monophosphate (CMP)–Neu5Ac] are transported to the endoplasmic reticulum and Golgi apparatus, where the stepwise biosynthesis of the glycans occurs (Figs. 4, 5) [6]. It is a complex process which involves many enzymes from different pathways. To date, about 700 glycan-related genes have been identified [13]. These genes code for the so-called glycosylation machinery

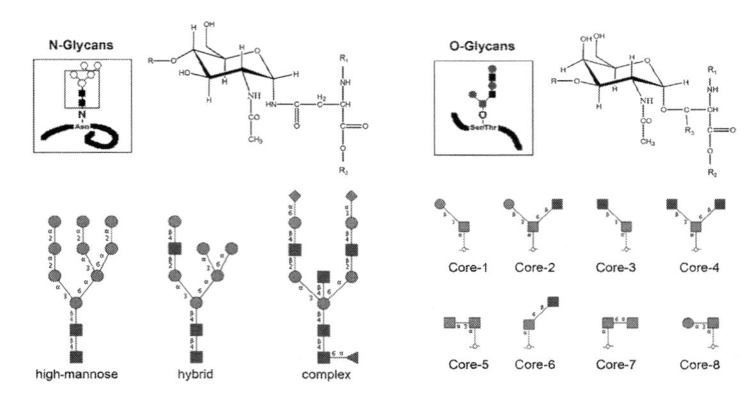

Fig. 3 Structure and linkage of *N*-glycans and *O*-glycans to the protein backbone. **a** *N*-Glycans linked to an asparagine residue of the polypeptide chain (the core structure is marked). The three types of *N*-glycans are shown below (high-mannose type, hybrid type, complex type). **b** *O*-Glycans linked to a serine or threonine residue of the polypeptide chain. For *O*-glycans, there is no common core structure, but eight different core structures known. *Green circles* mannose, *yellow circles* galactose, *blue squares* *N*-acetylglucosamine, *yellow squares* *N*-acetylgalactosamine, *red triangle* fucose, *purple diamonds* *N*-acetylneuraminic acid. R_1 and R_2 are polypeptide chains, R_3 is H (serine) or CH_3 (threonine)

such as kinases and epimerases in nucleotide biosynthesis, transporters, glycosyltransferases, glycosidases, glycan-modifying enzymes (e.g., glycan sulfation), and carbohydrate-binding proteins (lectins) [13]. The stepwise biosynthesis starts in the cytosol with the formation of a heptasaccharide on a lipid-linked precursor, Dol-P, consisting of two GlcNAc and five Man. After a "fliplike" mechanism from the cytosol into the endoplasmic reticulum lumen, the precursor is finalized to the common $Glc_3Man_9GlcNAc_2$ precursor and transferred via the oligosaccharide transferase complex to the polypeptide [14]. At this early stage, the correct folding undergoes a glycan-based quality control. Calnexin and calreticulin, two chaperone-like glycan-binding proteins, attach to and detach from proteins and recognize proper folding [15]. Once proteins are correctly folded, three Glc residues and one Man residue are cleaved by specific glycosidases and the newly formed glycoproteins enter the Golgi apparatus via vesicles [16] (Fig. 4). The glycan precursors are degraded to $Man_5GlcNAc_2$ structures. This deglycosylation is the starting point of the final glycoprotein processing, which is the provision and the transfer of UDP-GlcNAc, UDP-Gal, CMP-Neu5Ac, and GDP-Fuc residues by a subset of Golgi nucleotide transporters, glycosyltransferases, and glycosidases (Fig. 5) [17].

O-Glycan processing is initiated by the transfer of GalNAc to serine and threonine residues via a GalNAc transferase. Nascent *O*-glycan chains are further elongated by glycosyltransferases that transfer activated monosaccharides [18, 19].

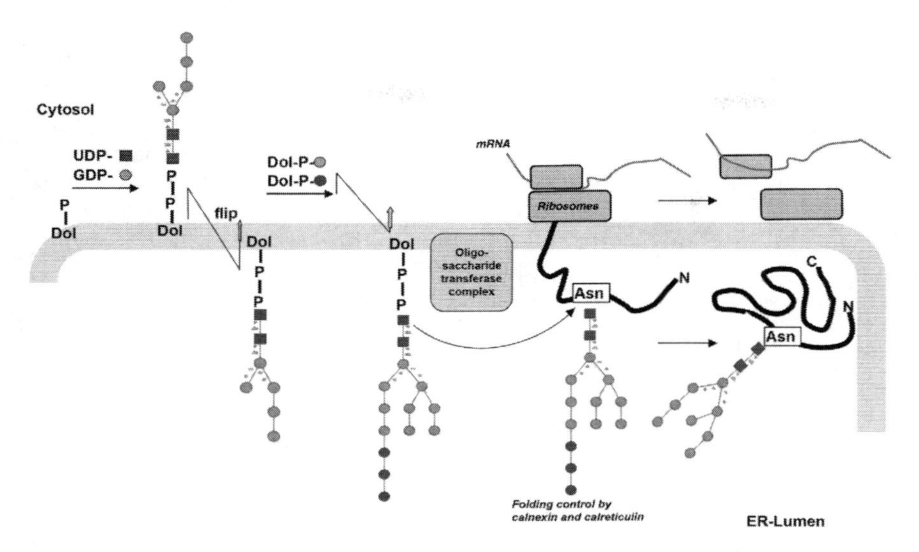

Fig. 4 Processing of the precursor for *N*-glycans in the endoplasmic reticulum. *Dol-P* dolichol phosphate, *ER* endoplasmic reticulum, *GDP* guanosine diphosphate, *mRNA* messenger RNA, *UDP* uridine diphosphate *green circles* mannose, *blue circles* glucose, *blue squares* *N*-acetylglucosamine

2.4 Sialic Acid Biosynthesis

Sialic acids, derived from neuraminic acid, consist of a backbone of nine carbons with an amino group at position C-5 and constitute the classic terminal acidic monosaccharide of glycoprotein glycans. They belong to a family of more than 50 members differing in the substitution types (e.g., acetyl, methyl, sulfate, phosphate) and positions (C-4, C-5, C-7, C-8, C-9) [20]. Sialic acids are characterized by a carboxyl group at position C-1 that confers strong acidity (p*K* 2.2) [21, 22]. The biosynthesis of sialic acids begins with UDP-GlcNAc, which enters the pathway by de novo synthesis starting with fructose 6-phosphate or by the salvage pathway via activation of GlcNAc from degraded glycoproteins [23]. UDP-GlcNAc is converted by the bifunctional enzyme UDP-*N*-acetylglucosamine 2-epimerase/*N*-acetylmannosamine kinase (GNE) into *N*-acetylmannosamine 6-phosphate. After condensation with phosphoenolpyrovate by Neu5Ac 9-phosphate synthase and dephosphorylation by Neu5Ac 9-phosphate phosphatase, free Neu5Ac is synthesized. Thus, Neu5Ac is the only monosaccharide, which is activated in the nucleus [24, 25]. After activation with cytidine triphosphate by CMP-Neu5Ac synthase, CMP-Neu5Ac is released in the cytosol. The activated neuraminic acids enter the Golgi apparatus, where they are transferred to the terminal position of glycoconjugates [26] or act as a negative-feedback inhibitor for GNE and consequently reduce the synthesis of neuraminic acids [27].

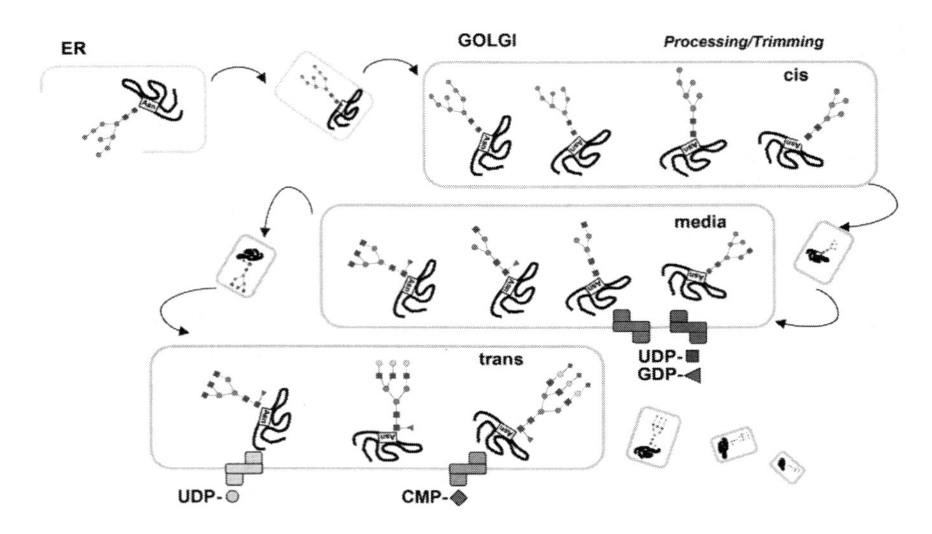

Fig. 5 *cis*-Golgi, *media*-Golgi, and *trans*-Golgi network with cytosolic UDP, GDP, and cytidine monophosphate (*CMP*) nucleotides and specific transmembrane transporters with the corresponding color code. Glycosyltransferases and glycosidases are not depicted. *Green circles* mannose, *yellow circles* galactose, *blue squares* *N*-acetylglucosamine, *red triangles* fucose, *purple diamonds* *N*-acetylneuraminic acid

3 Biological Impact of Protein Glycosylation

3.1 Stability and Serum Half-Life

The most obvious function of protein glycosylation is to facilitate protein solubility and stability. For instance, if fibrinogen and human granulocyte colony-stimulating factor are de-N-glycosylated and de-O-glycosylated, respectively, aggregates are formed, which results in biological inactivity [28, 29]. Glycosylation also ensures the protection of proteins against proteases by masking cleavage sites [30, 31]. Rudd et al. [32] suggested that the steric protection of the peptide moieties by the neighboring *N*-glycans is due to hydrogen bonding between the hydrophilic amino acids and glycans.

Another well-known function of sialylated glycans is to prolong circulation of glycoproteins in serum. When glycans of glycoproteins are terminated in Gal and not sialic acids, they are recognized by the asialoglycoprotein receptor (ASGPR), which results in a drastic reduction of serum half-life [33]. The ASGPR, located on the surface of hepatocytes [34, 35], is not able to recognize fully sialylated glycoproteins, but, during blood circulation, terminal sialic acids are cleaved off by unspecific sialidases. Subsequently, the ASGPR recognizes Gal and GalNAc, which are not capped anymore by sialic acids. Hence, glycoproteins are internalized and degraded [36–38].

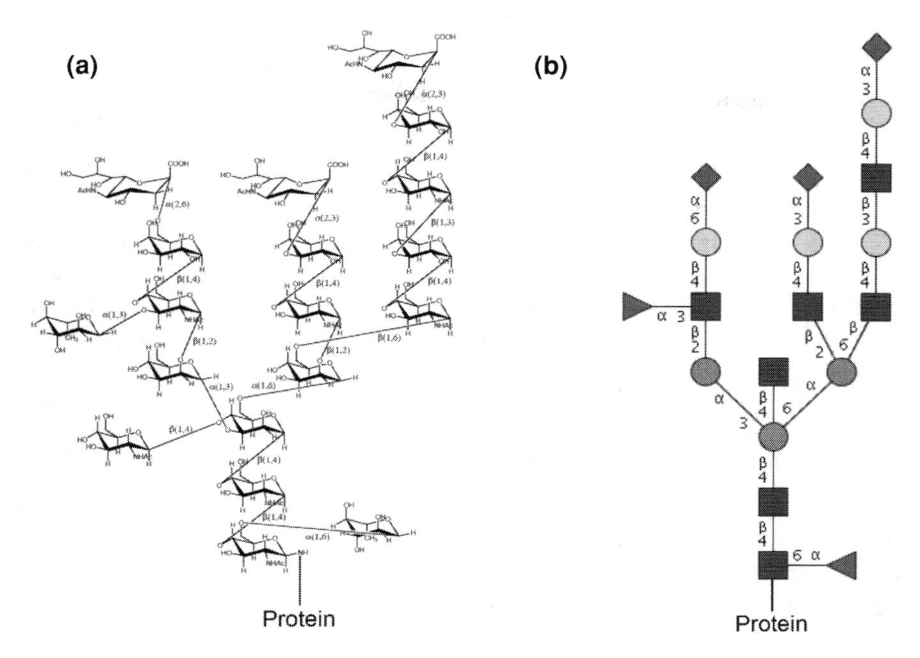

Fig. 6 **a** Chemical drawing with composition and linkage information, **b** Most frequently used simplified carbohydrate drawing (GlycoWorkbench) [119]. *Green circles* mannose, *yellow circles* galactose, *blue squares* *N*-acetylglucosamine, *red triangles* fucose, *purple diamonds* *N*-acetylneuraminic acid

3.2 Signal Transduction and Cell Adhesion

It has been established that bisecting GlcNAc, which is β1,4-linked to the Man residue located at the base of the trimannosyl core (Fig. 6), and core Fuc are involved in signal transduction and cell adhesion by regulating the function of glycoproteins. Wand et al. [39] and Saito et al. [40] showed that core fucosylation is essential for the binding of epidermal growth factor to its receptor, whereas bisecting GlcNAc favors the endocytosis of its receptor. The importance of bisecting GlcNAc and core Fuc was also established for recombinant antibodies that are used to treat various types of diseases such as cancer and autoimmune diseases [1]. It was shown that the absence of Fuc and the presence of bisecting GlcNAc at asparagine 297 in the Fc region enhance the effector functions of antibodies by up to 100-fold [41].

3.3 Immunogenicity

Human cells produce exclusively sialic acids of the Neu5Ac-type, whereas mammalian cell lines, used to produce biopharmaceuticals, express Neu5Ac as well as the non-human Neu5Gc. This monosaccharide is formed by CMP-Neu5Ac hydroxylase, which is absent in humans since a knockout mutation occurred about three million years ago [42]. As a consequence, Neu5Gc is immunogenic to humans [43] and recombinant glycoproteins from mammalian sources can bear Neu5Gc. Chinese hamster ovary (CHO) cells are the most widely used expression system for the production of FDA-approved recombinant therapeutics such as erythropoetin (EPO) (Epogen, Amgen) [44–46]. Glycoproteins expressed in CHO cells are usually highly sialylated and are decorated with α2,3-linked Neu5Ac as well as minor amounts of the immunogenic Neu5Gc (up to 3%) [47, 48]. Some human cells, such as stem cells, are grown with animal products such as serum or feeder layers during the culture [49]. The use of stem cells for regenerative therapies is therefore affected as well; the incorporation of Neu5Gc cannot be excluded and may result in immunological risks [50].

Mammals, with the exception of Old World monkeys, apes, and humans, express an α1,3-galactosyltransferase and accordingly add Gal residues to galactosylated glycans [51]. Humans only express a functional β1,4-galactosyltransferase, the α1,3-galactosyltransferase gene being a dysfunctional pseudogene [52]. As a consequence, glycans with α1,3 Gal residues are immunogenic to the human immune system, which prevents, for instance, xenotransplantations of pig organs [53, 54]. Murine NS0 or Sp2/0 cell lines used for the production of monoclonal antibodies (CD 20 antibody, ofatumumab, GlaxoSmithKline, IL-2R antibody, daclizumab, Hoffman-LaRoche) [55–57] may also contain traces of this epitope; therefore, the glycosylation of recombinant glycoproteins expressed in non-human systems, which may lead to hypersensitivity reactions when patients are injected with them, should particularly be controlled [58, 59].

4 Glycoengineering: Strategies to Influence Protein Glycosylation

More than half of the commercially available biopharmaceuticals that result from genetic engineering are glycoproteins [1]. Therefore, a major concern of biopharmaceutical laboratories is to monitor and tune glycosylation carefully. An optimal glycosylation is usually considered to be complete galactosylation (β1,4) and sialylation (α-linked Neu5Ac); in the following sections we review different "glycoengineering" or "glycodesign" approaches to influence glycan macroheterogeneity (site occupancy) as well as microheterogeneity (nature of glycans attached at a specific site) in order to modulate the degrees of galactosylation, fucosylation, and sialylation (Fig. 7).

4.1 Modifications of Glycan Biosynthetic Pathways

Each glycosyltransferase, glycosidase, and transporter involved in the biosynthetic pathway of the activated monosaccharides is a potential target to modulate the glycosylation machinery of a production cell line and therefore the glycosylation pattern of a biopharmaceutical.

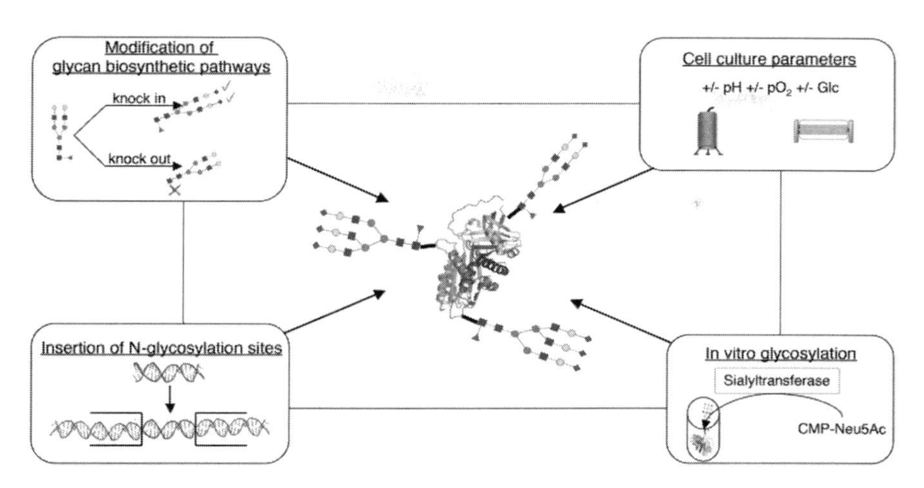

Fig. 7 Strategies to optimize glycoprotein glycosylation, so-called glycodesign. The choices regarding the expression system and parameters influence the resulting glycosylation. Various strategies, including in vitro glycosylation, modification of the biosynthetic pathways, and addition of N-glycosylation, are able to modulate glycoprotein glycosylation. *Glc* glucose, *Neu5Ac* *N*-acetylneuraminic acid

The key enzyme of the sialic acid pathway is the bifunctional GNE. The epimerase domain is regulated by a negative-feedback mechanism through the end product of the pathway, the activated sialic acid, CMP–sialic acid. A knockout of the epimerase domain results in a loss of the negative-feedback mechanism. Feeding the cell culture medium with *N*-acetylmannosamine, a sialic acid precursor, enhances sialylation via salvage pathways [60, 61]. On the basis of a pathological background in humans, it was shown that a mutant of GNE causes sialuria. Sialuria is a rare inborn disorder that is characterized by an excessive renal clearance of sialylated glycoproteins on the gram scale. This is due to a mutation within the epimerase domain, which results in a defective feedback inhibition process. This mutation has successfully been inserted in CHO cells and led to the production of highly sialylated recombinant EPO [62]. Another way to increase sialylation is to insert human α2,6-sialyltransferase in CHO cells as these cell lines produce α2,3-linked but no 2,6-linked Neu5Ac [63]. This insertion results in the production of humanized glycoproteins bearing both α2,3-linked and 2,6-linked Neu5Ac [64, 65]. A successful example of the knockout strategy is the reduction of the fucosylation by knocking out corresponding fucosyltransferases. In CHO cells, the FUT8 gene was knocked out and this resulted in the production of antibodies devoid of core Fuc that had a higher antibody-dependent cell-mediated cytotoxicity [41, 66]. An alternative defucosylation strategy is the decrease of the substrate availability, the reduction of GDP-Fuc. This is achieved by deflecting the Fuc de novo pathway using a highly effective prokaryotic enzyme [67].

A similar approach has been successfully established which combines a human-like glycosylation with high yields obtained using yeast and plant-based systems

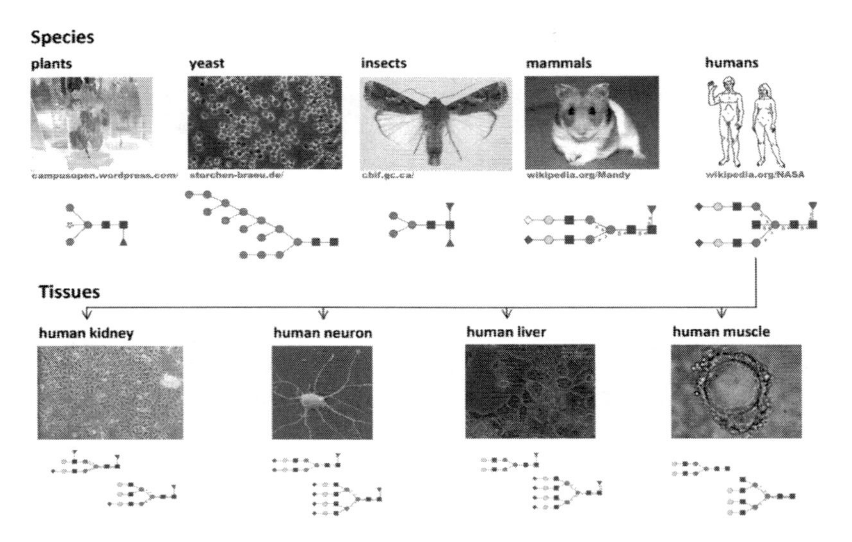

Fig. 8 Overview of different expression systems and their main types of glycosylation. Plant glycans contain xylose, which is antigenic for humans. Yeast glycoproteins bear exclusively high-mannose-type glycans and therefore recombinant products have a short half-life in serum. Insects produce only pauci-mannose structures, whereas the glycosylation machinery of mammals produces mainly complex glycans. Human cell lines express complex glycans containing *N*-acetylneuraminic acid but no *N*-glycolylneuraminic acid. Depending on the origin of the cell lines, their glycosylation machineries may be different (the different glycosylation patterns are shown below the type of tissue). *Green circles* mannose, *yellow circles* galactose, *blue squares* *N*-acetylglucosamine, *yellow squares* *N*-acetylgalactosamine, *red triangles* fucose, *purple diamond*, *N*-acetylneuraminic acid, *white diamond* *N*-glycolylneuraminic acid

("humanization" of the glycosylation machinery). Yeast glycoproteins are decorated with high-Man structures (Fig. 8), which are generally quickly recognized by the Man-binding receptor and removed from blood circulation [68]. In *Pichia pastoris*, nineteen yeast-specific enzymes were knocked out and glycosyltransferases from different biological sources were knocked in. This resulted in the production of antibodies having human-like sialylated biantennary structures [69, 70]. Plant glycosylation consists of trimannosyl chitobiose structures bearing two additional epitopes, namely, β1,2-xylose and core α1,3-fucose, that are immunogenic to mammals (Fig. 8). Plant glycosylation has recently been humanized by several research groups [71–73], resulting in the expression of diantennary digalactosylated *N*-glycan structures that are free from plant carbohydrate antigens [72].

4.2 Insertion of Additional N-Glycosylation Sites

An interesting approach to increase glycan macroheterogeneity is to raise the number of N-glycosylation sites of a given protein. The enhanced glycosylation and thus the increased sialylation should protect the biopharmaceutics against early degradation by the ASGPR. It is relatively easy to clone N-glycosylation motifs into the respective nucleic acid sequence. Generally, the Asn-X-Thr motif is more efficiently glycosylated than Asn-X-Ser. Studies have revealed that the occupancy of a particular glycosylation site additionally depends on the amino acid in the second position, the position of Asn-X-Ser/Thr in the three-dimensional structure, and the flanking structural confirmations [74]. Therefore, it is very important to locate the new glycosylation sites with the restrictions mentioned above and to avoid placing them in the functionally important domains of the protein. A very prominent and successful example is darbepoetin alfa (Amgen) [34, 75]. This genetically engineered EPO bears two additional N-glycosylation sites. Darbepoetin alfa is characterized by a 3 times prolonged serum half-life of 32 h compared with recombinant human EPO. Human alpha interferons are a family of cytokines that inhibit cell proliferation and viral infections. Recombinant human alpha interferon is an FDA-approved therapeutic used in the treatment of cancer and chronic viral diseases [76–78]. It is not glycosylated, which results in a short circulatory half-life in humans of about 4–8 h [79]. Four N-glycosylation sites were introduced by site-directed mutagenesis; the glycoengineered cytokine was posttranslationally modified with trisialylated and tetrasialylated *N*-glycans [80], resulting in a 25-fold increase in the half-life and a 20-fold decrease in the systemic clearance rate compared with the non-glycosylated cytokine [81]. The same strategy has been used for other recombinant glycoproteins, such as follicle-stimulating hormone [82]. In principle, this method can be used for all N-glycosylated glycoproteins and for non-glycosylated serum proteins as well. The location of the additional glycosylation sites ("design" strategy) is facilitated if information about the active site of the protein of interest is available (X-ray, nuclear magnetic resonance data). But its success depends on the quality of information available about the amino acids and domains that surround the new N-glycosylation sites during biosynthesis. Thus, effective N-glycosylation cannot be guaranteed because a proper protein folding is highly dependent on the first glycosylation steps in the endoplasmic reticulum. Proteins can be misfolded and degraded or additional glycosylation sites may not systematically modify the serum half-life.

4.3 Cell Culture Parameters

Cell culture parameters have been reported to influence significantly the glycan microheterogeneity of recombinant glycoproteins [83–87]. Temperature, pH, dissolved oxygen, and medium content such as ammonia content are paramount parameters to control in order to minimize charge-to-charge variations. The accumulation of ammonia has been correlated with significant loss of sialic acids on both *N*-glycans and *O*-glycans [86, 88]. Shear stress influences the glycosylation of recombinant glycoproteins [89]. Fucosylation was shown to increase with the percentage of dissolved oxygen during the production of EPO in CHO cells [84]. It was also demonstrated that pH variations (below 6.9 and above 8.2) lead to a decrease of the overall protein glycosylation [90]. Temperature variations may also result in altered glycosylation. Temperature decrease correlates with an increase in polylactosaminylation [91] and an increase in site occupancy [85], which may be due to the longer transit time of the nascent glycoproteins in the Golgi apparatus. Manganese and iron supplementation increases the site occupancy of human recombinant tissue plasminogen activator without interfering with cell growth or protein productivity [87].

4.4 In Vitro Glycosylation

In vitro glycosylation consists of the addition of carbohydrate moieties to the recombinant glycoproteins after the expression, which is performed either enzymatically or chemically. Raju et al. [92] extended the *N*-glycan chains of glycoproteins using β1,4-galactosyltransferase, α2,3-sialyltransferase, and the corresponding sugar nucleotides, which is time-consuming and quite costly. Fernandes et al. [93–95] chemically coupled polysialic acids to asparaginase and catalase, which enhanced their serum half-lives. Another example is the chemical coupling via oxime chemistry of mannose 6-phosphate to recombinant acid α-glucosidase, which is used in the treatment of Pompe disease [96]. The glycoengineered recombinant glycoprotein showed a higher affinity for the mannose 6-phosphate receptor, resulting in better uptake of the drug by muscle cells [97].

5 Glycoanalytics

5.1 Glycomics Compared with Genomics and Proteomics

If the sequence of a gene is elucidated, it is possible to predict the amino acid sequence of the resulting protein but not the glycans attached to it. DNA and proteins are linear molecules and, from an analytical point of view, are relatively easy to analyze compared with glycans, which are branched. Each hydroxyl group of a monosaccharide is potentially a new branching point of a glycosidic bond, which creates a new stereogenic center (Fig. 6). A peptide with three amino acids can build 3^3 (27) tripeptides. All peptides are linear and have the same type of linkage. Because of the structural diversity described above, three monosaccharides can theoretically result in 38,016 different trisaccharides calculated by [(permutation of sequence) × ring size × anomeric carbon atoms × linkages] or $E^n \times 2_r^n \times 2_a^n \times 4^{n-1}$ (linear forms) + E^n x $2_r^n \times 2_a^n$ x 6^{n-2} (branched forms) where E is the library of monosaccharides and n is the oligomeric size [98]. A calculation with the nine most common monosaccharides in a human system results in more than 15 million possible tetrasaccharides. If one relates this to hexasaccharides, there are 10^{15} theoretical possible structures from 20 monosaccharides compared with 20^6 hexapeptides from 20 proteinogenic amino acids and 4^6 possible hexanucleotides from four nucleotides [99, 100]. Such figures are quite high but nature does not synthesize all the possible combinations; therefore, glycan analysis is complex but not unmanageable.

Ongoing glycomic studies are interested in solving structure–function relationships between sets of glycans and in certain biological contexts. For that, national and international networks and research groups are coming together to unify the different carbohydrates syntaxes and to establish a public database for glycans which can be provided by data from different analytical methods, e.g., mass spectra and chromatograms, such as the Consortium for Functional Glycomics (USA; http://functionalglycomics.org), the Kyoto Encyclopedia of Genes and Genomes (Japan; http://www.genome.jp/kegg/glycan) and the European initiative EUROCarbDB (http://www.eurocarbdb.org). In 2003, the first data mining revealed 6,296 glycan structures [101]; in 2008, 23,118 distinct glycan structures were listed in the Complex Carbohydrate Structure Database (Complex Carbohydrate Research Center), which is the largest public glycan-related database [102]. This indicates that the calculated complexity of glycans does not match the analyzed structures and that glycan analysis is really sophisticated and difficult. In comparison with genomics and proteomics, about three billion base pairs and about 25,000 genes were sequenced and identified by the Human Genome Project during the same time period [103]. This discrepancy is due to the fact that glycan analysis is not as automated as genomics and proteomics are.

5.2 Glycan Analysis

As described in the previous sections, glycoengineering or "glycodesign" strategies as well as process parameters affect the glycan content of biopharmaceuticals. This may result in a modification of the efficacy of the end products. Therefore, international guidelines on the quality control of recombinant glycoproteins [104] recommend determining the glycan content of pharmaceuticals exhaustively. The methods used to analyze glycoproteins are part of the proteomics analysis repertoire and involve glycan-specific techniques to unravel structural complexity.

Clone screening can be performed using lectins. Lectins are (glyco-)proteins that bind specifically to monosaccharides or small carbohydrate domains mostly comprising disaccharides and/or trisaccharides [105]. They have been widely used to purify, enrich, or obtain a general overview of the glycosylation [106, 107]. They are useful for clone screening, but they are not used during the control of the quality of end products.

Each glycoprotein is unique with regard to its structural conformation, number of disulfide bridges, and sites of N- and O-glycosylation. This implies that a quantitative release of the glycans is always glycoprotein-dependent. As a

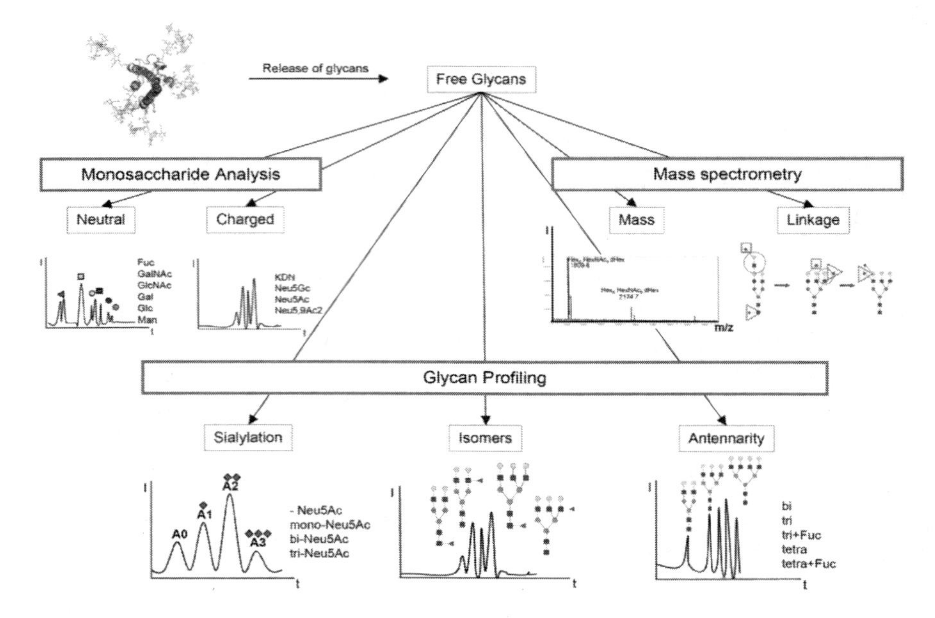

Fig. 9 Simplified overview of glycoanalytical methods

consequence, the so-called glycan release is the most difficult and critical step in a glycoanalytical route. Information about the protein sequence (potential protease cleavage sites), the host organism (bacterial, plant, mammalian, human), the nature of the sample (supernatant, kind of media), the biological constitution (purified supernatant, serum, tissue), the kind of glycosylation, the combination of N- and O-glycosylation, and finally the specific questioning are prerequisites to develop an analysis scheme. Owing to the different features of applied analytical methods, it is always advisable to combine several types of analyses to obtain consistent and reliable results. The broad methodical spectrum ranges from chromatographic and/or electrophoretic techniques to mass-spectrometric techniques (Fig. 9). *N*-Glycans and *O*-glycans are usually cleaved off the proteins, isolated, and finally characterized.

The nature and the total content of each carbohydrate constituent can be investigated by monosaccharide analysis, which provides general information about the type of glycans (high-Man, complex, hybrid). To this end, samples are hydrolyzed and the resulting monosaccharides are analyzed by high-performance anion-exchange chromatography coupled with pulsed amperometric detection (HPAEC-PAD) [67]. This technique is based on the separation of molecules according to their acidic properties. Monosaccharides, even neutral ones, are very weak acids and also weak anions in basic solutions. At pH 12, chromatographic separation of substances having very similar pK_a values, e.g., Glc (pK_a 12.28) and Gal (pK_a 12.39), can be achieved. Furthermore, HPAEC-PAD can also be used to profile and fractionate glycan pools [108]. This technique is very broadly used in the biopharmaceutical industry because PNGase F digests can be directly analyzed without any chemical derivatization. Another advantage is that isomer separation may be achieved in a single run.

The other techniques require chemical labeling of the reducing end for detection purposes. The well-established method of high-performance liquid chromatography (HPLC) is applied for the profiling and, if necessary, the fractionation of glycans. They are separated according to their antennarity (biantennary, triantennary, tetraantennary structures) or according to their charge (sialic acids, phosphorylation, sulfation) [109, 110]. Besides HPLC, a relatively recent method for the analysis of glycans is capillary electrophoresis (CE) [111, 112]. Both methods have the same time-consuming labeling step in common (2-aminobenzamide is used in HPLC, and 8-aminopyrene-1,3,6-trisulfonate is used in CE) but differ with respect to their time per run (20–30 min for CE, and 1–2 h for HPLC). CE, which separates glycans according to their charge to size ratio, is able to differentiate between structural isomers (core and antennary Fuc for instance). For migration purposes, 8-aminopyrene-1,3,6-trisulfonate, containing three negative charges, is the preferred method. Sialylated glycans, migrating too fast, are eluted almost simultaneously. Taking this technical principle into consideration, one obtains quantitative and fast CE results but loses information about the sialylation degree because of the necessary desialylation.

Mass spectrometry is one of the key tools for glycobiologists in the same way as it is in the field of proteomics [113–117]. The difference is that peptides are always ionized better than glycans; it is therefore necessary to separate glycans from peptides before performing analyses. To meet this challenge, each glycan preparation step, starting with denaturation and progressing to change of buffer conditions, desalting, enzymatic or chemical glycan cleavage, separation of peptides from glycans or glycopeptides, enrichment, and finally the purification of glycans, has to be performed very carefully to ensure the purity of glycan samples prior to mass-spectrometric analyses. The last and equally important working step is the interpretation of the resulting chromatograms, electropherograms, and spectra. As mentioned before, there is unfortunately no automated one-step analysis with online prediction of molecules. Semi-automatic tools are already available [118, 119] but most of the electrospray ionization data have to be assigned manually with a calculator.

6 Conclusion

Biotechnology is a relatively new branch in the pharmaceutical industry that has developed rapidly in the last three decades. As post-translational modifications have modulating effects on protein stability, prolonged half-life, and bioactivity, glycoengineering (or "glycodesign") tools have been developed to enhance the bioactivity and to suppress the potential immunogenicity of pharmaceuticals. In the field of glycan analysis, robust methods are now available, but automation is still being developed. Future advances will probably focus on the increase of productivity as well as the minimization of therapeutic doses in order to meet the growing demand.

References

1. Walsh G (2010) Nat Biotechnol 28(9):917
2. Goodman M (2009) Nat Rev Drug Discov 8(11):837
3. Hiller A (2009) Genet Eng Biotechnol News 29:153
4. Varki A (1993) Glycobiology 3(2):97
5. Laine RA (1997) In: Gabius H-J, Gabius S (eds) Glycosciences: status and perspectives. Chapman and Hall, London, pp 1–14
6. Ohtsubo K, Marth JD (2006) Cell 126(5):855
7. Apweiler R, Hermjakob H, Sharon N (1999) Biochim Biophys Acta 1473(1):4
8. von der Lieth CW, Freire AA, Blank D, Campbell MP, Ceroni A, Damerell DR, Dell A, Dwek RA, Ernst B, Fogh R, Frank M, Geyer H, Geyer R, Harrison MJ, Henrick K, Herget S, Hull WE, Ionides J, Joshi HJ, Kamerling JP, Leeflang BR, Lütteke T, Lundborg M, Maass K, Merry A, Ranzinger R, Rosen J, Royle L, Rudd PM, Schloissnig S, Stenutz R, Vranken WF, Widmalm G, Haslam SM (2010) Glycobiology 21(4):493
9. Julenius K, Mølgaard A, Gupta R, Brunak S (2005) Glycobiology 15(2):153
10. Varki A, Cummings R, Esko J, Freeze H, Stanley P, Bertozzi C, Hart G, Etzler M (2009) Essentials of glycobiology, 2nd edn. Cold Spring Harbor Laboratory Press, Cold Spring Harbor, New York
11. Hofsteenge J, Muller DR, de Beer T, Loffler A, Richter WJ, Vliegenthart JFG (1994) Biochemistry 33:13524
12. Haynes PA (1998) Glycobiology 8:1
13. Nairn AV, York WS, Harris K, Hall EM, Pierce JM, Moremen KW (2008) J Biol Chem 283(25):17298
14. Rush JS, Waechter CJ (1995) J Cell Biol 130:529
15. Molinari M, Eriksson KK, Calanca V, Galli C, Cresswell P, Michalak M, Helenius A (2004) Mol Cell 13(1):125
16. Hammond C, Braakman I, Helenius A (1994) Proc Natl Acad Sci U S A 91:913
17. Robbins PW, Hubbard SC, Turco SJ, Wirth DF (1977) Cell 12:893
18. Brockhausen I (1999) Biochim Biophys Acta 1473(1):67
19. Brockhausen I, Dowler T, Paulsen H (2009) Biochim Biophys Acta 1790(10):1244
20. Varki A, Schauer R (2009) In: Varki A, Cummings RD, Esko JD, Freeze HH, Stanley P, Bertozzi CR, Hart GW, Etzler ME (eds) Essentials of glycobiology, 2nd edn. Cold Spring Harbor Laboratory Press, Cold Spring Harbor, chap 14
21. Schauer R (2009) Curr Opin Struct Biol 19(5):507
22. Traving C, Schauer R (1998) Cell Mol Life Sci 54(12):1330
23. Hinderlich S, Berger M, Schwarzkopf M, Effertz K, Reutter W (2000) Eur J Biochem 267(11):3301
24. Kean EL (1970) J Biol Chem 245(9):2301
25. Kean EL, Munster-Kuhnel AK, Gerardy-Schahn R (2004) Biochimica Biophys Acta 1673(1–2):56
26. Harduin-Lepers A, Vallejo-Ruiz V, Krzewinski-Recchi MA, Samyn-Petit B, Julien S, Delannoy P (2001) Biochimie 83:727
27. Kornfeld S, Kornfeld R, Neufeld E, O'Brien PJ (1964) Proc Natl Acad Sci U S A 52:371
28. Langer BG, Weisel JW, Dinauer PA, Nagaswami C, Bell WR (1988) J Biol Chem 263: 15056

29. Oh-Eda M, Hasegawa M, Hattori K, Kuboniwa H, Kojima T, Orita T, Tomonou K, Yamazaki T, Ochi N (1990) J Biol Chem 265:11432
30. Cantell K, Hirvonen S, Sareneva T, Pirhonen J, Julkunen I (1992) J Interferon Res 12:177
31. Wittwer AJ, Howard SC (1990) Biochemistry 29:4175
32. Rudd PM, Joao HC, Coghill E, Fiten P, Saunders MR, Opdenakker G, Dwek RA (1994) Biochemistry 33(1):17
33. Ashwell G, Harford J (1982) Annu Rev Biochem 51:531
34. Egrie JC, Dwyer E, Browne JK, Hitz A, Lykos MA (2003) Exp Hematol 31(4):290
35. Ashwell G, Morell AG (1974) Adv Enzymol Relat Areas Mol Biol 41:99
36. Schwartz AL (1991) Target Diagn Ther 4:3
37. Chiu MH, Tamura T, Wadhwa MS, Rice KG (1994) J Biol Chem 269(23):16195
38. Yang Y, Thomas VH, Man S, Rice KG (2000) Glycobiology 10(12):1341
39. Wang X, Gu J, Ihara H, Miyoshi E, Honke K, Taniguchi N (2006) J Biol Chem 281(5):2572
40. Saito T, Kinoshita A, Yoshiura Ki, Makita Y, Wakui K, Honke K, Niikawa N, Taniguchi N (2001) J Biol Chem 276(15):11469
41. Shinkawa T, Nakamura K, Yamane N, Shoji-Hosaka E, Kanda Y, Sakurada M, Uchida K, Anazawa H, Satoh M, Yamasaki M, Hanai N, Shitara K (2003) J Biol Chem 278(5):3466
42. Hayakawa T, Aki I, Varki A, Satta Y, Takahata N (2006) Genetics 172(2):1139
43. Padler-Karavani V, Yu H, Cao H, Chokhawala H, Karp F, Varki N, Chen X, Varki A (2008) Glycobiology 18(10):818
44. Warner TG (1999) Glycobiology 9(9):841
45. Wurm FM (2004) Nat Biotechnol 22(11):1393
46. Chu L, Robinson DK (2001) Curr Opin Biotechnol 12(2):180
47. Hokke CH, Bergwerff AA, Van Dedem GW, Kamerling JP, Vliegenthart JFG (1995) Eur J Biochem 228(3):981
48. Bergwerff AA, van Oostrum J, Asselbergs FA, Bürgi R, Hokke CH, Kamerling JP, Vliegenthart JFG (1993) Eur J Biochem 212(3):639
49. Lanctot PM, Gage FH, Varki AP (2007) Curr Opin Chem Biol 11(4):373
50. Komoda H, Okura H, Lee CM, Sougawa N, Iwayama T, Hashikawa T, Saga A, Yamamoto A, Ichinose A, Murakami S, Sawa Y, Matsuyama A (2010) Tissue Eng Part A 16(4):1143
51. Galili U, Shohet SB, Kobrin E, Stults CL, Macher BA (1988) J Biol Chem 263(33):17755
52. Joziasse DH, Shaper JH, Jabs EW, Shaper NL (1991) J Biol Chem 266(11):6991
53. Oriol R, Ye Y, Koren E, Cooper DK (1993) Transplantation 56(6):1433
54. Ezzelarab M, Cooper DK (2005) Xenotransplantation 12(4):278
55. Barnes LM, Bentley CM, Dickson AJ (2000) Cytotechnology 32(2):109
56. Moran EB, McGowan ST, McGuire JM, Frankland JE, Oyebade IA, Waller W, Archer LC, Morris LO, Pandya J, Nathan SR, Smith L, Cadette ML, Michalowski JT (2000) Biotechnol Bioeng 69(3):242
57. Birch JR, Racher AJ (2006) Adv Drug Deliv Rev 58(5–6):671
58. Chung CH, Mirakhur B, Chan E, Le QT, Berlin J, Morse M, Murphy BA, Satinover SM, Hosen J, Mauro D, Slebos RJ, Zhou Q, Gold D, Hatley T, Hicklin DJ, Platts-Mills TA (2008) N Engl J Med 358(11):1109
59. Commins SP, Platts-Mills TA (2009) J Allergy Clin Immunol 124(4):652
60. Keppler OT, Horstkorte R, Pawlita M, Schmidt C, Reutter W (2001) Glycobiology 11(2):11R
61. Hinderlich S, Berger M, Keppler OT, Pawlita M, Reutter W (2001) Biol Chem 382(2):291
62. Bork K, Reutter W, Weidemann W, Horstkorte R (2007) FEBS Lett 581(22):4195
63. Grabenhorst E, Hoffmann A, Nimtz M, Zettlmeissl G, Conradt HS (1995) Eur J Biochem 232(3):718
64. Grabenhorst E, Schlenke P, Pohl S, Nimtz M, Conradt HS (1999) Glycoconj J 16(2):81
65. Bragonzi A, Distefano G, Buckberry LD, Acerbis G, Foglieni C, Lamotte D, Campi G, Marc A, Soria MR, Jenkins N, Monaco L (2000) Biochim Biophys Acta 1474(3):273
66. Yamane-Ohnuki N, Kinoshita S, Inoue-Urakubo M, Kusunoki M, Iida S, Nakano R, Wakitani M, Niwa R, Sakurada M, Uchida K, Shitara K, Satoh M (2004) Biotechnol Bioeng 87(5):614

67. von Horsten HH, Ogorek C, Blanchard V, Demmler C, Giese C, Winkler K, Kaup M, Berger M, Jordan I, Sandig V (2010) Glycobiology 20(12):1607
68. Mistry PK, Wraight EP, Cox TM (1996) Lancet 348(9041):1555
69. Hamilton SR, Gerngross TU (2007) Curr Opin Biotechnol 18(5):387
70. Hamilton SR, Davidson RC, Sethuraman N, Nett JH, Jiang Y, Rios S, Bobrowicz P, Stadheim TA, Li H, Choi BK, Hopkins D, Wischnewski H, Roser J, Mitchell T, Strawbridge RR, Hoopes J, Wildt S, Gerngross TU (2006) Science 313(5792):1441
71. Bakker H, Rouwendal GJ, Karnoup AS, Florack DE, Stoopen GM, Helsper JP, van Ree R, van Die I, Bosch D (2006) Proc Natl Acad Sci U S A 103(20):7577
72. Schähs M, Strasser R, Stadlmann J, Kunert R, Rademacher T, Steinkellner H (2007) Plant Biotechnol J 5(5):657
73. Jacobs PP, Callewaert N (2009) Curr Mol Med 9(7):774
74. Jones J, Krag SS, Betenbaugh MJ (2005) Biochim Biophys Acta 1726(2):121
75. Elliott S, Lorenzini T, Asher S, Aoki K, Brankow D, Buck L (2003) Nat Biotechnol 21:414
76. Neumann AU, Lam NP, Dahari H, Gretch DR, Wiley TE, Layden TJ, Perelson AS (1998) Science 282(5386):103
77. Goldstein D, Laszlo J (1986) Cancer Res 46:4315
78. Borden EC, Lindner D, Dreicer R, Hussein M, Peereboom D (2000) Semin Cancer Biol 10:125
79. Chatelut E, Rostaing L, Grégoire N, Payen JL, Pujol A, Izopet J, Houin G, Canal PA (1999) Br J Clin Pharmacol 47(4):365
80. Ceaglio N, Etcheverrigaray M, Kratje R, Oggero M (2010) Biochimie 92(8):971
81. Ceaglio N, Etcheverrigaray M, Kratje R, Oggero M (2008) Biochimie 90(3):437
82. Perlman S, van den Hazel B, Christiansen J, Gram-Nielsen S, Jeppesen CB, Andersen KV, Halkier T, Okkels S, Schambye HT (2003) J Clin Endocrinol Metab 88(7):3227
83. Gu X, Wang DI (1998) Biotechnol Bioeng 58(6):642
84. Restelli V, Wang MD, Huzel N, Ethier M, Perreault H, Butler M (2006) Biotechnol Bioeng 94(3):481
85. Andersen DC, Bridges T, Gawlitzek M, Hoy C (2000) Biotechnol Bioeng 70(1):25
86. Yang M, Butler M (2000) Biotechnol Bioeng 68(4):370
87. Gawlitzek M, Estacio M, Fürch T, Kiss R (2009) Biotechnol Bioeng 103(6):1164
88. Andersen DC, Goochee CF (1995) Biotechnol Bioeng 47(1):96
89. Senger RS, Karim MN (2003) Biotechnol Prog 19(4):1199
90. Borys MC, Linzer DI, Papoutsakis ET (1993) Biotechnology (N Y) 11(6):720
91. Nabi IR, Dennis JW (1998) Glycobiology 8(9):947
92. Raju TS, Briggs JB, Chamow SM, Winkler ME, Jones AJ (2001) Biochemistry 40(30):8868
93. Fernandes AI, Gregoriadis G (1997) Biochim Biophys Acta 1341(1):26
94. Fernandes AI, Gregoriadis G (1996) Biochim Biophys Acta 1293(1):90
95. Gregoriadis G, Fernandes A, Mital M, McCormack B (2000) Cell Mol Life Sci 57(13–14): 1964
96. Zhu Y, Li X, McVie-Wylie A, Jiang C, Thurberg BL, Raben N, Mattaliano RJ, Cheng SH (2005) Biochem J 389(3):619
97. Zhu Y, Jiang JL, Gumlaw NK, Zhang J, Bercury SD, Ziegler RJ, Lee K, Kudo M, Canfield WM, Edmunds T, Jiang C, Mattaliano RJ, Cheng SH (2009) Mol Ther 17(6):954
98. Laine RA (1994) Glycobiology 4:759
99. Dove A (2001) Nat Biotechnol 19:913
100. Laine RA (1997) In: Gabius H-J, Gabius S (eds) Glycosciences: status and perspectives. Chapman and Hall, London, pp 1–14
101. Aoki KF, Yamaguchi A, Okuno Y, Akutsu T, Ueda N, Kanehisa M, Mamitsuka H (2003) Genome Inform 14:134
102. Haslam SM, Julien S, Burchell JM, Monk CR, Ceroni A, Garden OA, Dell A (2008) Immunol Cell Biol 86:564
103. Venter JC (2001) Science 291(5507):1304

104. Committee for Proprietary Medicinal Products (2003) Guideline on comparability of medicinal products containing biotechnology-derived proteins as active substance: quality issues. EMEA/CPMP/BWP/3207/00/Rev1 2003. European Medicines Agency, London
105. Rosenfeld R, Bangio H, Gerwig GJ, Rosenberg R, Aloni R, Cohen Y, Amor Y, Plaschkes I, Kamerling JP, Maya RB (2007) J Biochem Biophys Methods 70(3):415
106. Merkle RK, Cummings RD (1987) Methods Enzymol 138:232
107. Yamamoto K, Tsuji T, Osawa T (1993) Methods Mol Biol 14:17
108. Gohlke M, Blanchard V (2008) Methods Mol Biol 446:239
109. Royle L, Dwek RA, Rudd PM (2006) Curr Protoc Protein Sci 12:12.6
110. Rudd PM, Dwek RA (1997) Curr Opin Biotechnol 8(4):48
111. Callewaert N, Geysens S, Molemans F, Contreras R (2001) Glycobiology 11(4):275
112. Schwarzer J, Rapp E, Reichl U (2008) Electrophoresis 29(20):4203
113. Dell A, Morris HR (2001) Science 291(5512):2351
114. Geyer H, Geyer R (2006) Biochim Biophys Acta 1764(12):1853
115. Pabst M, Altmann F (2011) Proteomics 11(4):631
116. Wuhrer M, Deelder AM, Hokke CH (2005) J Chromatogr B Anal Technol Biomed Life Sci 825(2):12
117. Morelle W, Canis K, Chirat F, Faid V, Michalski JC (2006) Proteomics 6(14):3993
118. Deshpande N, Jensen PH, Packer NH, Kolarich D (2010) J Proteome Res 9(2):1063
119. Ceroni A, Maass K, Geyer H, Geyer R, Dell A, Haslam SM (2008) J Proteome Res 7(4):1650

Adv Biochem Engin/Biotechnol (2012) 127: 187–219
DOI: 10.1007/10_2011_113

Published Online: 20 October 2011

Protein Glycosylation Control in Mammalian Cell Culture: Past Precedents and Contemporary Prospects

Patrick Hossler

Abstract Protein glycosylation is a post-translational modification of paramount importance for the function, immunogenicity, and efficacy of recombinant glycoprotein therapeutics. Within the repertoire of post-translational modifications, glycosylation stands out as having the most significant proven role towards affecting pharmacokinetics and protein physiochemical characteristics. In mammalian cell culture, the understanding and controllability of the glycosylation metabolic pathway has achieved numerous successes. However, there is still much that we do not know about the regulation of the pathway. One of the frequent conclusions regarding protein glycosylation control is that it needs to be studied on a case-by-case basis since there are often conflicting results with respect to a control variable and the resulting glycosylation. In attempts to obtain a more multivariate interpretation of these potentially controlling variables, gene expression analysis and systems biology have been used to study protein glycosylation in mammalian cell culture. Gene expression analysis has provided information on how glycosylation pathway genes both respond to culture environmental cues, and potentially facilitate changes in the final glycoform profile. Systems biology has allowed researchers to model the pathway as well-defined, inter-connected systems, allowing for the *in silico* testing of pathway parameters that would be difficult to test experimentally. Both approaches have facilitated a macroscopic and microscopic perspective on protein glycosylation control. These tools have and will continue to enhance our understanding and capability of producing optimal glycoform profiles on a consistent basis.

Keywords Genomics · Mammalian cell culture · Protein glycosylation · Systems biology

Abbreviations

ADCC Antibody-dependent cellular cytotoxicity
BHK Baby hamster kidney cells

P. Hossler (✉)
Abbott Laboratories, Abbott Bioresearch Center, Worcester, MA 01605, USA
e-mail: patrick.hossler@abbott.com

CDC	Complement-dependent cytotoxicity
CHO	Chinese hamster ovary cells
CMP-NeuAc	Cytosine monophosphate-N-acetylneuraminic acid
DO	Dissolved oxygen
EPO	Erythropoietin
ER	Endoplasmic reticulum
Fuc	Fucose
FucT	Glycoprotein 6-α-L-fucosyltransferase
Gal	Galactose
GalNAc	N-Acetylgalactosamine
GalT	β-N-Acetylglucosaminyl glycopeptide β-1,4-galactosyltransferase
GDP-Fuc	Guanidine diphosphate-fucose
GlcNAc	N-Acetylglucosamine
GnT I	α-1,3-Mannosyl-glycoprotein 2-β-N-acetylglucosaminyltransferase
GnT II	α-1,6-Mannosyl-glycoprotein 2-β-N-acetylglucosaminyltransferase
GnT III	β-1,4-Mannosyl-glycoprotein 4-β-N-acetylglucosaminyltransferase
GnT IV	α-1,3-Mannosyl-glycoprotein 4-β-N-acetylglucosaminyltransferase
GnT V	α-1,6-Mannosyl-glycoprotein 4-β-N-acetylglucosaminyltransferase
GS	Glutamine synthetase
IFN-γ	Interferon-gamma
Man	Mannose
Man I	Mannosyl-oligosaccharide 1,2-α-mannosidase
Man II	Mannosyl-oligosaccharide 1,3-1,6-α-mannosidase
ManNAc	N-Acetylmannosamine
NeuAc	N-Acetylneuraminic acid (sialic acid)
NeuGc	N-Glycolylneuraminic acid
PK	Pharmacokinetics
PCA	Principal component analysis
tPA	Tissue plasminogen activator
SiaT	β-Galactoside α-2,3/6-sialyltransferase
UDP-Gal	Uridine diphosphate-galactose
UDP-GlcNAc	Uridine diphosphate-N-acetylglucosamine
UPR	Unfolded protein response

Contents

1 Introduction

It has been deduced that more than half of all catalogued proteins are actually glycoproteins, with varying levels of glycosylation characterization [1]. Due to their ubiquity in nature and in recombinant protein expression, protein glycosylation is a post-translational modification that is of paramount interest to the biopharmaceutical industry. Glycosylation has been proven to affect various physiochemical properties of glycoprotein therapeutics, including protein folding [2], solubility [3], binding [4], and stability [5]. The contributing factors that help shape the final glycoform profile have been reviewed [6], and include the cell expression system, the glycoprotein structure itself, the activity and specificity of enzymes involved in the metabolic pathway, as well as the various cell culture environmental variables. Understanding the dependence of the final glycoform profile on these parameters is of ongoing interest to the biopharmaceutical industry.

Due to the nature of the metabolic pathway and its inherent dependence on a multitude of factors, there is typically a diverse array of asparagine linked (N-) and serine/threonine linked (O-) glycans across the various glycosylation sites on the protein product. The heterogeneous glycan structures that are observed at a particular glycosylation site on different glycoprotein molecules is termed microheterogeneity, whereas the differences in glycan site-occupancy at different glycosylation sites across different glycoprotein molecules is termed macroheterogeneity. Microheterogeneity exists for a variety of reasons, with reports showing a direct relationship to cell culture conditions, and to the particular cell line. Macroheterogeneity also exists for a variety of reasons, including protein folding. Since hundreds of different N- and O-glycan species have been described to date, numerous nomenclature systems have been devised to facilitate better classification and communication. Though no completely standard naming convention has been adopted, a common convention is highlighted in Butler et al. [7] as well as the convention established by the Consortium for Functional Glycomics [8] (Fig. 1). In order to make the pathway more computer-representation-friendly, there have been efforts towards development of an XML notation for glycan structures as well [9,10].

An extensive survey of the various glycan structures, their structural and functional roles on the molecule, and the biological processes they modulate has

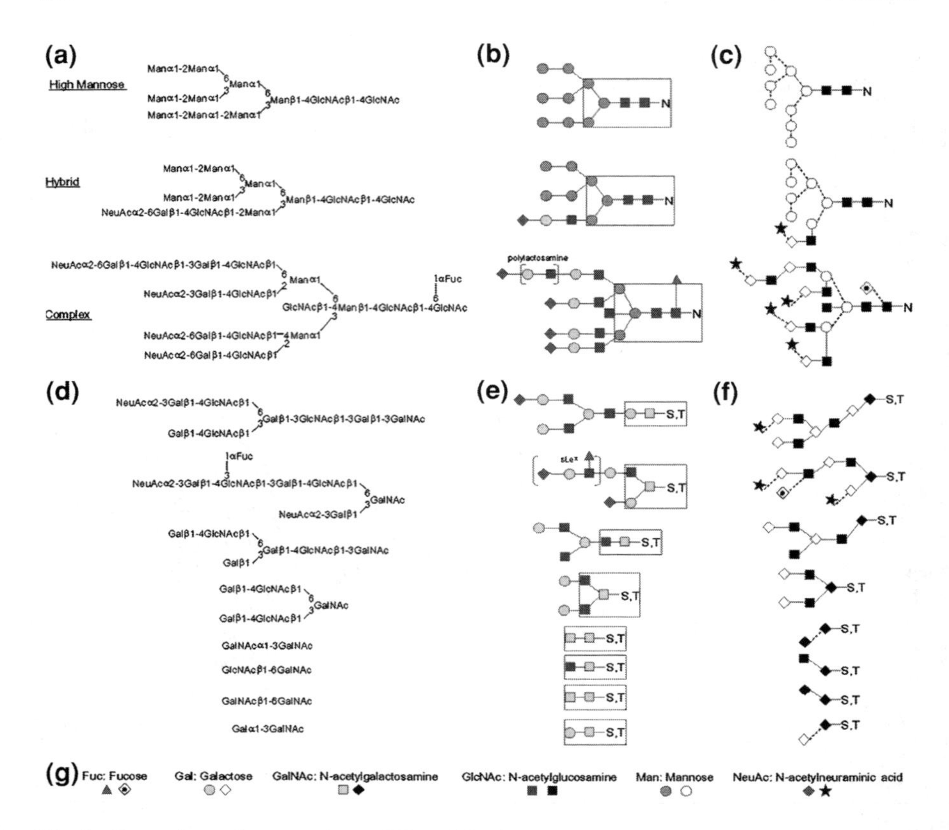

Fig. 1 Comparison of commonly used N- and O-glycan abbreviated nomenclature. **a** Consortium for functional glycomics (CFG) text nomenclature for N-glycans; principal types of N-glycans are shown for representative purposes. **b** CFG symbolic nomenclature for N-glycans; common core pentasaccharide structure are shown in boxes. **c** Symbolic nomenclature of N-glycans consistent with Butler et al. [7]. **d** CFG text nomenclature for O-glycans; eight principal types of O-glycans are shown for representative purposes. **e** CFG symbolic nomenclature for O-glycans; eight different core structures are shown in boxes. **f** Symbolic nomenclature of O-glycans consistent with Butler et al. [7]. **g** Monosaccharide symbol legend

been documented [11,12]. The presence of individual N- and O-glycan structures has been shown to have important roles in the immune system [13], and can affect clearance rate [14] and antigenicity [15] of the glycoprotein therapeutic. For example, the heavy chains of monoclonal antibodies typically contain at least one N-linked glycosylation site, which facilitates two N-glycans per fully assembled antibody molecule. These glycans have been shown to be essential for both interactions with Fc receptors (FcR) and their accompanying effector functions, including antibody-dependent cellular cytotoxicity (ADCC) and complement-dependent cytotoxicity (CDC) [16,17].

Despite a typically diverse glycoform profile, there is no fundamentally ideal glycoform profile ubiquitous for all glycoproteins. Some glycans have been found to be more ideal depending upon the function of the glycoprotein they are attached to. In other cases, there is a relative insensitivity towards the particular glycan on the resulting glycoprotein function [18]. Thus, there is sufficient evidence in the literature and other sources to necessitate at least the evaluation of this potential effect and for biopharmaceutical companies to adequately characterize the expressed glycoform profile to ensure it lies within the assigned acceptability limits. Typically, a consistent spectrum of product glycans is targeted in industry, since in most cases it is not possible to fully characterize or predict the in vivo effect of every particular glycosylation pattern [19]. To achieve consistency during manufacturing, all efforts in upstream and downstream process development are controlled as tightly as possible.

Since there is a relatively large repertoire of product glycans, controlling their relative amounts in production cell culture often proves difficult to understand. Reports of glycosylation control have centered on either the cell culture process environment, or at the genetic level through overexpression or suppression of key glycosylation enzymes. Two additional tools that have been applied recently towards the understanding and control of protein glycosylation include genomics and systems biology. The gene expression studies conducted to date have highlighted the diverse responses towards cell culture environments which have helped elucidate the regulation of this important metabolic pathway. Systems biology has facilitated our understanding of the protein glycosylation pathway as a system with direct and indirect relationships whose ultimate controllability is dependent upon our understanding of these relationships, and their inherently complex nature. Through continual use of these tools, a more thorough understanding of the regulation and control of protein glycosylation should continue to be realized.

2 Metabolic Pathway

Protein glycosylation is characterized by two main forms depending on the amino acid the oligosaccharides are added onto, including asparagine-linked glycans (N-glycans) and serine/threonine-linked glycans (O-glycans). The processing of N- and O-glycans occurs through a series of monosaccharide removal and chain elongation reactions within the endoplasmic reticulum (ER) and Golgi. This sequential series of reactions forms a metabolic pathway that is linear in some stages, in that each reactant produces only one glycan product. In other stages the pathway is branched, in which a glycan reactant is capable of producing multiple glycan products (Fig. 2). As a result, the glycan processing pathways can diverge at multiple processing steps, which eventually become increasingly convergent as the terminal glycan processing steps (galactosylation and sialylation) are met.

N-glycosylation is initiated by the en bloc transfer through an oligosaccharyltransferase enzyme (OST) from a pre-formed 9-mannose glycan attached to dolichol on asparagine residues of glycoprotein molecules with an Asn–X–Thr/Ser (where X represents any amino acid but proline). This processing step occurs

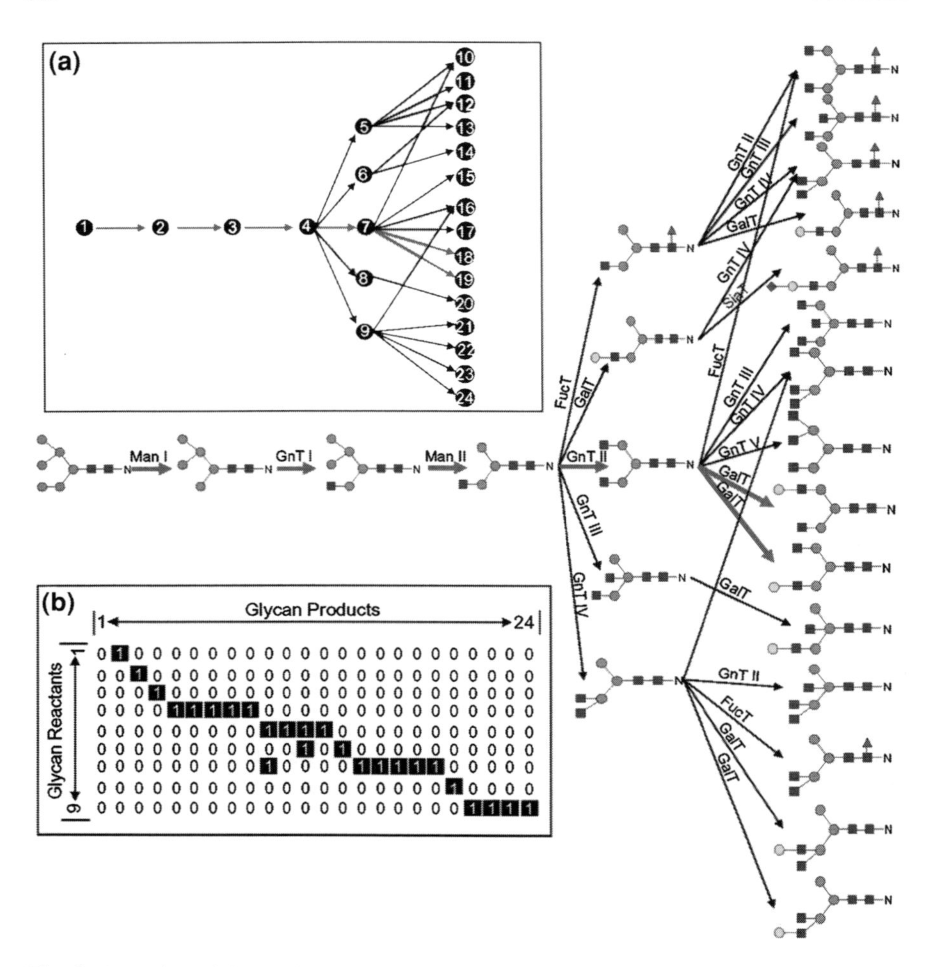

Fig. 2 A portion of the N-glycan biosynthetic pathway with hypothetical preferential pathway highlighted in *red*. Subset A: Metabolic network representation of the glycosylation pathway highlighting a systems level perspective with arbitrary glycan numbering. Subset B: Adjacency matrix representation of glycosylation network; allowed reactions between reactant and product N-glycans are assigned a value of 1

shortly around the time of initial protein folding in the lumen of the ER. A trimer of glucose sugars on this N-glycan are removed in a step-wise fashion, while concomitantly binding to the calnexin and calreticulin lectins, which ensure proper protein folding and serve as a quality control step within the ER before transport to the Golgi [20]. Should improper folding or attenuated removal of these glucose monosaccharides exist, a signal is provided that directs the glycoprotein towards proteasomal degradation. The high-fidelity mechanism of glycan processing that exists inside the ER ensures the N-glycans are approximately uniform. However, a similar mechanism does not exist inside the Golgi. After some initial mannose removal reactions in the ER, the N-glycans are transported to the Golgi apparatus where a series of additional mannose removal steps precedes a highly diverse series of step-wise additions of a small number of monosaccharides that facilitates a potentially large number of product glycan structures. The innate substrate specificity of the enzymes towards their substrate glycans, as well as the relative concentrations of enzymes and nucleotide-sugars across the various cisternae of the Golgi, control the final glycoform profile. The diverse array of contributing factors towards protein glycosylation have been documented [6]. As a result of differences in the extent of Golgi processing, three N-glycan types have been categorized. High mannose glycans are the least processed, complex type glycans are the most processed, and hybrid type glycans are intermediately processed with structural attributes of both high mannose and complex glycans (Fig. 1a–c).

In contrast to N-glycosylation, O-glycosylation begins with the transfer of a single monosaccharide (GalNAc), compared to a pre-formed oligosaccharide attached to dolichol. GalNAc is transferred from its cognate sugar-nucleotide UDP-GalNAc onto the hydroxyl group of Ser or Thr residues of glycoproteins. They are then further modified by sialic acid, fucose, galactose, xylose, and/or polylactosamine additions (Fig. 1a–c). The cellular compartment in which the pathway is initiated has been found to be either the ER or Golgi [21]. In contrast to N-glycans, which have a common core pentasaccharide structure, O-glycans have up to eight different core structures which make their study and experimental measurement more difficult (Fig. 1d–f). O-glycans have also been shown to have an intricate role in the physiochemical properties of glycoproteins. A good overview of the metabolic pathway and the effects of O-glycans on proteins can be found in Van den Steen et al. [22].

The monosaccharide source for addition onto N- and O-glycans comes from a common pool of nucleotide-sugars transported into the lumen of the ER and Golgi organelles either from de-novo biosynthesis, or from the recycling of monosaccharides through proteins degraded in the lysosome. All nucleotide-sugars are generated in the cytosol, except CMP-sialic acid which is generated in the nucleus. An excellent review highlighting the details of nucleotide-sugar transporters is given in Ishida and Kawakita [23].

3 Established Methods for Protein Glycosylation Control

3.1 Expression Host

Glycosylation varies frequently between and even within each species. This is especially true in immunoglobulin G (IgG), which is the backbone molecule of most of today's monoclonal antibodies [24]. A comparative study of IgG between various species (human, rhesus, dog, cow, guinea pig, sheep, goal, horse rat, mouse, rabbit, cat, and chicken) revealed differences in neutral oligosaccharides and NeuAc/NeuGc content, as well as isomer differences, suggesting not only differences in glycan processing, but also enzymatic specificities [25]. Even with knowledge of human glycoform profiles for a particular glycoprotein, matching those attributes on the recombinantly expressed biologic is not trivial. Human serum erythropoietin (EPO) has been shown to be significantly different from recombinantly expressed EPO [26]. It is due to these potential differences that the choice of host cell expression system is very important, and one of the principal reasons for the biopharmaceutical industry's focus on a few established mammalian cell lines. The choice of host cell expression system and the resulting protein glycosylation has been extensively reviewed [27, 28].

Some monosaccharides present on N- and O-glycans from mammalian cells are undesirable. Human cells, for example, do not normally express α(1, 3)GalT due to a frameshift mutation [29], and are able to mount an immune response to the attached α(1, 3)Gal. Numerous mouse expression systems, including hybridomas producing humanized antibodies, do add them on secreted glycoproteins. However, other commonly used expression systems, such as NS0 and CHO cells, do not add this monosaccharide and as a result there is usually not a problem with the expressed proteins [28]. N-Glycolylneuraminic acid (NeuGc) is another monosaccharide that is generally undesirable for recombinant therapeutic glycoproteins. Adult human cells typically have terminal NeuAc on their glycans. However, other mammals sometimes have terminal NeuGc, which differs from NeuAc by the incorporation of one additional hydroxyl group through the action of the CMP-NeuAc hydroxylase enzyme. Thus recombinant proteins expressed by rodent species, including mouse and hamster, can have terminal NeuGc within the glycoform profile. NeuGc has been detected on the glycans attached to glycoproteins derived from NS0 [30], as well as CHO [31].

Frequently the choice of mammalian expression host has an effect on the resulting glycoform profile due to a different complement of functionally expressed protein glycosylation enzymes. For example, α(2, 6)SiaT activity varies 50–100-fold in various rat tissues [32], but cell lines such as CHO and BHK both lack a functional α(2, 6) SiaT, and make exclusively α(2,3) linked NeuAc. In another example, CHO cells do not typically have detectable levels of GnT III, while human and other mammalian cell lines do [33].

Mammalian cells glycosylate proteins that are on average more consistent with humans than non-mammalian cells. Some of the more common non-mammalian host cell expression systems for recombinant glycoproteins include insect, plant, and yeast systems. These cells all have been shown to glycosylate expressed proteins; however, in the majority of the cases their glycoform profiles are either different from that observed in human, or are not optimal. Insect cell lines have been used as host cells for glycoproteins, including monoclonal antibodies. However, the glycoproteins expressed are typically hybrid, high-mannose, or paucimannose glycans [34]. Some insect cell lines have also been shown to add non-human α(1,3) Fuc, which is potentially immunogenic. Plant cell lines have also been used as expression systems for glycoproteins, but typically lack Gal and NeuAc, and instead add xylose and α(1,3) Fuc. Yeast expression systems have become more common for glycoproteins, especially *Pichia pastoris* because of robust expression as well as scalable fermentation. However, yeast systems also hypermannosylate the expressed proteins. As a result of this, efforts have been made to successfully humanize glycosylation in these cells, as well as to incorporate human glycosylation characteristics, including sialylation [35]. Although there have been a few very successful attempts to "humanize" glycosylation patterns in these cells, in the majority of the cases mammalian cell lines are still the preferred host expression system towards the production of recombinant glycoprotein therapeutics with human-like glycosylation characteristics.

3.2 Cell Culture Process Conditions

The effect of cell culture environmental conditions on the resulting protein glycosylation has been well-documented. Different culture conditions frequently facilitate differences in the resulting protein glycosylation. The choice of bioreactor production mode, process control variables, as well as culture media and respective nutrient levels have all been documented to elicit an effect on the resulting protein glycosylation. Previous reviews have described in detail these potential effectors of glycoform profiles [19].

Different production bioreactor modes and vessels have been shown to cause differences in protein glycosylation using the same cell line and culture media. In Kunkel et al. [36], even with the same cell line expressed under the same process conditions, but with different bioreactor controller platforms, differences were observed in the final glycoform profile. In another study it was shown that with CHO-derived human IFN-γ cultured in a perfused, fluidized-bed bioreactor attached to macroporous microcarriers there was an increase in the proportion of tri- and tetrasialylated N-glycans, along with a decrease in monosialylated and neutral N-glycans after 210 h in culture [37]. In contrast, cells that were cultured in a stirred tank bioreactor showed similar sialylation time profiles until late-stage, where a decrease in overall sialylation was observed as the cells began to die, potentially due to extracellular degradation. In another study the three main glycoforms from a cytokine fusion protein were expressed from BHK cells, and showed subtle differences across different culture platforms, and at different times in a fixed bed bioreactor with continuous perfusion [38].

Process variables that are typically monitored and/or controlled during the course of a typical mammalian cell culture process have been documented to have a major role in modulating the final glycoform profile. Dissolved oxygen (DO) levels, bioreactor pH, and culture temperature are three process parameters commonly controlled in any cell culture process. All three variables have been shown to have an appreciable effect on protein glycosylation in at least some instances [19]. Butler [6] provides a very good review of the relationship between process control variables and the resulting protein glycosylation. The results published to date all point to the absolute necessity of mapping the acceptable operating ranges (AORs) for each of these variables to ensure glycosylation consistency in a good manufacturing practice production environment.

Dissolved oxygen levels have a proven role in modulating glycoform profiles. In one study, it was reported that high dissolved oxygen levels were correlated with a higher NeuAc content on recombinant follicle simulating hormone (FSH) [39]. In another study, low levels of DO caused little change in the glycosylation of tPA from CHO cells [40]; however, a change was observed in the case of the glycosylation of an IgG1 antibody in mouse hybridomas which demonstrated decreased galactosylation with decreases in DO levels [41]. The above results suggest that DO does play a role in glycosylation, but the effect is likely cell- and protein-specific, and relationships need to be investigated on a case-by-case basis.

Process temperature has also shown mixed results with respect to protein glycosylation. Shifts in temperature setpoints during production bioreactor cultures is a common practice for increasing final product titers. In numerous studies, temperature shifts have been shown to preserve product quality compared to non-shifted cultures. A fourfold increase in specific productivity was elicited by a temperature shift from 37 C to 33 C from CHO-derived EPO [42]. The sialic acid profiles from both process conditions were in fact comparable, with at most a few percent differences in composition. Reducing the culture temperature from 37 C to 33 C in a CHO cell line expressing human granulocyte macrophage colony stimulating factor saw little change in protein glycosylation and terminal sialylation, but did enhance specific productivity [43]. Interestingly, even the glycosidase enzymes between temperature-shifted and non-temperature-shifted cultures were reported to be similar. A temperature shift from 37 C to 34 C over an 8-day time period caused intracellular sialidase activity to spike immediately after inoculation for both cultures, followed by a decay over time that was roughly similar to that of a non-temperature-shifted culture [44]. Despite these results, other researchers have reported differences in the final glycoform profile on shifting temperature. In one investigation, EPO-Fc expressed by CHO cells facilitated a 20–40% decrease in sialylation with a shift to a lower temperature [45]. In yet another study, it was found that a temperature shift at a later stage of a CHO cell culture facilitated a 59% decrease in NeuGc levels compared to when the temperature shift occurred earlier [31]. These results suggest that the commonly used temperature shift strategy to enhance productivity is generally a safe practice from a protein glycosylation perspective, but glycoform profile changes have been reported in at least some instances.

pH is also an important process control variable with a proven role in controlling glycoform profiles. In one study, different pH levels were shown to affect the final glycoform profile of a monoclonal antibody in hybridoma cultures [46]. At lower pH levels (6.9 and 7.2), relatively higher levels of agalacto and monogalacto complex type N-glycans were measured, compared to higher pH levels (7.4) which facilitated the highest degree of galactosylation as well as the highest NeuAc/NeuGc ratio. Typically, base is added into a production bioreactor to control pH. In one study it was found that the base type can have an important role on the resulting protein glycosylation [31]. In CHO cell cultures expressing glycoprotein B1 it was found that using sodium hydroxide as the base instead of sodium carbonate facilitated a 33% decrease in NeuGc levels.

The harvest criteria for a production bioprocess is frequently determined by cell culture viability, which is monitored throughout the duration of the culture. The fractional amount of non-viable cells is an important process variable because of the proven role of extracellular glycosidases which can accumulate in the media from lysed cells, and step-wise remove monosaccarides from the glycan. Prominent amongst these is sialidase, which has been shown to accumulate in cell culture media, and is active at neutral pH [47, 48]. Other glycosidases found to date include β-galactosidase, β-hexosaminidase, and fucosidase found in 293, NS0, and hybridoma cells. Interestingly, the sialidase activity found in cultures of these other cells is much lower than that found in CHO cell culture [49].

Metabolic waste products have been shown to affect the resulting protein glycosylation in mammalian cells. In particular, the effect of ammonia on the resulting glycosylation has been well documented. Ammonium chloride has been shown to prevent terminal glycosylation of immunoglobulins in plasma cells without affecting secretion [50]. Ammonia has also been reported to decrease terminal sialylation in both N- and O-glycans. For example, ammonium ion concentrations above 2 mM resulted in reduced α(2, 6) NeuAc attached to O-glycans on granulocyte colony stimulating factor (GCSF) produced by recombinant CHO cells [51], and excess ammonia supplemented to CHO cells expressing EPO also caused a decrease in terminal sialylation [52]. The extent of glycosylation has also been shown to decrease over time in batch culture [53], potentially due to the lack of sufficient nutrient availability in late-stage culture, or the accumulation of toxic metabolites such as ammonia. Indeed, this was proven by Borys et al. [54], where increased ammonium ions reduced the extent of recombinant placental lactogen N-glycosylation in CHO cells. The reason for these effects of ammonia is not completely known, but it has been observed that ammonium chloride causes an increase in intracellular pH [55].

3.3 Cell Culture Media

Cell culture "core" media components (glucose, trace metals, and amino acids) have been shown to modulate the resulting protein glycosylation profile. The glucose concentration in culture, in particular, has been shown to affect the glycosylation of monoclonal antibodies produced by human hybridomas in batch culture and of IFN-γ produced by CHO cells in continuous culture [56–58]. Studies have also shown mixed results with respect to low glucose levels and the resulting impact on glycosylation. There was reported to be little change observed with continuous cultures of BHK-21 cells producing a recombinant IgG-IL2 fusion protein, where nutrient limitations (glucose, glutamine) caused microheterogeneity to be largely unchanged [59]. However, in another report it was found that the fraction of fully processed N-glycans attached to CHO-expressed IFN-γ was higher during culture periods of glucose excess [56]. A potential reason for these changes in protein glycosylation as a result of low nutrient levels may be explained through another study where glucose- and glutamine-limited CHO cell cultures led to a significantly lower intracellular pool of UDP-GlcNAc and UDP-GalNAc [60].

Metal ions have also been shown to have an important role towards the resulting protein glycoform profile. Many glycosylation enzymes are dependent on metal ions for activity. Manganese in particular has been shown to have a prominent role in increasing galactosylation [61]. CMP-sialic acid synthetase (the enzyme responsible for CMP-sialic acid biosynthesis) has been shown to be dependent on metal ions for activity, including manganese which facilitated optimal activity at neutral pH, and magnesium which facilitated optimal activity at pH 9.5. Interestingly, lower activity levels for this enzyme were demonstrated with the addition of copper, and no effect on activity levels were observed upon the addition of iron, zinc, and strontium, suggesting that the increase in activity is likely to be specific [62].

Amino acids have also been shown to affect the resulting protein glycosylation profile. One study showed how the supplementation of key amino acids (cysteine, isoleucine, leucine, tryptophan, valine, asparagines, aspartate, and glutamate) facilitated an increase in a lower sialylated fraction of recombinant EPO expressed in CHO [61]. Whether or not this was due to the resulting increase in productivity elicited as a result of the amino acid addition which exceeded the reactive capability of the glycosylation enzymes remains to be seen.

Commonly added media supplements have also been shown to modulate the resulting protein glycosylation. Although serum is used less nowadays in production of biologics, researchers have evaluated the effect of fetal calf serum on the oligosaccharide profile on IL-Mu6 glycoproteins secreted by BHK cells in both suspension and microcarrier cultures [63]. Decreases in sialylation levels in both cultures upon the incorporation of 2% serum were observed, as well as large changes in the relative amounts of complex type N-glycans. In another study, researchers supplemented CHO batch and fed-batch cultures expressing IFN-γ with Primatone RL tissue hydrolysate and found decreased sialylation at each of the two glycosylation sites [64]. There have been a few case studies that have explored the effect of sodium butyrate addition on the resulting glycosylation. In one study, CHO cells expressing tPA were cultured under conditions of decreased growth rate by supplementing with sodium butyrate and/or quinidine which resulted in a glycoform profile with increased site occupancy of Asn-184 [65]. In another report, it was found that the addition of sodium butyrate decreased the NeuGc levels on glycoprotein B1 expressed by CHO cells by 50–62% [31].

The addition of nucleotide-sugar precursors to cell culture media is an established method of modulating the final glycoform profile. In one published study, it was found that supplementing ManNAc into a CHO cell culture expressing IFN-γ resulted in an increase in sialylation on oneN-glycan site associated with a 30-fold increase in the intracellular levels of CMP-sialic acid [66]. In another study, Baker et al. [30] supplemented glucosamine, uridine, and ManNAc to GS-CHO and GS-NS0 cells expressing tissue inhibitor of metalloproteinases 1 (TIMP-1).

The glucosamine- and uridine-supplemented cultures increased N-glycan antennarity in CHO, but not NS0. Sialylation also decreased in this case. In the case of ManNAc addition, there was no change in the resulting sialylation in either CHO or NS0 despite an increase in intracellular CMP-sialic acid levels. In another study, researchers investigated the effect of intracellular glycosylation on various feeding strategies of nucleotide-sugars in shake flask cultures of CHO cells expressing IFN-γ [67]. The addition of galactose $\pm$ uridine, glucosamine $\pm$ uridine, and ManNAc $\pm$ cytidine enabled a 12%, 28%, and 32% increase in sialylation, respectively, compared to untreated control cultures. These results were partially attributable to the $\sim$20-fold increase in UDP-hexose, 6–15-fold increase in UDP-HexNAc, and 30–120-fold increase in CMP-NeuAc observed between the above conditions. An overall conclusion made by the researchers was that the addition of nucleotide-sugar precursors to the cultures had a prolonged effect on the resulting increased levels of intracellular nucleotide-sugars with the resulting behavior observed up to 48 h post-supplementation.

The collective information of the effects of cell culture media components and supplements on the resulting protein glycosylation suggests that studies need to be conducted on an individual glycoprotein and process basis to ascertain any potential role in glycomodulation. This had led some to attempt to incorporate high-throughput screening technologies to study the effect of these cell culture conditions, as well as the incorporation of high-throughput glycan measurements via lectin arrays [68], and high-throughput HPLC [69].

3.4 Cell Line Engineering and Selection

Innate differences in glycoprotein processing inherent in the use of different mammalian cell lines is typically associated with a different repertoire of active glycosylation enzymes inside the ER and Golgi. In addition, these enzymes have overlapping specificities for multiple glycans as potential substrates. Relative gene expression amongst the glycosyltransferases and glycosidases are furthermore not similar for all mammalian cells. Thus, potential changes in the glycoform profile, either through biosynthesis or degradation, are expected, and are dependent upon the relative enzyme amounts directly or peripherally related to protein glycosylation. Researchers have attempted to utilize this by increasing or decreasing the gene expression profile of various genes through cell line engineering. To date there has been a considerable amount of work on the cell line engineering of individual protein glycosylation enzymes. The information gleaned from these studies suggests that glycosylation of proteins expressed in mammalian cell lines does typically respond to changes in gene expression of glycosylation enzymes and nucleotide-sugar transporters.

There have been many successful attempts at modulating the protein glycosylation profile through the genetic manipulation of individual protein glycosylation genes. These attempts have included the introduction of new active enzymes, as well as the targeted increase and decrease in enzyme expression. Many research attempts have targeted maximizing NeuAc content, through either maximizing the addition onto N-glycans, or precluding its removal. A high NeuAc, low NeuGc content on glycoproteins is frequently targeted, since it has been found to promote a longer circulatory half-life [70]. In Weikert et al. [71], α(2, 3) SiaT was overexpressed in CHO cells expressing a TNFR-IgG fusion protein and tPA resulting in a significant increase in NeuAc content, and a significant decrease in N-glycan microheterogeneity. α(2, 6) SiaT is not typically expressed in hamster cells such as CHO and BHK; however, researchers have been successful in stably transfecting this enzyme into both cell types, with the EPO protein having α(2, 6) NeuAc content [72, 73]. In Ferrari et al. [74], researchers took the opposite approach and attempted to maximize sialic acid content by reducing its removal. The researchers used antisense RNA technology to significantly knock-down expression of sialidase in CHO cells expressing Dnase, which resulted in a higher sialic acid content since its removal was attenuated. Another strategy towards increasing NeuAc content is to increase the supply of CMP-NeuAc, the nucleotide-sugar donor for NeuAc addition onto N-glycans. In Wong et al. [75], researchers reported the transfection of a hamster CMP-NeuAc transporter. Transfection and overexpression of the transporter in a CHO cell line producing recombinant human IFN-γ resulted in single clones that had 2–20-fold increase in total CMP-NeuAc transporter expression at the transcript level and 1.8–2.8-fold increase in CMP-SiaT at the protein level when compared to untransfected CHO IFN-γ-expressing cells. The overexpression of this transporter facilitated a 4–16% increase in site sialylation of IFN-γ. As mentioned previously, NeuGc is the sialic acid type that is generally undesirable within the final glycoform profile. Cell line engineering studies have also targeted the enzyme responsible for its formation. In one report, CHO cells were transfected with antisense CMP-sialic acid hydroxylase, which resulted in an 80% reduction in hydroxylase activity [76]. The targeted maximization of NeuAc and minimization of NeuGc will likely be of continued interest in the years ahead.

Additional protein N- and O-glycosylation enzymes have been successfully overexpressed or knocked-down in mammalian cell lines. GnT III adds a bisecting GlcNAc on N-glycans (Fig. 2) in many mammals, but not CHO [33]. Umana et al. [77] successfully introduced this enzyme in CHO cells expressing a chimeric IgG_1antibody. At various levels of expression, the researchers found that the enzyme could support a final glycoform profile with approximately 50% bisected, complex-type N-glycans. Additional studies have also shown the successful introduction of this gene into CHO cells expressing an IgG [78]. Other notable cell line glycoengineering studies include Mori et al. [79], who used siRNA against FucT VIII in CHO cells, which facilitated an increase in ADCC, and Potvin et al. [80], who transfected α(1, 3) FucT into CHO and found activation of the previously silent host enzyme which supplemented the activity of the newly transfected enzyme. The targeted cell line glycoengineering of mammalian cell lines has indeed had many success stories.

Mammalian cell lines with targeted enzyme knock-outs as well as mutations that allow for non-functional protein glycosylation genes have been reported in the literature. In particular, there have been numerous reports on the isolation of CHO mutants defective in various components of the protein glycosylation pathway whose characterization has significantly increased our knowledge of the protein glycosylation pathway in general, and in CHO cells in particular [25]. In Yamane-Ohnuki et al. [81], a FucT VIII CHO knockout was generated, resulting in a significant decrease in fucose content and a significantly higher ADCC. In another report, researchers reported a CHO mutant line, MAR-11, which was selected using lectin affinity for decreased levels of cell surface sialic acid relative to wild-type CHO cells [82]. These particular cells expressed a truncated version of the CMP-NeuAc transporter which facilitated recombinantly expressed IFN-γ devoid of sialic acid.

The introduction, suppression, or knockout of protein glycosylation genes is a proven method for protein glycosylation control in mammalian cells. The studies to date have shown an appreciable sensitivity of the final glycoform profile to the corresponding gene expression levels of the enzymes and nucleotide-sugar transporters involved. However, the genetic perturbation of these enzymes on an individual basis does underscore the need for better understanding of the regulation of the pathway, how it can be optimized further, and also to better answer the degree to which the pathway can be further controlled. Genomics studies on protein glycosylation have been attempted, and the information gleaned from these studies has begun to provide a more comprehensive perspective of the pathway, its regulation, and the cellular response to different culture environments.

4 Contemporary Prospects for Understanding and Enhancing Protein Glycosylation Control

4.1 Genomics

One central theme throughout the established methods for protein glycosylation control is that these control points often have mixed results, and that a case-by-case study is generally required for understanding glycosylation control. In addition, most of these studies are univariate in nature, in that they have targeted one process variable, media component, or protein glycosylation enzyme at a time. Through the advent of genomics and large-scale gene expression analysis tools, researchers have been able to take a more multivariate approach to the understanding of the protein glycosylation pathway, and its relationship to other biological pathways. These results highlight some interesting conclusions that have in fact advanced our understanding of the protein glycosylation pathway, its relationship to other metabolic pathways, and its relative sensitivity to the systems investigated. The information generated through these case studies has begun to permit the transition of the current status of protein glycosylation control from a purely heuristic exercise to that of better process and cellular understanding.

The effect of elevated metabolic waste products on the resulting protein glycosylation gene expression has been evaluated. In Chen and Harcum [83], the effect of elevated ammonia was evaluated via quantitative real-time reverse-transcriptase polymerase chain reaction (qRT-PCR) on 12 genes in CHO cells. In general, the authors found that numerous cytosol- and ER-localized genes associated with early glycosylation steps were insensitive to ammonia levels, in contrast to the later glycan processing steps in the Golgi which were sensitive. Specifically, the authors found that $\alpha(1, 3)$ FucT and sialidase did not change over time with elevated ammonium levels. In contrast, the UDP-Gal transporter, $\beta(1,4)$ GalT, and $\alpha(2, 3)$ SiaT genes were found to have lower expression over time in those cultures with elevated ammonium levels. CMP-NeuAc transporter and UDP-glucose pyrophosphorylase were found to be sensitive towards ammonium levels, but not culture time. Only one gene, the UDP-Gal transporter, was found to be expressed at a higher level in the higher ammonium cell culture condition. Since the gene expression profile for $\alpha(2, 3)$ SiaT was significantly lower in the elevated ammonium-treated cultures, and the CMP-NeuAc transporter profile was modestly lower, the authors concluded that this effect on gene expression elicited by elevated ammonium is at least partially responsible for the commonly observed decrease in NeuAc content under elevated ammonium. Sialidase levels were not significantly different regardless of ammonium levels in the culture media, suggesting that NeuAc addition was attenuated, and there was no elevated NeuAc removal from the tPA glycoprotein analyzed. This case study is a fine example of how glycosylation gene expression information complements the understanding of protein glycosylation control.

The effect of cell culture media components and supplement levels has also been evaluated on glycosylation gene expression. As mentioned previously, sodium butyrate is a commonly added supplement to target increases in product titer levels, as it is a known modulator of chromatin structure and has been extensively studied for its role in increasing cellular specific productivity [84, 85]. In an interesting study, researchers added sodium butyrate to both hybridoma and CHO cell cultures and measured the gene expression relative to control cultures via mouse and CHO cDNA microarrays [86]. The addition of sodium butyrate was shown to cause differential expression of a modest number of protein glycosylation genes in both the hybridoma and CHO cell cultures. These results suggest that commonly used cell culture media supplements can have a significant role in modulating the protein glycosylation gene expression profile. Although differential expression was observed, it is uncertain whether the glycoform profiles from these cultures changed. However, it was shown in other reports that the addition of sodium butyrate does not appreciably change the glycoform profile of expressed proteins [87].

In another study highlighting the effect of culture media conditions on gene expression levels, Korke et al. [88] cultured hybridoma cells in continuous bioreactor cultures in the presence of controlled low glucose levels which facilitated low lactate production. Using a mouse cDNA and oligonucleotide microarray, the researchers evaluated the gene expression profiles at different metabolic states. In particular, they reported no differentially expressed protein glycosylation genes in the low-lactate-producing cells. In another study, researchers cultured primary fibroblasts with tunicamycin as well as glucose-limited cultures and measured the resulting relative gene expression [89]. The authors found a much higher fold change in expression (>2-fold) with cells cultured with tunicamycin compared to glucose-limited cultures. Hence, in these two particular cases it appears that low nutrient levels do not cause a significant change in protein glycosylation gene expression. The effect of cell culture media conditions on the resulting protein glycosylation gene expression profiles will likely be of continued interest in the years ahead.

The supplementation of nucleotide-sugar precursors in cell culture media is an established method for protein glycosylation control. Researchers have utilized a microarray with glycosylation-pathway-specific genes and qRT-PCR to analyze 79 protein glycosylation-related genes to study in-depth the effect on intracellular glycosylation of feeding various nucleotide-sugars in shake flask cultures of CHO cells expressing IFN-γ [67]. This analysis enabled both an intracellular and extracellular perspective on protein glycosylation control. Gene expression results indicated a significant upregulation of genes involved in CMP-NeuAc biosynthesis for the cultures supplemented with galactose and uridine. In each of the conditions evaluated, however, there was no significant change in the expression of the nucleotide-sugar transporters. Similar to previously discussed studies, genes involved in early glycan processing were not significantly changed in the various cultures; however, those involved in glycan branching and terminal processing (galactosylation and sialylation) were. GnT II and GnT IV were upregulated for those cultures with glucosamine $\pm$ uridine and N-acetylmannosamine $\pm$ cytidine. Lysosomal sialidase and cytosolic sialidase were upregulated for the glucosamine $\pm$ uridine and N-acetylmannosamine $\pm$ cytidine supplemented cultures. Another interesting result from this study was that the addition of the nucleotide-sugar precursors had a consistent response in the downregulation of protein-folding genes, including chaperones (BiP), as well as calnexin, calreticulin, and protein disulfide isomerase. Hence, the relationship between protein glycosylation and the other protein-processing steps in the secretory pathway is indeed a connected one.

In an investigation evaluating the effect of bioreactor culture time on the resulting protein glycosylation, researchers targeted 24 N-glycosylation genes for expression analysis through qRT-PCR during the exponential, stationary, and death phases of a fed-batch culture of CHO cells expressing IFN-γ [90]. The genes investigated included those involved in N-glycan chain extension, glycan branching, terminal sialylation, nucleotide-sugar synthesis and transport, and N-glycan degradation. The researchers found that 21 of the 24 genes evaluated were either up- or downregulated throughout the time course of the experiment. It was observed that as the culture progressed there was a decreased expression of CMP-NeuAc biosynthesis genes (UDP-GlcNAc-2-epimerase), as well as increased expression of sialidases which correlated to the decrease in sialic acid measured on the secreted product (~ 13% increase in asialo glycan structures). Sialyltransferase gene expression increased in expression throughout the culture, indicating that abrogated biosynthesis of CMP-sialic acid, as well as the increase in sialic acid removal, were more likely the reason for the decrease in sialic acid attached to the protein. Nucleotide-sugar transporter gene expression did not significantly change over time. Biantennary glycans were found to increase over time at the expense of tri- and tetra-antennary glycans, which correlated to the relatively higher GnT II expression measured. The authors further analyzed the gene expression of low-glutamine controlled CHO fed-batch bioreactor cultures and found that GnT V was upregulated and GnT IV was downregulated, which was consistent with the N-glycan measurements which reflected a decrease in tetra-antennary glycans, and an increase in tri-antennary glycans. At low glutamine levels there was also a reduction in sialylation levels, which was consistent with the measured decrease in expression of UDP-GlcNAc-2 epimerase (a ManNAc biosynthesis gene required for CMP-sialic acid biosynthesis), as well as α(2, 3) SiaT genes. The above results highlight the more in-depth understanding afforded through the use of microarrays and gene expression applications in general, since the relative expression results and the resulting glycosylation results were in alignment in this particular study.

Maintaining high cell-specific productivity conditions is certainly the goal of any production bioreactor. A supporting goal, however, is to ensure that this does not come at the expense of product quality. A case study on the effect of 11 different GS-NS0 cell clones operating at different IgG productivity levels on the resulting differential gene expression has been recently described [91]. In this study, differential expression was evaluated using an oligonucleotide mouse microarray. The extensive statistical analysis revealed significant differential expression of numerous genes involved with post-translational protein processing as a direct result of the different productivity levels. It would be interesting to see what if any glycoform profile differences were observed amongst the cell lines investigated. Whether the glycosylation machinery compensates for increases in specific productivity with concomitant increases in expression, or alternatively the fraction of non-fully processed N- and O-glycans increases as a result of expression and/or enzyme activity not compensating fast enough, remains an open question. However, reports have shown both direct and inverse relationships between protein synthesis rates and the resulting protein glycosylation. In one study, lowering the protein synthesis rate by cycloheximide improved the glycan site occupancy of recombinant prolactin from C127 cells [92]. Similarly, decreasing culture temperature to 21 C in HL-60 cells to decrease secretory pathway activity resulted in a 30–50% increase in the presence of N-acetyllactosamine repeats [93]. However studies on tPA synthesis in CHO cells suggest that the protein synthesis rate has little effect on protein glycosylation [94]. In a study using BHK and CHO cells producing glycoproteins, similar glycan profiles were observed even after a 200-fold increase in productivity [95].

The glycosylation gene expression results from cell culture studies to date suggest that the terminal glycan-processing enzymes are more sensitive to culture environment changes compared to earlier glycan-processing enzymes.

Comelli et al. [96] probed the relative gene expression profile of glycosylation-related genes using mammalian tissues and have confirmed this. The researchers developed an oligonucleotide microarray with 436 genes to focus on glycosylation-related genes, including human and murine glycosyltransferases, nucleotide sugar transporters, nucleotide synthesis, glycosidases, proteoglycans, and glycan-binding proteins. The relative expression profiles of nine murine tissues were investigated. The biosynthetic enzymes responsible for the core regions of N- and O-linked oligosaccharides were ubiquitously expressed in the tissues evaluated; however, the enzymes for the addition of terminal sugars were expressed in a more tissue-specific manner. These enzymes included families of sialyltransferases and fucosytransferases. These gene expression results show that most of the diversity at the gene expression level is more typically associated with the terminal sugar additions, that is, those that are localized in the Golgi, rather than at the initial stages of protein glycosylation in the ER.

How a mammalian cell line responds to changes in glycosylation or protein production rates is not fully understood. However, in yeast the cellular response has been shown to involve internal cellular signaling networks that may provide clues for mammalian cells. Several genomics studies have been reported for the characterization of the cellular response to abrogation of glycosylation reactions [89, 97] or analyzing mutants partially defective for N-linked glycosylation [98]. In the latter case, a good example can be seen in Cullen et al. [99]. In this study the researchers used a yeast glycosylation mutant with a partial loss of function of PMI40, an enzyme required for conversion of fructose-6-phosphate to mannose 6-phosphate (an early step in protein glycosylation). The major class of differentially expressed genes could be considered part of the starvation response. These included genes involved in the tricarboxylic acid and glyoxylate cycles, and protein and amino acid biosynthesis. Expression profiling generated from this study demonstrated that genes encoding structural proteins of two mitogen-activated (MAP) kinase pathways, and most of the genes that play a role in the unfolded protein response (UPR), were also induced in response to this glycosylation defect. Whether such a coordinated cellular response occurs in mammalian cell lines is not clear at this time. Also not clear is what occurs when the glycosylation rate is enhanced, rather than attenuated as the above references all documented.

In another study with yeast, researchers studied the unfolded protein response (UPR), which was found to have a direct interaction with the protein glycosylation pathway [97]. The UPR is activated from the accumulation of unfolded proteins in the ER, and the UPR facilitates changes which allows the cell to alleviate the stress. The authors found multiple targets in protein glycosylation that changed as a result. These included 17 genes in the ER and six genes in the Golgi, in addition to numerous other genes involved in protein transport. The genes upregulated as a response by the UPR included those involved in core oligosaccharide synthesis, glycosyltransferases, glycosylphosphatidylinisotol anchoring, and O-linked glycosylation, as well as vesicle trafficking in both anterograde and retrograde fashion. It is uncertain whether a similar response in mammalian cells would exhibit these changes on the gene expression level as a result of a corresponding change in the UPR. But it does show that the glycosylation pathway is readily changed at the gene expression level as a result of other glycoprotein-processing steps in secretory pathway organelles.

The protein glycosylation pathway has a proven role in interacting with numerous other metabolic pathways. In a case study using yeast, Klebl et al. [98] studied a mutant deficient in the synthesis of the lipid-linked oligosaccharide precursor (the initiation step for protein glycosylation). Their results indicated that this deficiency and the resulting perturbation of the N-glycosylation network caused a dramatic reprogramming of various signaling and metabolic networks. Some of these included genes involved in the mitogen-activated protein kinase (MAPK) cascades, as well as phosphate, amino acid, carbohydrate, mitochondrial, and ATP metabolism. This genomics study helped elucidate the protein targets with which glycosylation interacts in peripheral networks, as well as revealing the metabolic and cellular signaling and regulatory elements with which the pathway has an interaction. More such studies are needed in mammalian cell lines to understand better the intricate relationship between the protein glycosylation pathway and other closely related pathways.

4.2 Systems Biology

The complex relationship between cell, process, and metabolic pathway parameters shaping the final glycoform profile suggests that a more systems-level perspective is required to achieve a better understanding of glycosylation control. As discussed earlier, the protein glycosylation pathways are highly complex, with multiple routes possible towards the generation of each of the measured product glycans. With additional glycan site-occupancy variability on the protein, and an intricate relationship with nucleotide-sugar biosynthesis pathways, one can certainly appreciate the complexity of the system. The particular pathway(s) traversed in generating a given glycoform profile is dependent on the relative concentrations of the enzymes involved, the glycan accessibility on the protein to the various enzymes, the spatial localization of these enzymes across the ER and Golgi compartments, the relative nucleotide-sugar concentrations, the substrate specificity for each of the reactant glycans and their subsequent glycan products, as well as the many other factors that have been shown to have some interaction with the protein N- and O-glycosylation pathways. The above sections highlight some of the studies that have identified the cell culture process variables which modulate protein glycosylation, as well as the attempts to control the pathway into a more idealized state.

The interpretation of how these parameters help modulate micro- and macro-heterogeneity is difficult and indirect. Understanding how they interact with the variables that have a direct role on the pathway is especially important. In addition, for many of these successful attempts at modulating the glycosylation pathway, there are typically similar studies that have shown little to no affect on the resulting glycosylation. As a result of the very interactive and case-by-case nature of these variables, the non-linearity in the response and the complexity of the pathway itself has led some to incorporate a more mathematical interpretation of the protein glycosylation pathway and its relationship to the final glycoform profile. Attempts have been made towards the systematic understanding of protein N- and O-glycosylation with systems biology approaches and metabolic modeling. Numerous systems biology approaches have been attempted to understand the protein N- and O-glycosylation pathways. Though not typically viewed as such, protein glycosylation has begun to be looked at from a more network-based perspective (Fig. 2a, b) that has typically been associated with other models of metabolism such as central energy metabolism [100]. The information gleaned from these case studies has reinforced our awareness of the pathway complexity, and has helped increase our understanding of glycoform profile sensitivity to the various controlling parameters.

Some of the mathematical and statistical tools that are commonly used in systems biology include metabolic kinetic models, sensitivity analysis, numerical optimization, and principal component analysis. Sensitivity analysis evaluates the measured or calculated response of a particular parameter as a result of a change in an interacting variable. This change can be large or small, though small changes are typically targeted. Numerical optimization attempts to determine the set of target conditions that best allow for convergence of the response to a specified criteria. In this analysis, an objective function of the various control variables is defined in which the optimization routine attempts to find a local minimum or maximum. Also required is an initial first guess at a particular set of target conditions, since multiple convergent values are possible. Principal component analysis (PCA) provides a quantitative measure of determining the relative importance of control parameters towards a particular measured response. In this procedure, the factors important for a response are correlated into a smaller number of uncorrelated variables that account for the variability in the response data. These mathematical and statistical tools have been used for the analysis of metabolic pathways including the protein N- and O-glycosylation pathways.

In an early systems biology approach to protein glycosylation, Umana and Bailey [101] developed a metabolic model of N-glycan microheterogeneity for 33 different N-glycans, including high-mannose, hybrid, and complex-type oligosaccharides. Using kinetic parameters obtained from the literature, the authors were able to simulate the effect of perturbing the various model parameters on the resulting glycoform profile. Included in this model were four discretely localized Golgi compartments, differential glycosylation enzyme localization across these four compartments, and intra-Golgi transport. The model was essentially a reaction-transport system, where the ten Golgi protein glycosylation enzymes must react upon the substrate glycan(s) to form the subsequent product glycan(s) before transport to the next compartment occurs. Simulating the effects of increasing specific productivity, the model predicted decreases in tri-antennary glycans, and at sufficiently high levels decreases in bi-antennary glycans and increases in non-fully processed high-mannose-type glycans. The reason for this is intuitive in that without concomitant increases in the levels of these enzymes to compensate for the increase in protein traversing the system, a decrease in the degree of glycan processing is to be expected. The researchers also simulated the effect of overexpressing the GnT III enzyme, which due to its overlap in glycan substrate specificity led to the preferential channeling of glycan fluxes towards bisected-hybrid-type glycans and away from bisected-complex-type glycans. The formation of bisected-complex glycans could be rescued through the mislocalization of GnT III, or overexpression of GnT II and Man II so as to change the sequence of enzymatic reactions to ensure the complex-type glycans are not restricted from being formed. The model developed by the researchers provided an early glimpse of the detailed glycosylation pathway understanding that could be achieved through the use of systems biology.

Later work expanded this model by incorporating additional reactions involving galactosylation, fucosylation, sialylation, and the incorporation of N-acetyllactosamine repeats (see Fig. 1b), as well as the consideration of nucleotide-sugars in the Golgi compartments [102]. Additional N-glycan substrate specificities were included in this model. The final system considered 7,565 N-glycans involved in 22,871 enzymatic reactions. The researchers successfully applied nonlinear optimization in their model for the estimation of enzyme concentrations requisite for the matching of the predicted N-glycan profile with that of published data from recombinant human TPO [103]. Simulating the effect of increasing specific productivity, the researchers saw a steady decline in the degree of sialylation. The researchers later applied their model to the prediction of enzyme concentrations and UDP-GlcNAc levels needed for returning the predicted N-glycan profile for a four-fold increase in specific productivity back to the glycoform profile observed with control levels of specific productivity. Information gleaned from use of the model supports its utility as a guidance tool for better understanding of glycosylation, as well as an experimental guide for re-directing intracellular glycan fluxes.

Hossler et al. [104] studied the protein N-glycosylation pathway in the Golgi, and constructed two different metabolic models to describe the potential biosynthesis of 341 N-glycans. One model described the Golgi maturation view of intra-Golgi transport, and the other described the vesicular transport view. Which of these two views is the most valid has been debated in the literature, with studies providing evidence for both [105,106]. In the Golgi maturation model, the Golgi cisternae have been described as relatively mobile and mature as they traverse geographically within the cell along with their protein contents. In the vesicular transport model, the cisternae are conceived as being relatively stationary, while Golgi vesicles transport the protein contents in both directions through anterograde and retrograde transport. Mathematically, these two systems were analyzed using a system of 341 nonlinear ordinary differential equations which were solved simultaneously for each of the reactant and product glycans. Reported glycan substrate specificities for each of the ten enzymes considered were incorporated into the respective models, as were the additional variables of preferential enzyme localization across the Golgi cisternae [107], nucleotide-sugar levels, and intra-Golgi transport. The metabolic fluxes through the pathway and the glycan distribution were visualized directly on the pathway itself through the GlycoVis visualization program [108], previously written for the purposes of analyzing glycan biosynthetic fluxes, as well as deducing preferential glycan biosynthetic pathway specificities (Fig. 3).

Simulations using the Golgi maturation model revealed that the resulting terminally processed glycan profile was particularly sensitive to a few key enzymes, especially GalT which, due to the addition of Gal at particular key branching steps, could have potentially dramatic effects on the resulting N-glycan profile. The use of nonlinear optimization revealed that, although many terminally processed glycans could be predicted to be formed in significantly high abundance, not all of them could do so. That is, glycan uniformity could be achieved for some glycans, but not all, due to the nature of the metabolic pathway and the location of key branching points in the network. Controlling the order of monosaccharide addition is thus an important variable in determining the resulting glycoform profile. In addition, it was also found that, on incorporating a multi-compartmental system in this Golgi maturation model, the appearance of early and intermediately processed N-glycans was observed in the final glycoform profile. These results agree well with reports from the literature where the appearance of high-mannose-type glycans is frequently reported [109].

Simulations using the vesicular transport model revealed a wider spectrum of intermediately processed N-glycans, rather than an over-representation of a few glycans in abundance. This suggested to the authors that a purely vesicular transport system as simulated using the chosen model parameters cannot completely predict experimentally measured glycan profiles without the incorporation of additional model variables, including perhaps imperfect mixing of the enzymes in each of the Golgi compartments, or selective transport perhaps through some form of sorting mechanism. The results also pointed to the Golgi maturation model as having a higher likelihood for modeling intra-Golgi transport, or perhaps a hybrid mechanism involving both maturation and vesicular transport, a mechanism whose possibility has been discussed elsewhere [110]. However, both models did highlight a shared importance of the particular order of monosaccharide addition in determining the final N-glycan profile. Using the compartmentalized Golgi maturation model, glycan uniformity could be generated for each of the terminally processed N-glycans (i.e., those not capable of reacting further). This could only be achieved by mislocalizing the glycosylation enzymes so as to ensure the correct sequence of processing occurred, to prevent enzymatic glycan substrate specificities from channeling glycan fluxes away from the desired product N-glycans. Glycosylation enzyme mislocalization has been documented in the literature, suggesting that this glycoengineering strategy is a viable means of glycomodulation [111].

In another systems biology approach to the study of microheterogeneity, researchers derived an equation for the analysis of the degree of sialylation on a generalized glycoprotein [112]. Assuming Michaelis–Menten-type kinetics for a sialyltransferase enzyme with glycan substrate, the authors found that at

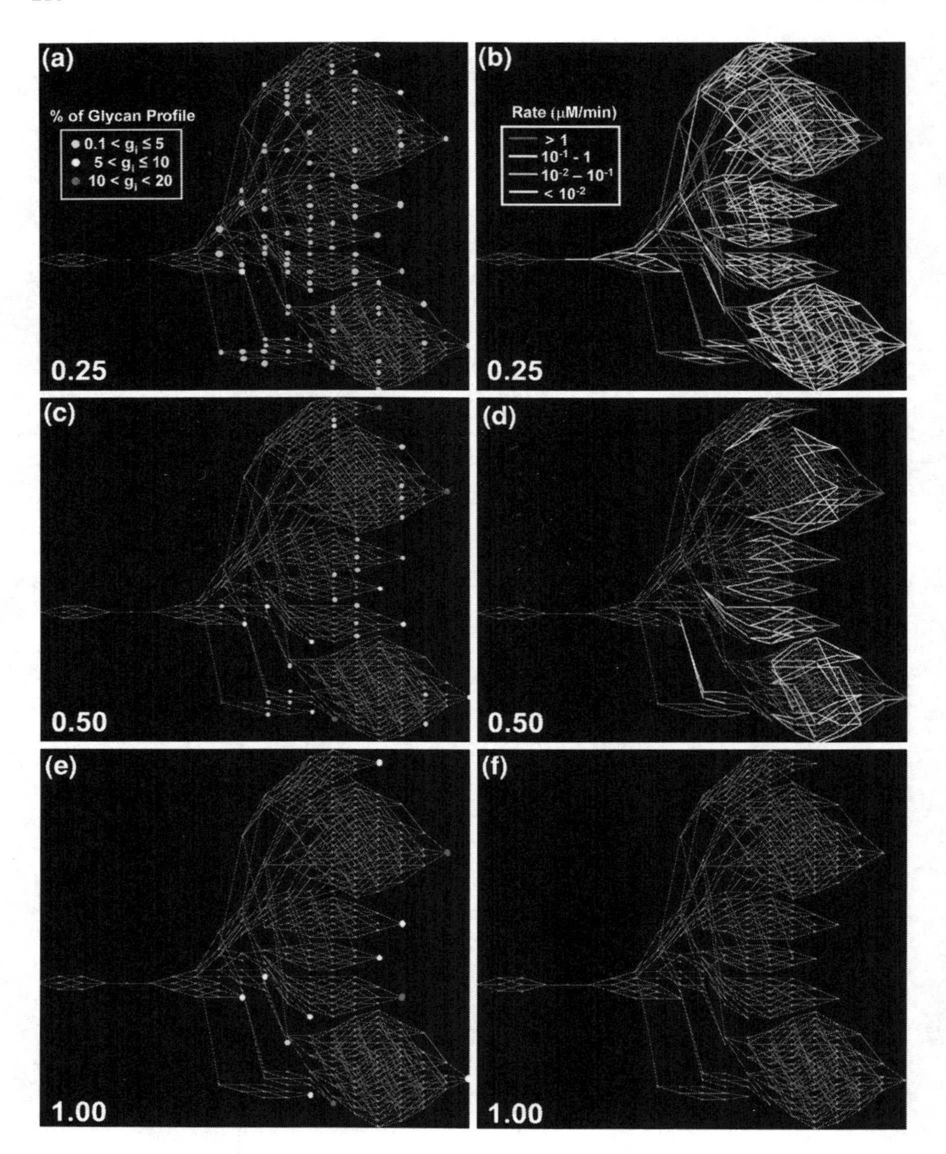

Fig. 3 *In silico* mapping of the glycoform profile (**a**, **c**, **e**) and glycan reaction rates (**b**, **d**, **f**) through the protein N-glycosylation pathway in a simulated Golgi maturation model. Numbers reflect normalized distance across a Golgi organelle at 25% (**a**, **b**), 50% (**c**, **d**), and 100% (**e**, **f**) traversal. [reprinted from Hossler et al. [104], originally published under the license terms of The Public Library of Science (PLoS) and the Creative Commons Attribution License (CCAL)]

sufficiently high concentrations of glycans, the amount of sialylation was dependent on the enzyme concentration, enzymatic kinetic parameters, and the protein residence time within the reaction compartment. However, under non-saturation conditions of glycan substrate, the degree of sialylation was actually independent of the amount of protein in the trans-Golgi network (TGN), and the reactant glycans attached to it. Although this study concentrated on sialylation alone, these mathematical results do point out to us a potential reason for the strong dependence of the final glycoform profile on the many cell culture parameters reported in the aforementioned sections. In a typical production bioprocess using a relatively high-expressing cell line, the recombinant protein is either likely to be transported through the secretory pathway relatively fast, the amount of glycoprotein being transported is relatively high, or a combination of both. In each of these scenarios it is expected that the cumulative glycan throughput is in an elevated state, in which case the strong dependence on proven pathway control parameters, such as relative enzyme amounts, is consistent with this study.

Macroheterogeneity of N-glycans has also been modeled using a systems approach [113]. In this particular model the initiation of protein glycosylation in the ER in the timeframe and space immediately after translation in the ER was analyzed. The mathematical framework established provides some interesting insights into the origins of glycan site-occupancy and macroheterogeneity. The central hypothesis of this work is that the initial glycosylation event catalyzed by the oligosaccharyltransferase (OST) enzyme takes place in a very defined region in space in the ER lumen, which also competes with other processing events, including protein folding. The translocation into the ER lumen proceeds at the same rate as protein translation, which is modeled as a linear velocity parameter. By incorporating a mass balance, the authors were able to derive various dimensionless groups which describe their system as a function of the glycosylation event, as well as the effects of incorporating protein folding. The main conclusion from this work was that due to the competing nature of the first initial glycosylation event, and the protein folding that occurs in the ER of mammalian cells, there is an optimal mRNA elongation rate one could expect to observe for achieving maximum site occupancy. Though no study testing this experimentally could be found in the literature, the modeling results do suggest that specific productivity can indeed have a role in helping determine the glycan site occupancy.

The O-glycan biosynthetic pathway has also been examined through a systems-level approach. In Liu et al. [114], researchers analyzed the formation of sialyl Lewis-X (sLe^{x}) structures (see Fig. 1e) on P-selectin glycoprotein ligand-1 (PSGL-1), which has 71 potential O-glycan sites. Though not typically seen on glycoproteins constituitively secreted from mammalian cells, this glycan attached to the cell surface of leukocytes does have prominent biological roles, and the approach and results from this study are nonetheless instructive in the analysis of O-glycan pathways in general. The researchers' goal was to combine various modeling and analysis methods for the *in silico* determination of enzymatic reaction rate constants which could be compared to experimental measurements for comparison purposes. Lysates from human leukocyte (HL-60) cells were used as the source material for the experimentally measured rate constants. Five glycosyltransferases were included in their model, as well as the known product glycans that had been reported elsewhere to generate a model of the diverse pathways potentially traversed from an initial reactant glycan to give rise to the spectrum of product glycans. Subset pathways were subsequently focused on, due to their more prominent role connecting the substrate and known product glycans. Focusing on these subset pathways, the researchers applied a nonlinear optimization approach for the determination of the rate constants which facilitated the optimal agreement between the predicted versus experimental O-glycan profile. The various subset pathways which facilitated close agreement to the final profile were analyzed using hierarchical clustering followed by principal component analysis (PCA) to assign relative importances to the individual network reactions. Assessing the relative importance of each of the individual glycosylation reactions facilitated the determination of the most likely pathway traversed for the two sialyl Lewis-X structures of interest. Sensitivity analysis was also used to find that final sLe^x levels had a direct relationship with the expression of one particular sialyltransferase (α(2,3)ST3Gal-IV), and an inverse relationship with that of another (α(2,3)ST3Gal-I/II). Through this very detailed and systematic approach to O-glycan biosynthesis, the authors were able to conclude that the final O-glycan profile is determined principally through individual enzyme activity, as well as competition of the enzymes for the O-glycan substrates.

In a complementary case study, researchers combined gene expression measurements with those of glycosyltransferase activity in an attempt to predict the resulting O-glycoform of PSGL-1 expressed by HL-60 cells and neutrophils on their cell surface [115]. Although distinct differences were observed by the researchers in the enzyme activity between neutrophils and HL-60 cells, overall there was a relatively close agreement between measured enzyme activities and the corresponding enzyme gene expression (although the agreement was not strictly linear). Differences observed between activity and expression results were hypothesized to be due to gene expression changes manifesting rather quickly compared to the intracellular enzyme levels, which have a more pronounced turnover time. Whether or not protein glycosylation enzymes in mammalian cells have slow intracellular turnover times in general remains to be seen, however. The authors further found that enzyme activity measurements were more directly correlated with the resulting O-glycan profile than with gene expression measurements. For more accurate prediction of the final O-glycan profile utilizing gene expression data, additional information including preferential enzyme localization in the Golgi and additional glycan substrate specificities would need to be considered. This conclusion certainly points to the validity and utility of adopting systems-level perspectives in predicting final glycoform profiles.

5 Discussion of Protein Glycosylation Control

Protein glycosylation is a fascinating pathway comprising both divergent and convergent pathways characterized by a relatively small number of enzymes, but with a potentially larger number of substrate and product glycans. The fact that not all these glycans are measured in a typical cell culture process suggests that a great deal of control is manifested. There have been various reports in the literature of researchers utilizing these controlling mechanisms for glycomodulation, either directly through cell line engineering, or indirectly by controlling cell culture process and media conditions. Many cell culture process and media conditions have shown mixed results on glycosylation, with specific results sometimes being observed, and sometimes not. These results suggest that understanding the effect of cell culture process and media conditions requires a case-by-case analysis. Attempts to control protein glycosylation through cell line engineering have also had many successes. The N- and O-glycoform profiles from CHO cells appear to be particularly responsive to these attempts. Many enzymes have been over-expressed, silenced, or knocked-out completely in CHO and other industrially relevant mammalian cell lines. The cumulative results of these studies indicates that in most cases the targeted genetic manipulation of these cells does indeed have the expected effect on the resulting N- and O-glycan profiles. Thus cell line engineering remains a very viable means of attempting glycomodulation in cell culture bioprocesses. It is likely that the effect of cell culture media components and process conditions on protein glycosylation, as well as cell line engineering, will be of interest to the biopharmaceutical industry for some time to come. Despite the limitations in understanding from a glycosylation pathway perspective, the reason for these potential changes has become more apparent.

The proven sensitivity of the protein glycosylation pathway to the various parameters mentioned in this review highlights the difficulty in controlling this pathway. The dependence of the pathway on cell culture process conditions, relative enzyme and nucleotide sugar expression levels, and the expression host is well-established. What is not understood as well in many cases, however, is the reason for these changes at the glycosylation pathway level. Through the use of tools such as genomics and systems biology, there is now even more appreciation for the complexity of protein glycosylation. Gene expression analysis and systems biology have provided additional information on the understanding of the pathway as a whole, and its relationship to other pathways, in particular protein secretion.

Gene expression results have shown that the pathway is particularly sensitive to the expression of the terminal processing enzymes, including galactosylation and sialylation, as well as the nucleotide-sugar transporters responsible for nucleotide-sugar import into the ER and Golgi organelles. Additional gene expression results have shown that the resulting measured glycoform profile does not always correspond exactly to the relative gene expression levels. This further suggests that the enzyme expression and relative enzyme amounts and activities in the ER and Golgi need to be investigated in parallel to provide a more comprehensive understanding.

Systems biology studies of N- and O-glycan micro- and macroheterogeneity have attempted to incorporate these various pathway considerations and have revealed much about the protein glycosylation pathway as a result. *In-silico* prediction of preferential pathway specificities needed for the prediction of glycoform profiles that match those of experimentally measured glycans has shown that there is significant pathway redundancy even for the theoretical prediction of uniform protein glycosylation. Enzyme concentration levels are necessary but not sufficient for achieving uniform glycosylation for some terminally processed glycans. In these particular cases, controlling the order of monosaccharide addition onto an N- or O-glycan is required through the preferential localization of enzymes across the Golgi compartments. The application of mathematical and statistical tools, including optimization, PCA, and clustering have also permitted the theoretical prediction of pathway specificities for the formation of experimentally measured glycan profiles. Indeed, these systems-level approaches have allowed for the theoretical analysis of the pathway requirements for producing uniform and/or desired glycoform profiles. The use of these tools will continue to supplement and guide experimental efforts at glycomodulation in the future.

The aforementioned case studies of genomics and system biology on protein glycosylation in mammalian cells have certainly increased our awareness of the complexity of the protein glycosylation pathway. We do not yet have the capability to fine-tune at will this pathway in mammalian cells. With time and continued incorporation of these relatively newer tools in mammalian cell culture, that capability, as well as a clearer picture of protein glycosylation control, should continue to emerge. Indeed these approaches will continue to contribute towards our ultimate goal of producing optimal glycoproteins on a consistent basis in mammalian cell culture.

Acknowledgments Content review and helpful suggestions from Sean McDermott and Christopher Racicot is gratefully acknowledged.

References

1. Apweiler R, Hermjakob H, Sharon N (1999) On the frequency of protein glycosylation, as deduced from analysis of the SWISS-PROT database. Biochim Biophys Acta 1473(1):4–8
2. Walsh MT et al (1990) Effect of the carbohydrate moiety on the secondary structure of beta 2-glycoprotein. I. Implications for the biosynthesis and folding of glycoproteins. Biochemistry 29(26):6250–6257
3. Leavitt R, Schlesinger S, Kornfeld S (1977) Impaired intracellular migration and altered solubility of nonglycosylated glycoproteins of vesicular Stomatitis virus and Sindbis virus. J Biol Chem 252(24):9018–9023
4. Wallick SC, Kabat EA, Morrison SL (1988) Glycosylation of a VH residue of a monoclonal antibody against alpha (1–6) dextran increases its affinity for antigen. J Exp Med 168(3): 1099–109
5. Wyss DF, Wagner G (1996) The structural role of sugars in glycoproteins. Curr Opin Biotechnol 7(4):409–416
6. Butler M (2006) Optimisation of the cellular metabolism of glycosylation for recombinant proteins produced by mammalian cell systems. Cytotechnology 50(1–3):57–76
7. Butler M et al (2003) Detailed glycan analysis of serum glycoproteins of patients with congenital disorders of glycosylation indicates the specific defective glycan processing step and provides an insight into pathogenesis. Glycobiology 13(9):601–622
8. Raman R et al (2005) Glycomics: an integrated systems approach to structure-function relationships of glycans. Nat Methods 2(11):817–824

9. Kikuchi N et al (2005) The carbohydrate sequence markup language (CabosML): an XML description of carbohydrate structures. Bioinformatics 21(8):1717–1718
10. Sahoo SS et al (2005) GLYDE—an expressive XML standard for the representation of glycan structure. Carbohydr Res 340(18):2802–2807
11. Lowe JB, Marth JD (2003) A genetic approach to mammalian glycan function. Annu Rev Biochem 72:643–691
12. Lis H, Sharon N (1993) Protein glycosylation. Structural and functional aspects. Eur J Biochem 218(1):1–27
13. Rudd PM et al (2001) Glycosylation and the immune system. Science 291(5512): 2370–2376
14. Gross V et al (1988) Involvement of various organs in the initial plasma clearance of differently glycosylated rat liver secretory proteins. Eur J Biochem 173(3):653–659
15. Sathyamoorthy N et al (1991) Evidence that specific high mannose structures directly regulate multiple cellular activities. Mol Cell Biochem 102(2):139–147
16. Lifely MR et al (1995) Glycosylation and biological activity of CAMPATH-1H expressed in different cell lines and grown under different culture conditions. Glycobiology 5(8): 813–822
17. Shields RL et al (2002) Lack of fucose on human IgG1 N-linked oligosaccharide improves binding to human Fcgamma RIII and antibody-dependent cellular toxicity. J Biol Chem 277(30):26733–26740
18. Millward TA et al (2008) Effect of constant and variable domain glycosylation on pharmacokinetics of therapeutic antibodies in mice. Biologicals 36(1):41–47
19. Hossler P, Khattak SF, Li ZJ (2009) Optimal and consistent protein glycosylation in mammalian cell culture. Glycobiology 19(9):936–949
20. Helenius A, Aebi M (2001) Intracellular functions of N-linked glycans. Science 291(5512):2364–2369
21. Carraway KL, Hull SR (1989) O-glycosylation pathway for mucin-type glycoproteins. Bioessays 10(4):117–121
22. Van den Steen P et al (1998) Concepts and principles of O-linked glycosylation. Crit Rev Biochem Mol Biol 33(3):151–208
23. Ishida N, Kawakita M (2004) Molecular physiology and pathology of the nucleotide sugar transporter family (SLC35). Pflugers Arch 447(5):768–775
24. Rademacher TW (1993) Glycosylation as a factor affecting product consistency. Biologicals 21(2):103–104
25. Stanley P, Raju TS, Bhaumik M (1996) CHO cells provide access to novel N-glycans and developmentally regulated glycosyltransferases. Glycobiology 6(7):695–699
26. Skibeli V, Nissen-Lie G, Torjesen P (2001) Sugar profiling proves that human serum erythropoietin differs from recombinant human erythropoietin. Blood 98(13):3626–3634
27. Durocher Y, Butler M (2009) Expression systems for therapeutic glycoprotein production. Curr Opin Biotechnol 20(6):700–707
28. Jenkins N, Parekh RB, James DC (1996) Getting the glycosylation right: implications for the biotechnology industry. Nat Biotechnol 14(8):975–981
29. Larsen RD et al (1990) Frameshift and nonsense mutations in a human genomic sequence homologous to a murine UDP-Gal:beta-D-Gal(1,4)-D-GlcNAc alpha(1,3)-galactosyltransferase cDNA. J Biol Chem 265(12):7055–7061
30. Baker KN et al (2001) Metabolic control of recombinant protein N-glycan processing in NS0 and CHO cells. Biotechnol Bioeng 73(3):188–202
31. Borys MC et al (2010) Effects of culture conditions on N-glycolylneuraminic acid (Neu5Gc) content of a recombinant fusion protein produced in CHO cells. Biotechnol Bioeng 105(6):1048–1057
32. Lee EU, Roth J, Paulson JC (1989) Alteration of terminal glycosylation sequences on N-linked oligosaccharides of Chinese hamster ovary cells by expression of beta-galactoside alpha 2, 6-sialyltransferase. J Biol Chem 264(23):13848–13855

33. Campbell C, Stanley P (1984) A dominant mutation to ricin resistance in Chinese hamster ovary cells induces UDP-GlcNAc:glycopeptide beta-4-N-acetylglucosaminyltransferase III activity. J Biol Chem 259(21):13370–13378
34. Kim YK et al (2005) Production and N-glycan analysis of secreted human erythropoietin glycoprotein in stably transfected Drosophila S2 cells. Biotechnol Bioeng 92(4):452–461
35. Hamilton SR et al (2006) Humanization of yeast to produce complex terminally sialylated glycoproteins. Science 313(5792):1441–1443
36. Kunkel JP et al (2000) Comparisons of the glycosylation of a monoclonal antibody produced under nominally identical cell culture conditions in two different bioreactors. Biotechnol Prog 16(3):462–470
37. Goldman MH et al (1998) Monitoring recombinant human interferon-gamma N-glycosylation during perfused fluidized-bed and stirred-tank batch culture of CHO cells. Biotechnol Bioeng 60(5):596–607
38. Burger C et al (1999) An integrated strategy for the process development of a recombinant antibody-cytokine fusion protein expressed in BHK cells. Appl Microbiol Biotechnol 52(3):345–353
39. Chotigeat W et al (1994) Role of environmental conditions on the expression levels, glycoform pattern and levels of sialyltransferase for hFSH produced by recombinant CHO cells. Cytotechnology 15(1–3):217–221
40. Lin AA, Kimura R, Miller WM (1993) Production of tPA in recombinant CHO cells under oxygen-limited conditions. Biotechnol Bioeng 42(3):339–350
41. Kunkel JP et al (1998) Dissolved oxygen concentration in serum-free continuous culture affects N-linked glycosylation of a monoclonal antibody. J Biotechnol 62(1):55–71
42. Yoon SK, Song JY, Lee GM (2003) Effect of low culture temperature on specific productivity, transcription level, and heterogeneity of erythropoietin in Chinese hamster ovary cells. Biotechnol Bioeng 82(3):289–298
43. Bollati-Fogolin M et al (2005) Temperature reduction in cultures of hGM-CSF-expressing CHO cells: effect on productivity and product quality. Biotechnol Prog 21(1):17–21
44. Clark KJ, Chaplin FW, Harcum SW (2004) Temperature effects on product-quality-related enzymes in batch CHO cell cultures producing recombinant tPA. Biotechnol Prog 20(6):1888–1892
45. Trummer E et al (2006) Process parameter shifting: part I. Effect of DOT, pH, and temperature on the performance of Epo-Fc expressing CHO cells cultivated in controlled batch bioreactors. Biotechnol Bioeng 94(6):1033–1044
46. Muthing J et al (2003) Effects of buffering conditions and culture pH on production rates and glycosylation of clinical phase I anti-melanoma mouse IgG3 monoclonal antibody R24. Biotechnol Bioeng 83(3):321–334
47. Gramer MJ, Goochee CF (1993) Glycosidase activities in Chinese hamster ovary cell lysate and cell culture supernatant. Biotechnol Prog 9(4):366–373
48. Gramer MJ et al (1995) Removal of sialic acid from a glycoprotein in CHO cell culture supernatant by action of an extracellular CHO cell sialidase. Biotechnology (N Y) 13(7):692–698
49. Gramer MJ, Goochee CF (1994) Glycosidase activities of the 293 and NS0 cell lines, and of an antibody-producing hybridoma cell line. Biotechnol Bioeng 43(5):423–428
50. Thorens B, Vassalli P (1986) Chloroquine and ammonium chloride prevent terminal glycosylation of immunoglobulins in plasma cells without affecting secretion. Nature 321(6070):618–620
51. Andersen DC, Goochee CF (1995) The effect of ammonia on the O-linked glycosylation of granulocyte colony-stimulating factor produced by chinese hamster ovary cells. Biotechnol Bioeng 47(1):96–105
52. Yang M, Butler M (2000) Effects of ammonia on CHO cell growth, erythropoietin production, and glycosylation. Biotechnol Bioeng 68(4):370–380
53. Curling EM et al (1990) Recombinant human interferon-gamma. Differences in glycosylation and proteolytic processing lead to heterogeneity in batch culture. Biochem J 272(2):333–337

54. Borys MC, Linzer DI, Papoutsakis ET (1994) Ammonia affects the glycosylation patterns of recombinant mouse placental lactogen-I by chinese hamster ovary cells in a pH-dependent manner. Biotechnol Bioeng 43(6):505–514
55. Grammatikos SI et al (1998) Intracellular UDP-N-acetylhexosamine pool affects N-glycan complexity: a mechanism of ammonium action on protein glycosylation. Biotechnol Prog 14(3):410–419
56. Hayter PM et al (1992) Glucose-limited chemostat culture of Chinese hamster ovary cells producing recombinant human interferon-gamma. Biotechnol Bioeng 39(3):327–335
57. Hayter PM et al (1993) The effect of the dilution rate on CHO cell physiology and recombinant interferon-gamma production in glucose-limited chemostat culture. Biotechnol Bioeng 42(9):1077–1085
58. Tachibana H et al (1994) Changes of monosaccharide availability of human hybridoma lead to alteration of biological properties of human monoclonal antibody. Cytotechnology 16(3):151–157
59. Cruz HJ et al (2000) Metabolic shifts do not influence the glycosylation patterns of a recombinant fusion protein expressed in BHK cells. Biotechnol Bioeng 69(2):129–139
60. Nyberg GB et al (1999) Metabolic effects on recombinant interferon-gamma glycosylation in continuous culture of Chinese hamster ovary cells. Biotechnol Bioeng 62(3):336–347
61. Crowell CK et al (2007) Amino acid and manganese supplementation modulates the glycosylation state of erythropoietin in a CHO culture system. Biotechnol Bioeng 96(3):538–549
62. Higa HH, Paulson JC (1985) Sialylation of glycoprotein oligosaccharides with N-acetyl-, N-glycolyl-, and N-O-diacetylneuraminic acids. J Biol Chem 260(15):8838–8849
63. Gawlitzek M et al (1995) Characterization of changes in the glycosylation pattern of recombinant proteins from BHK-21 cells due to different culture conditions. J Biotechnol 42(2):117–131
64. Gu X et al (1997) Influence of Primatone RL supplementation on sialylation of recombinant human interferon-gamma produced by Chinese hamster ovary cell culture using serum-free media. Biotechnol Bioeng 56(4):353–360
65. Andersen DC et al (2000) Multiple cell culture factors can affect the glycosylation of Asn-184 in CHO-produced tissue-type plasminogen activator. Biotechnol Bioeng 70(1):25–31
66. Gu X, Wang DI (1998) Improvement of interferon-gamma sialylation in Chinese hamster ovary cell culture by feeding of N-acetylmannosamine. Biotechnol Bioeng 58(6):642–648
67. Wong NS et al (2010) An investigation of intracellular glycosylation activities in CHO cells: effects of nucleotide sugar precursor feeding. Biotechnol Bioeng 107(2):321–336
68. Rosenfeld R et al (2007) A lectin array-based methodology for the analysis of protein glycosylation. J Biochem Biophys Methods 70(3):415–426
69. Lauc G et al (2010) Complex genetic regulation of protein glycosylation. Mol Biosyst 6(2):329–335
70. Ashwell G, Kawasaki T (1978) A protein from mammalian liver that specifically binds galactose-terminated glycoproteins. Methods Enzymol 50:287–288
71. Weikert S et al (1999) Engineering Chinese hamster ovary cells to maximize sialic acid content of recombinant glycoproteins. Nat Biotechnol 17(11):1116–1121
72. Schlenke P et al (1999) Construction and characterization of stably transfected BHK-21 cells with human-type sialylation characteristic. Cytotechnology 30(1–3):17–25
73. Zhang X, Lok SH, Kon OL (1998) Stable expression of human alpha-2,6-sialyltransferase in Chinese hamster ovary cells: functional consequences for human erythropoietin expression and bioactivity. Biochim Biophys Acta 1425(3):441–452
74. Ferrari J et al (1998) Chinese hamster ovary cells with constitutively expressed sialidase antisense RNA produce recombinant DNase in batch culture with increased sialic acid. Biotechnol Bioeng 60(5):589–595
75. Wong NS, Yap MG, Wang DI (2006) Enhancing recombinant glycoprotein sialylation through CMP-sialic acid transporter over expression in Chinese hamster ovary cells. Biotechnol Bioeng 93(5):1005–1016

76. Chenu S et al (2003) Reduction of CMP-N-acetylneuraminic acid hydroxylase activity in engineered Chinese hamster ovary cells using an antisense-RNA strategy. Biochim Biophys Acta 1622(2):133–144
77. Umana P et al (1999) Engineered glycoforms of an antineuroblastoma IgG1 with optimized antibody-dependent cellular cytotoxic activity. Nat Biotechnol 17(2):176–180
78. Davies J et al (2001) Expression of GnTIII in a recombinant anti-CD20 CHO production cell line: expression of antibodies with altered glycoforms leads to an increase in ADCC through higher affinity for FC gamma RIII. Biotechnol Bioeng 74(4):288–294
79. Mori K et al (2004) Engineering Chinese hamster ovary cells to maximize effector function of produced antibodies using FUT8 siRNA. Biotechnol Bioeng 88(7):901–908
80. Potvin B et al (1990) Transfection of a human alpha-(1,3)fucosyltransferase gene into Chinese hamster ovary cells. Complications arise from activation of endogenous alpha-(1,3)fucosyltransferases. J Biol Chem 265(3):1615–1622
81. Yamane-Ohnuki N et al (2004) Establishment of FUT8 knockout Chinese hamster ovary cells: an ideal host cell line for producing completely defucosylated antibodies with enhanced antibody-dependent cellular cytotoxicity. Biotechnol Bioeng 87(5):614–622
82. Lim SF et al (2008) The Golgi CMP-sialic acid transporter: a new CHO mutant provides functional insights. Glycobiology 18(11):851–860
83. Chen P, Harcum SW (2006) Effects of elevated ammonium on glycosylation gene expression in CHO cells. Metab Eng 8(2):123–132
84. Oh SK et al (1993) Substantial overproduction of antibodies by applying osmotic pressure and sodium butyrate. Biotechnol Bioeng 42(5):601–610
85. Palermo DP et al (1991) Production of analytical quantities of recombinant proteins in Chinese hamster ovary cells using sodium butyrate to elevate gene expression. J Biotechnol 19(1):35–47
86. De Leon Gatti M et al (2007) Comparative transcriptional analysis of mouse hybridoma and recombinant Chinese hamster ovary cells undergoing butyrate treatment. J Biosci Bioeng 103(1):82–91
87. Mimura Y et al (2001) Butyrate increases production of human chimeric IgG in CHO-K1 cells whilst maintaining function and glycoform profile. J Immunol Methods 247(1–2): 205–216
88. Korke R et al (2004) Large scale gene expression profiling of metabolic shift of mammalian cells in culture. J Biotechnol 107(1):1–17
89. Lecca MR et al (2005) Genome-wide analysis of the unfolded protein response in fibroblasts from congenital disorders of glycosylation type-I patients. FASEB J 19(2):240–242
90. Wong DC et al (2010) Profiling of N-glycosylation gene expression in CHO cell fed-batch cultures. Biotechnol Bioeng 107(3):516–528
91. Seth G et al (2007) Molecular portrait of high productivity in recombinant NS0 cells. Biotechnol Bioeng 97(4):933–951
92. Shelikoff M, Sinskey AJ, Stephanopoulos G (1994) The effect of protein synthesis inhibitors on the glycosylation site occupancy of recombinant human prolactin. Cytotechnology 15(1–3):195–208
93. Wang WC et al (1991) The poly-N-acetyllactosamines attached to lysosomal membrane glycoproteins are increased by the prolonged association with the Golgi complex. J Biol Chem 266(34):23185–23190
94. Bulleid NJ et al (1992) Cell-free synthesis of enzymically active tissue-type plasminogen activator. Protein folding determines the extent of N-linked glycosylation. Biochem J 286(Pt 1): 275–280
95. Grabenhorst E et al (1999) Genetic engineering of recombinant glycoproteins and the glycosylation pathway in mammalian host cells. Glycoconj J 16(2):81–97
96. Comelli EM et al (2006) A focused microarray approach to functional glycomics: transcriptional regulation of the glycome. Glycobiology 16(2):117–131
97. Travers KJ et al (2000) Functional and genomic analyses reveal an essential coordination between the unfolded protein response and ER-associated degradation. Cell 101(3):249–258

98. Klebl B et al (2001) A comprehensive analysis of gene expression profiles in a yeast N-glycosylation mutant. Biochem Biophys Res Commun 286(4):714–720
99. Cullen PJ et al (2006) Genome-wide analysis of the response to protein glycosylation deficiency in yeast. FEMS Yeast Res 6(8):1264–1273
100. Gerdtzen ZP, Daoutidis P, Hu WS (2004) Non-linear reduction for kinetic models of metabolic reaction networks. Metab Eng 6(2):140–154
101. Umana P, Bailey JE (1997) A mathematical model of N-linked glycoform biosynthesis. Biotechnol Bioeng 55(6):890–908
102. Krambeck FJ, Betenbaugh MJ (2005) A mathematical model of N-linked glycosylation. Biotechnol Bioeng 92(6):711–728
103. Inoue N et al (1999) Asn-linked sugar chain structures of recombinant human thrombopoietin produced in Chinese hamster ovary cells. Glycoconj J 16(11):707–718
104. Hossler P, Mulukutla BC, Hu WS (2007) Systems analysis of N-glycan processing in mammalian cells. PLoS One 2(1):e713
105. Elsner M, Hashimoto H, Nilsson T (2003) Cisternal maturation and vesicle transport: join the band wagon!. Mol Membr Biol 20(3):221–229
106. Marsh BJ, Howell KE (2002) The mammalian Golgi-complex debates. Nat Rev Mol Cell Biol 3(10):789–795
107. Colley KJ (1997) Golgi localization of glycosyltransferases: more questions than answers. Glycobiology 7(1):1–13
108. Hossler P et al (2006) GlycoVis: visualizing glycan distribution in the protein N-glycosylation pathway in mammalian cells. Biotechnol Bioeng 95(5):946–960
109. Spellman MW et al (1989) Carbohydrate structures of human tissue plasminogen activator expressed in Chinese hamster ovary cells. J Biol Chem 264(24):14100–14111
110. Mironov A Jr, Luini A, Mironov A (1998) A synthetic model of intra-Golgi traffic. FASEB J 12(2):249–252
111. Ferrara C et al (2006) Modulation of therapeutic antibody effector functions by glycosylation engineering: influence of Golgi enzyme localization domain and co-expression of heterologous beta1, 4-N-acetylglucosaminyltransferase III and Golgi alpha-mannosidase II. Biotechnol Bioeng 93(5):851–861
112. Monica TJ, Andersen DC, Goochee CF (1997) A mathematical model of sialylation of N-linked oligosaccharides in the trans-Golgi network. Glycobiology 7(4):515–521
113. Shelikoff M, Sinskey AJ, Stephanopoulos G (1996) A modeling framework for the study of protein glycosylation. Biotechnol Bioeng 50(1):73–90
114. Liu G et al (2008) Systems-level modeling of cellular glycosylation reaction networks: O-linked glycan formation on natural selectin ligands. Bioinformatics 24(23):2740–2747
115. Marathe DD et al (2008) Systems-level studies of glycosyltransferase gene expression and enzyme activity that are associated with the selectin binding function of human leukocytes. FASEB J 22(12):4154–4167

Adv Biochem Engin/Biotechnol (2012) 127: 221–249
DOI: 10.1007/10_2011_104

Published Online: 31 December 2011

Modeling of Intracellular Transport and Compartmentation

Uwe Jandt and An-Ping Zeng

Abstract The complexity and internal organization of mammalian cells as well as the regulation of intracellular transport processes has increasingly moved into the focus of investigation during the past two decades. Advanced staining and microscopy techniques help to shed light onto spatial cellular compartmentation and regulation, increasing the demand for improved modeling techniques. In this chapter, we summarize recent developments in the field of quantitative simulation approaches and frameworks for the description of intracellular transport processes. Special focus is therefore laid on compartmented and spatiotemporally resolved simulation approaches. The processes considered include free and facilitated diffusion of molecules, active transport via the microtubule and actin filament network, vesicle distribution, membrane transport, cell cycle-dependent cell growth and morphology variation, and protein production. Commercially and freely available simulation packages are summarized as well as model data exchange and harmonization issues.

Keywords Spatiotemporal modeling · Stochastic modeling · Cell culture · Model exchange · Modeling packages

Contents

U. Jandt (✉) · A.-P. Zeng
Institute of Bioprocess and Biosystems Engineering,
Hamburg University of Technology,
Denickestreet 15, D-21071 Hamburg, Germany
e-mail: uwe.jandt@tu-harburg.de

A.-P. Zeng
e-mail: aze@tu-harburg.de

1 Introduction

Within the past two decades, numerous improvements have been made in the in-vivo and in-situ visualization and characterization of transport processes in cells. Techniques such as confocal microscopy, stimulated emission depletion (STED) microscopy [50, 118], dual color localization microscopy (2CLM) [42], spectral position determination microscopy (SPDM) [76] and enhanced staining [45, 108] allow for detailed and statistical analysis of translocation trajectories and distribution of intracellular organelles or marked enzymes and enzyme complexes down to the nanometer scale [26]. The number of spatiotemporally resolvable and quantifiable intracellular transport processes is thus constantly increasing and the complexity of the resulting network of cooperating and competing transportation events is rising. Interaction between different cells such as clustering and quorum sensing adds a further dimension of complexity.

The sheer complexity of such intra- and intercellular biological processes increasingly drives the need for the development of sophisticated modeling techniques in order to understand causalities and provide tools for the determination of bottlenecks and optimization of biotechnological processes. The increasing amount of available data combined with still exponentially growing computation power allows speculation about the possibility of whole-cell simulations, i.e. the construction of virtual cells in silico, in the not-so-remote future, an idea that has been especially in vogue at the beginning of the millenium [130]. Computational implications of whole-cell simulation with respect to the complexity of intracellular proteome and metabolome diversity, as analyzed in [119] and [6], suggest that it may in principle be possible to perform a cell cycle of, e.g., *E. coli* within a couple of weeks on powerful computer clusters, provided that nanoscale (i.e. molecular) effects are widely neglected. However, this holds only true if parallelization and therefore high scalability can be provided, which is a non-trivial problem [6, 92].

In recent years, a multitude of specialized models for well-defined biological questions have been published. Meanwhile, more generalized and multi-purpose software packages have been developed in order to ease handling for non-specialists. This chapter will address both aspects, with the focus on transportation processes within cells. We start with simple free-diffusion compartmented models in Sect. 2, then proceed to active intracellular transport (Sect. 3), to quantitative descriptions of vesicle distribution (Sect. 4), and membrane transport (Sect. 5). We consider the influence of cell morphology fluctuations and cell growth, as well as the impact of transport variations during the cell cycle in unsynchronized cultures (Sect. 6), and the dynamics of protein translation (Sect. 7). In the following section, the issue of standardized model data interchange will be addressed, and finally various general-use software packages for cell and/or biochemical reaction network simulation will be introduced.

2 Free Diffusion

Biochemical models, in their simplest form, assume that every compartment under consideration is perfectly mixed and hence neglect transport processes exhibited by the contributors to the biochemical reactions modeled . This can be a reasonable approximation if transport and diffusion processes are considered to be fast compared to the actual reactions. In many cases, however, due to the relatively large volume of eukaryotic cells and their complex compartmentation, the spatiotemporal motion dynamics of substrates, reaction products, co-factors, vesicles or even the reaction sites themselves (e.g. fluctuating membrane-bound protein complexes) cannot be neglected. For this purpose, it is necessary to incorporate a reasonably resolved three-dimensional view of the cells and/or compartments under investigation, e.g. obtained using confocal electron or light microscopy. The numerical simulation of diffusion processes (Wiener process, Fick's law) in predefined three-dimensional geometries is within the scope of standard cell simulation packages, which will be summarized in Sect. 9.

The limiting factor in free diffusion processes is denoted by the diffusion coefficient in the cytosol, D_{cyto}, which can be given either in absolute units (e.g. $\mu\mathrm{m}^2/\mathrm{s}$) or as relative diffusion coefficient compared to water $D_{\mathrm{cyto}}/D_{\mathrm{water}}$. The diffusion coefficient strongly depends on the properties of the solute and solvent. In general, this is defined by the Navier–Stokes equation:

$$D = \frac{k_B T}{6\pi\eta R_0}, \tag{1}$$

with k_B denoting the Boltzmann constant ($k_B = 1.3806504 \times 10^{-23}$J/K[93]), T the temperature (in K), η the dynamic viscosity of the solvent, and R_0 the hydrodynamic radius, or Stokes radius. In real processes, the Stokes radius is roughly correlated with the molar mass [10], leading to generally rapid diffusion of small solutes compared to solutes of high molecular weight. Due to their high diffusion rates, small molecules or ions are therefore suitable to act as messengers in intracellular signalling networks, which have been reviewed in [71].

The diffusion coefficients for numerous substances have been determined empirically, usually by application of fluorescence recovery after photobleaching (FRAP) [112, 129]. Experimentally derived diffusion coefficients of substrates of different type and molar weight in numerous compartments have been reviewed in [126]. It is noted that small solutes (BCECF, MW $= 550 - 820\,\mathrm{Da}$ [61]) diffuse approximately four times slower in the cytosol than in water, a finding similar to that of FITC-dextrans and FITC-Ficolls up to a MW $\approx 500\,\mathrm{kDa}$, green fluorescent protein (GFP)(MW $\approx 27\,\mathrm{kDa}$, $D_{\mathrm{GFP,cyto}} = 2.5 - 3 \times 10^{-7}\mathrm{cm}^2 s^{-1}$) and very small DNA fragments of not more than $\approx 100\,\mathrm{bp}$, i.e. MW $\approx 70\,\mathrm{kDa}$. In any case, whenever the limiting molar mass of a specific substrate is exceeded, the diffusion rate decreases in a highly non-linear fashion to almost zero [83], an effect which is presumably mainly determined by the influence of the actin cytoskeleton [19]. Diffusion measurements of unconjugated GFP in mitochondria ($D_{\mathrm{GFP,mito}} = 2 - 3 \times 10^{-7}\,\mathrm{cm}^2\,\mathrm{s}^{-1}$) [106] indicate similar free diffusion characteristics, while the apparent diffusion in the endoplasmic reticulum (ER) is decreased approximately three-fold ($D_{\mathrm{GFP,ER}} = 0.5 - 1 \times 10^{-7}\,\mathrm{cm}^2\,\mathrm{s}^{-1}$)[21], which is suspected to be caused mainly by the complex geometry of ER.

The a-priori prediction of free protein diffusion kinetics in aqueous solutions according to their structure and conformation has been investigated in the past. A numerical prediction model for protein diffusion coefficients in water was proposed in [11]. It can be adapted to a wide range of protein surfaces. The model was compared to conventional approaches of Einstein and [36] and yielded comparably superior results with respect to experimental data of lysozyme and tobacco mosaic virus. A more recent model based on the Yukawa potential was published in [43].

The estimation of protein diffusion on charged membranes was the focus of the work published in [70]. This model considers lipid lateral reorganization and demixing when adsorbed charged macromolecules diffuse on membranes in a time scale of microseconds, which is several orders of magnitudes faster than conventional molecular dynamics (MD) approaches.

3 Active Transport

In eukaryotic cells, a significant portion of the intracellular transport of particles and vesicles is facilitated by active transport along microtubules (MT) and/or actin filaments (AF). This is necessary due to the size of eukaryotic cells and their complicated compartmentation, which would make particle transport by pure diffusion very inefficient. The resulting process has similarities with "facilitated diffusion" which can be found in other biological processes such as oxygen transport by myoglobin [142, 143] and hemoglobin [88], rapid association of DNA binding proteins with specific DNA sequences [47, 66], or ribosomal subunit translocation [65, 69]. MT are polymers that are mostly formed by nucleation starting on microtubule organization centers (MTOC) [82]. They exhibit a characteristic fluctuation pattern throughout the cell cycle [100], usually distinguished by growth, shrinking, and pause phases; and additionally occasional "catastrophe" events [16].

Simple compartmented models as described in Sect. 2 assume force-free diffusion of particles and substances within each compartment. Hence, directed transport is neglected. In recent years, increasing effort has been put into the mathematical description of intracellular active transport processes. Special attention has been laid on the non-continuous, but saltatory active movement of particles ("random walk"), i.e. the fact that particles are usually transported for distances of up to $5 - 20\,\mu\,\mathrm{m}$ at more or less constant velocities, with intermediate pauses of more than 1 s and even occasional reverse transport, depending on the region within the cell [72]. An early publication [128] assumed different transport velocities (fast anterograde, slow anterograde and retrograde), which were later questioned [3]. The work of [123], based on improved intracellular movement classification approaches [139], describes an analytical mathematical and one-dimensional model of motor-assisted transport via MTs or AFs of intracellular particles. It distinguishes between an unidirectional model with all filaments having the same polarity, and a bi-directional model. It incorporates simple kinetics for the association and dissociation of particles to and from MTs and/or AFs. It allows for free diffusion of unattached particles and constant motion of attached particles based on partial differential equations. The motion of particles satisfies the one-dimensional transport equation

$$
\begin{aligned}
\frac{\partial n_0(x,t)}{\partial t} - D\frac{\partial^2 n_0(x,t)}{\partial x^2} &= -(k_+ + k_-)n_0 + k'_+ n_+ + k'_- n_- \\
\frac{\partial n_\pm(x,t)}{\partial t} + v_\pm \frac{\partial n_\pm(x,t)}{\partial x} &= k_\pm n_0 - k'_\pm n_\pm,
\end{aligned}
\tag{2}
$$

with $n_0(x,t)$ denoting the density of free particles at distance x along the filament at time t, and $N_\pm\,(x,t)$ the densities on right- and left-directed filaments. k_+ and k_- are the first-order rate constants for binding to filaments, and k'_+ and k'_- for detachment, respectively. The motor velocities are denoted by v_+ and v_-. The particle flux $J(x,t)$ consequently reduces to

$$
J(x,t) = -D\frac{\partial n_0(x,t)}{\partial x} + n_+(x,t)v_+ + n_-(x,t)v_-. \tag{3}
$$

From these equations, one can derive the mean free-diffusion lifetime $\tau_{\mathrm{off}} = (k_+ + k_-)^{-1}$, the mean free-diffusion path length $l_{\mathrm{off}} = \sqrt{D/(k_+ + k_-)}$, the mean active transport lifetime $\tau_\pm = 1/k'_\pm$, and the mean active transport path length $l_\pm = |v_\pm|/k'_\pm$. Under the assumption that free diffusion is absent, the mean drift velocity due to active transport is equal to

$$\overline{v} = \frac{K_+ v_+ + K_- v_-}{K_+ + K_- + 1}, \tag{4}$$

with $K_\pm \equiv k_\pm / k'_\pm$. The effective diffusion constant yields

$$D_* = \frac{D + K_+ (v_+ - \overline{v})^2 / k'_+ + K_- (v_- - \overline{v})^2 / k'_-}{K_+ + K_- + 1}. \tag{5}$$

The model of [123] considered diffusion as a one-dimensional process along each MT/AF. If a rotationally symmetric distribution of MTs/AFs in the cell is assumed, the extension to three dimensions is relatively straightforward. In the work of [34], the model has been extended with respect to three-dimensional diffusion in a cylindrical neighborhood of each MT or AF.

This, however, seems to show that pure analytical mathematical models are not appropriate to cope with complex (i.e. non-symmetric and non-periodic) cell, MT, and AF geometries, since the number of equations to solve would be hard to handle when dealing with variable MT and AT construction and destruction kinetics, non-linear MT and AF geometries, arbitrarily-shaped compartments etc. Thus, numerical and hence spatiotemporally quantized models based on Monte Carlo methods are becoming more prevalent in order to obtain more flexibility with respect to the aforementioned model properties.

A computational model for quantification and analysis of the switching process between MT and AF transport has been published by [122]. This model simulates pigment transport in melanophores. Here, radial particle motion is classified as three discrete states—movement to the MT plus end (P_+), to the MT minus end (P_-), and pause(P_0), with the corresponding transition rate constants k_{1-6}. Transport on AFs is considered as two-dimensional diffusion, and the transition rates for the switching of a particle from MTs to AFs or AFs to MTs are denoted by k_{MA} and k_{AM}, respectively. The cell geometry is idealized to a 2D circular shape and MTs are considered to form an ideal radial array. The resulting set of linear differential equations has been solved numerically, since analytical treatment turned out to be impractical due to the fact that the unknown parameters enter both the equations and boundary conditions. The numerical solution has been obtained using the VirtualCell framework [121] (see also the introduction of generalized cell simulation toolboxes in Sect. 9) on a mesh size of 0.5–0.55 μm. The parameter fits found in this study ($k_{\mathrm{AM}} \ll k_{\mathrm{MA}}$ during dispersion) suggest that the transfer from MTs to AFs is almost irreversible during dispersion.

Intracellular transport processes for synthetic, i.e. non-viral, gene delivery during transient gene transfection processes have been the subject of a simulation approach published in [23], which is based on preceding works [24, 103, 123]. It utilizes stochastic modeling in order to obtain the highest flexibility with respect to arbitrary cell and nuclear morphology and intracellular organization. No model fitting has been incorporated.

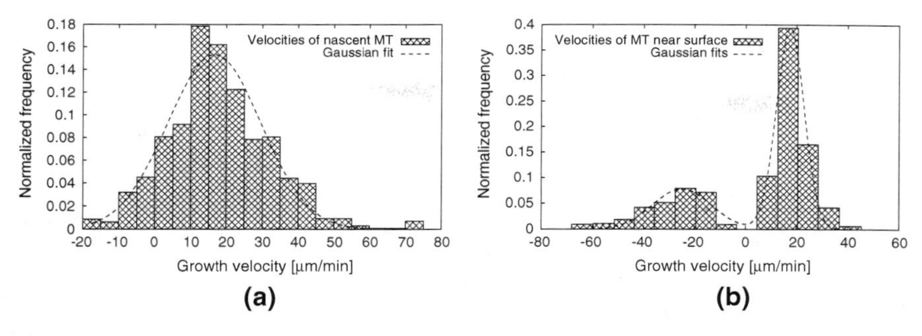

Fig. 1 Adaptation of microtubule (MT) dynamics for a two-stage model and Gaussian fits. **a** For nascent MTs. **b** For MTs near the cell membrane (data adapted from [72])

A non-linear and saltatory MT dynamics model is included that utilizes frequent growth and shrinkage cycles, as well as rare catastrophe, i.e. complete MT destruction, events. In this model, the transfected plasmids, mediated by polyplexes such as polyethylenimine (PEI), are internalized via endocytosis. The endosomes generated can be subject to either diffusion or active transport along MTs in the positive or negative direction. The contained plasmids can unpack from the polyplex, underlie degradation, escape to the cytosol or transfer to lysosomes. After leaving vesicles, the plasmids and/or polyplexes can only move passively by diffusion. Cell division and mitosis are neglected, as well as cell growth. The simulation results are compared to measured data of human skin normal fibroblasts.

It has been proven by simulation that the results of in-vitro transfection optimization cannot easily be transferred to in-vivo conditions. For example, the transfection efficiency is highly dependent on the endosomal escape rate of plasmid vectors and/or polyplexes; overshooting the maximum can lead to an up to 1,000-fold reduction in delivery efficiency. It can be concluded that since cell morphology plays a significant role in the transfection process, the stochastic simulation scheme presented, though containing several simplifications, provides a flexible and consistent quantitative description of intracellular transport processes during gene delivery.

More detailed analyses of MT dynamics and transport have been performed in [2], finding that the growth and shrinking behavior of freshly assembled MT differs significantly from those close to the cell membrane. Based on this data, we derived a detailed MT dynamics model that has been implemented in a stochastic model developed in our group for the quantitative analysis of transient gene transfection processes in mammalian cells. The MT model is two-fold: a strongly growth-favoring variant for nascent MTs (phase I) with a length of less than 85% of the cell diameter is implemented as a Gaussian distribution fitted to measured growth velocites (Fig. 1a, with the mean velocity $v_{\mu}^{I} = 16.6\,\mu\text{m}\,\text{min}^{-1}$ and standard deviation $v_{\sigma}^{I} = 13.0\,\mu\text{m}\,\text{min}^{-1}$. For MTs in phase II, i.e. those which have exceeded 85% of the cell diameter, a different distribution can be derived, as shown in Fig. 1b. The number of

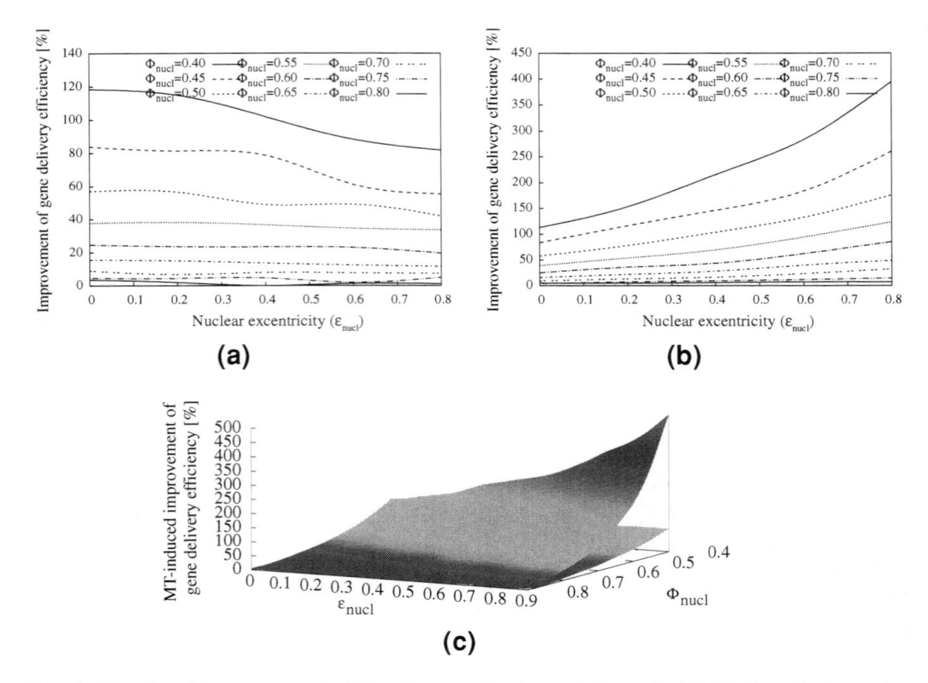

Fig. 2 Simulated improvement of lipoplex transfection efficiency in HEK293s cells by active transport via MT, depending on nucleus-to-cell size ratio, off-centering of the nucleus and clustering of cells. **a** Relative increase of with active transport based on MT (without cell clustering). **b** With cell clustering. **c** *Bottom* plane: without clustering; *top* plane: with cell clustering

shrinking MTs is significantly higher, thus two combined Gaussian fits are employed with $v^{II}_{\alpha,\mu} = -24.9\,\mu\text{m}\,\text{min}^{-1}$, $v^{II}_{\alpha,\sigma} = 11.4\,\mu\text{m}\,\text{min}^{-1}$(shrinking), and $v^{II}_{\beta,\mu} = 17.4\mu\text{m}\,\text{min}^{-1}$, $v^{II}_{\beta,\sigma} = 5.4\mu\text{m}\,\text{min}^{-1}$. The fraction of MTs in rest, which has been measured independently, is denoted by k^{I}_{0} for phase I and k^{II}_{0} for phase II. In rare cases, MTs can collapse completely (catastrophe), an event defined by the mean collapse frequency f_c.

In a simulation framework describing transient transfection dynamics in mammalian cells [63], the impact of active microtubule-facilitated polyplex transport compared to pure passive transport can be simulated depending on the cell geometry (nucleus-to-cell diameter ratio Φ_{nucl} and eccentricity of the nucleus ϵ_{nucl}), and on clustering of the cells, as visualized in Fig. 2. The transfection efficiency increases by approx 9–10% in the range of $\Phi_{\text{nucl}} = 0.7$ and centered nucleus ($\epsilon_{\text{nucl}} = 0$), no matter whether cells are clustered or not. With the nucleus being more off-center ($\epsilon_{\text{nucl}} \rightarrow 1$), the improvement decreases to almost 0% in the case of non-clustered cells, but increases to 37% in the case of clustered cells. This effect is stronger for smaller nuclei sizes.

Further complications when considering active MT transport arise from the fact that MTs usually do not really grow linearly and radially, as usually assumed.

A detailed analysis of MT bending characteristics within the microtubule array in 3T3 fibroblasts has been performed in [2], yielding the result that, although the bending angle does usually not exceed $10-15^{\circ}$ within $5\,\mu$m, the amount of strongly bent MTs rises with distance from the MTOC (60% at $10\,\mu$m distance from the MTOC, and 75% at $20\,\mu$m.) Another issue to consider is the de-novo formation of non-centrosomal MTs, which are evident in neurons [136], but have also been discovered, e.g., in A498 cells [144].

4 Dynamics of Vesicle Distribution

The spatiotemporal distribution of vesicles, especially transport vesicles, in eukaryotic cells is subject to a complex regulation system that is still not completely understood. A generalized theory of spatial patterns of intracellular organelles has been reported in [22]. It describes the flow of organelles within cells, which is mediated by three transport processes, driven by kinesin, dynein, and myosin. Interactions between organelles are not considered.

The transport processes are denoted by s, with $s=0$ for free diffusion, $s=-1$ and $s=+1$ for microtubule (MT)-mediated transport in the minus or plus direction, respectively, and $s=2$ for actin filament (AF) transport. The corresponding attachment and detachment rates are k_s and k_s', respectively, with the affinity constant being $K_s = k_s/k_s'$. The organelle density at time t at radial position r is denoted by $\tilde{c}_s(r,t)$.After nondimensionalization and approximation over all states $\tilde{c}(r,t)=\sum \tilde{c}_s(r,t)$, the model yields

$$\Pi \frac{\partial c}{d\tau} \approx \Phi \frac{1}{\xi}\frac{\partial\,(c\xi)}{\partial \xi} + \Omega \frac{1}{\xi}\frac{\partial^2\,(c\xi)}{\partial \xi^2} + \Delta \frac{1}{\xi}\frac{\partial}{\partial\xi}\left[\xi \frac{\partial c}{\partial \xi}\right], \tag{6}$$

with

$$\xi = r/R_c, \xi_N = R_N/R_C, \tau = tV/R_C, c = /C_0,$$
$$\Pi = 1 + K_{+1} + K_{-1} + K_2, \Phi = K_{-1} - K_{+1}, \Delta = \tilde{D}_2 K_2 + \tilde{D}_0,$$
$$\Omega = \frac{V}{R_C}\left(\frac{K_{+1}}{k'_{+1}}\left(1+\frac{\Phi}{\Pi}\right)^2 + \frac{K_{-1}}{k'_{-1}}\left(1-\frac{\Phi}{\Pi}\right)^2 + \frac{K_2}{k'_2}\left(\frac{\Phi}{\Pi}\right)^2\right),$$
$$\tilde{D}_2 = D_2/VR_C, \text{and}\, \tilde{D}_0/VR_c.$$

Directed motions on MTs are represented by $\Phi > 0$ (toward nucleus) and $\Phi < 0$ (away from nucleus), dispersive radial motions along MTs are represented by Ω, and dispersive motions along the cell surface as a combination of cytosolic diffusion and AF movement are denoted by Δ. The authors define two Péclet numbers to quantify the relative contributions of each type of motion, leading to a certain equilibrium spatial distribution. First, the one-dimensional Péclet number $P_{e,1D} = \Phi/\Omega$ compares directed and diffusive motions on MTs. Secondly, the two-dimensional Péclet number $P_{e,2D} = \Phi/\Delta$ describes the the ratio of directed and radial motion along MTs to diffusive motion along the cell surface. After solving Eq. 6, four distinct limiting patterns can be determined: (1) &aggregation near the cell center; (2) hyperdispersion which denotes the concentation of organelles near the cell periphery; (3) areal dispersion, i.e. uniform distribution over the cell surface area; and (4) radial dispersion, i.e. uniform radial distribution. A regime map can be derived that provides a correlation between the two relevant transport variables, $P_{e,2D}$ and $P_{e,2D}/P_{e,1D} = \Omega/\Delta$, and the resulting equilibrium spatial distribution. The computed patterns have been confirmed by experimental data [22].

5 Membrane Transport

Biomembranes are lipid bilayers with embedded proteins that serve as selectively passable barriers between either the cell's interior and its exterior or between different compartments within the cell. The transport of solutes through biomembranes is facilitated by one or more of three distinct processes: passive diffusion, facilitated diffusion and/or active transport. The composition of membranes and embedded proteins is however highly specific and dynamic, a necessary prerequisite for the correct trafficking of nutrients, lipids etc. in and out of the cell and into and from the corresponding organelles. For an overview of the functional diversity of membrane transport, the reader is referred to excellent reviews [48, 89].

5.1 Passive Diffusion

Pure lipid bilayers, i.e. synthetic membranes without proteins, enable small molecules to diffuse passively through the membrane. This spontaneous and concentration-balancing process is strongly dependent on molecule size and polarization. Ions are virtually incapable of diffusion through membranes. The passive diffusion velocity is proportional to the concentration deviation, hence the diffusion can be described by a solute-dependent permeability coefficient P [cm/s].

The diffusion rate can therefore be expressed as

$$\frac{\partial n}{\partial t} = PA\,\frac{\partial C}{\partial x}, \tag{7}$$

with $\partial n/\partial t$ denoting the diffusion rate in $\mathrm{mol}\, s^{-1}$, A the considered membrane surface area, and $\partial C/\partial x$ the concentration gradient.

Typical permeability coefficients P vary in the range of, e.g., $4.0 \times 10^{-6}\mathrm{cm\,s^{-1}}$ for urea, $3.4 \times 10^{-3}\mathrm{cm\,s^{-1}}$ for H_2O, and $2.9\,\mathrm{cm\,s^{-1}}$ for hydrochloric acid [137].

5.2 Channels and Facilitated Diffusion

In free diffusion processes, the diffusion of small lipophobic or hydrophilic molecules and ions through the membrane is enabled by membrane-bound channel or port proteins with a hydrophilic interior, forming transmembrane pores. This can either be performed without binding of the channel protein to the substrate, resulting in high transport capacities of $\gg 10^6$ molecules or ions per second per channel (free diffusion) or with binding and substrate-triggered conformation change of the porter protein. Since the transport is passive, its direction is always from high to low substrate concentration, resulting in a decrease of concentration gradient.

Unlike in passive diffusion as expressed in Eq. 7, the transport rate does not increase linearly with increasing concentration gradient, but reaches a saturation level determined by $v_{\max}$ in $\mathrm{mol}/(\mathrm{cm}^3\,\mathrm{s})$:

$$\frac{\partial n}{\partial t} = \frac{v_{\max}}{1 + K/\dfrac{\partial C}{\partial x}}, \tag{8}$$

with K (in $\mathrm{mol\,cm}^{-4}$) determining the speed of saturation, i.e. the time elapsed until $1/2\, v_{\max}$ is reached.

Crucial characteristics of channels and ports are their substrate selectivity and their regulation properties, i.e. how the flux is maintained or interrupted depending on external signals. These signals can be mediated by signal molecules, especially ions, phosphorylation/dephosphorylation, voltage variations etc.

5.2.1 Aquaporins

AQP are special channel proteins dedicated to the active transport of water through membranes [12]. They reach a transport capacity of up to 3×10^9 water molecules per seconds per channel [1] while preventing protons from moving through the channel along the water molecule network using the Grotthuss mechanism [86], which is essential for the conservation of electrochemical gradients. The single channel water permeability of AQP1 is $\approx 4.6 - 11 \times 10^{-14} \mathrm{cm}^3/\mathrm{s}$ [114, 145].

The related sub-family of aquaglyceroporins are responsible for transport of other small molecules, such as glycerol, ammonia and urea. Furthermore, aquaporins seem to play a pivotal role in cell migration, e.g. of tumor cells [104, 135].

5.3 Active Transport

During active transport processes, the substrate molecules are transported against the concentration gradient, thus consuming energy that is usually derived from the cell's metabolism. The transport is enabled by carrier proteins. The transportation process can be suppressed via either competitive inhibitors that bind to the transportation site of the carrier enzyme instead of the substrate, or via allosteric inhibitors that bind to elsewhere than the active site of the carrier, forcing it to change its conformation.

If the transport of single substrate molecules is performed without the need for any second substrate (primary transport), the carrier is described as a uniporter. In this case, due to the need for active binding of the transporter protein or cofactors to the substrate and their limited number, the transport velocity characteristics is analogous to that of facilitated transport as denoted in Eq. 9. In the case of coupled transport in coordination with a second substrate, the carrier is referred to as a co-transporter. The coupled transport can either take place in the same direction (symporter) or in the opposite direction (antiporter). In either case, the passive transport of one substrate along its concentration gradient increases entropy and therefore yields energy, which is in consequence used for the active transport of the second substrate against the concentration gradient (secondary active transport; see also [48]). For coupled transports, the flux can be approximated by the sigmoidal function

$$\frac{\partial n}{\partial t} = \frac{v_{\max}}{1 + \left(K / \frac{\partial C}{\partial x}\right)^{h}}, \tag{9}$$

with $h > 1$ denoting the degree of (positive) cooperativity (see also [73]).

5.4 Microdomains Within Membranes

In recent years, the view of the distribution of functional proteins such as transporters along the membrane (especially cell membrane and ER) surface has changed. Formerly regarded as a homogeneous fluid, it is now believed that there are well organized sections of approx 50 nm lateral size with specialized functions, so-called functional rafts [120]. They are formed in a highly dynamic manner with contributions from sphingolipid and cholesterol [46, 78, 97], presumably in coordination with the underlying actin-based cytoskeletal network [111]. Functional rafts are believed to play a pivotal role in protein sorting in the endoplasmic reticulum (ER) and Golgi apparatus [51].

A common model assumption in the morphology of rafts is that they form self-assembling dynamic lipid shells [4, 87], able to bind specific proteins. The characterization of the motility of membrane-bound proteins has moved increasingly into the focus of investigation, featuring confocal and two-photon microscopy as well as single particle tracking (SPT) with sophisticated tracking algorithms (reviewed in [62, 77]). The “hop-diffusion” motion characteristics of transmembrane proteins when shuffled between microdomains has been reviewed in [20] and [111]. A common but not yet fully conclusive finding is that the proteins are seemingly able to diffuse freely within a small area and sporadically exhibit a “hop” to another region. A corresponding simulation approach therefore implements a two-stage transmembrane protein diffusion model that involves reduced long-range diffusion by introducing barriers while short-range diffusion remains unchanged [134].

6 Cell Cycle and Cell Morphology Dynamics

Most current intracellular transport modeling approaches neglect the effect of transport variations during the progress of the cell cycle and the impact of cell growth or cell morphology changes during the cell cycle. Since it is clear that cell and compartment morphology as well as intracellular organization have a strong influence on the individual transport processes and their balance [23], it is worthwhile increasing model accuracy by including cell cycle-induced effects.

The dynamic implications of complex cell cycle regulation have been evaluated based on general physiological properties of the most important regulatory modules of the CdK network regulating the cell division cycle, which have been described in a generic computational model [18]. This model aims at a generic representation of yeast, frog cell and human cell cycle regulation. It contains numerous different control loops, which have been tentatively adapted to yeast and mammalian cells by simplification of sub-networks and clustering. By means of bifurcation analysis, the typical oscillatory behavior of the regulating enzymes can be detected and analyzed. The increase in cell mass (and therefore cell volume) during the cell cycle is presumably connected to the cell cycle control [25],[1]which is incorporated in this model by defining the cytosolic synthesis rate of the regulatory proteins A, B, and E as proportional to the cell mass. The mass increase during the cell cycle is assumed to be proportional to the cell size, leading to exponential growth behavior, i.e. $m(t) = m_0\, e^{t/\mu}$ with $0 \leq t < \mu \ln(2)$. This implies an almost linear diameter growth $d(t) = d_0\, e^{t/(3\mu)} \approx d_0\,(1 + 0.3654t)$, assuming that growth is isotropic. However, growth is usually non-constant during the progress of the cell cycle (reviewed in [40]), but can at least be considered linear and isotropic in G1 phase.

A specialized and empirical model for the advance of the cell cycle in myeloma cells has been proposed in [79]. If the cell cycle is denoted by ${}^c\bar{t}_\phi = (t - {}^c t_\delta)/{}^c\tau$, with t representing the simulated time point in s, ${}^c t_\delta$ the time point of the last division of cell c, and ${}^c\tau$ the cell's cell cycle duration, the cell cycle phase Ω with increasing ${}^c\bar{t}_\phi$ according to [79] is defined as

$$\Omega = \begin{cases} G1; & 0 \leq {}^c\bar{t}_\phi < 0.50 \\ S; & 0.50 \leq {}^c\bar{t}_\phi < 0.82 \\ G2M; & 0.82 \leq {}^c\bar{t}_\phi < 1 \end{cases}. \tag{10}$$

[1] Controversial publications note that this may not hold true for every kind of eukaryotic cells; e.g., proliferating Schwann cells seem to not need cell-size checkpoints [17]—further discussed in [41]—a phenomenon which the authors speculate is valid for many mammalian cells as well.

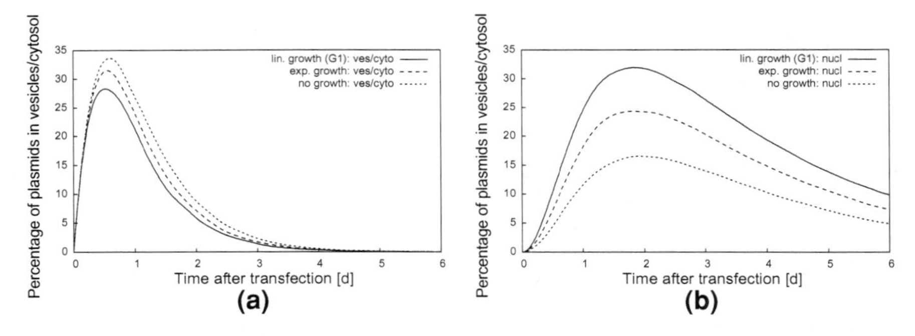

Fig. 3 Influence of growth on transfection dynamics in HEK293s cells. Continuous lines with volumetrically linear growth in G1 phases, dotted lines without growth and fixed averaged size. **a** Percentage of plasmids, both complexed and pure, in cells outside of nucleus (accumulated for vesicles and cytosol). **b** Percentage of plasmids in nucleus

Using the aforementioned dynamics of the cell cycle and cell growth, the impact and spatiotemporal transport processes can be estimated. In the transient transfection simulation model described in [63], the retention times of transfected plasmids in different compartments of HEK293s cells can be estimated depending on cultivation and transport parameters, with and without growth (but with same average volume). In this model, the volumetric growth is assumed to be exponential during the G1 phase, and linear in S/G2M phases. Figure 3denotes the corresponding trajectories in the different compartments (cytosol and vesicles combined; nucleus). The (peak) number of plasmids delivered to the nucleus reduces by approx 48% when no growth is assumed.

6.1 Cell-Cycle-Dependent Transport Variations

It has been well known for a long time [113] that significant intracellular transport rate variations can occur during the progress of the cell cycle. Many of them may be related to actin polarization as cells enter the S phase. For example, plasmid uptake rates via endocytosis in synchronized D407 cells can vary by up to 60% [85]. In addition, the activity of nuclear import facilitated by the nuclear pore complex (NPC) is altered drastically [84, 132]. Subsequent transport rate variations of macromolecules between cytoplasm and nucleus have been reviewed in [15, 138].

7 Translation and Protein Transport

The protein translocation and secretion process in eukaryotic cells is highly complex, selective and non-linear [74, 107] and is thus a worthwhile subject for detailed quantitative analysis, especially for optimization of protein expression in pharmaceutical protein production cell lines. However, the intracellular post-translational transport of proteins in eukaryotes has not yet been the subject of detailed mechanistic and spatiotemporally resolved models, unlike in bacteria, for which protein traffic is quantitatively better understood [29]. It is thus desirable to obtain a deeper mechanistic and quantitative understanding of translation and &post-translational processes in order shed light on biological questions such as quantification of the impact of mRNA degradation, protein folding and selection, post-translational modification, glycosylation and the secretory pathway on the expression level.

7.1 Overview and Limiting Factors

The protein transport and targeting during and after translation consists of multiple steps which will be briefly mentioned in this section, with each of them potentially influencing the overall protein production ability. For more detailed insight into the numerous processes contributing to protein translocation, targeting and secretion, the reader is referred to excellent reviews [44, 52, 64, 67, 109].

After transcription and mRNA export from the nucleus, the majority of proteins are synthesized in the cytosol with the help of ribosomes which are either suspended in the cytosol or bound to the rough endoplasmic reticulum (ER) or to the nuclear envelope. The translocation into the ER is conducted in either co-translational (mammalian cells) or post-translational fashion (mostly in yeast). The protein remains in an unfolded or loosely folded state and is transferred across the ER membrane through the translocon either completely (soluble proteins with usually amino-terminal, cleavable signal sequences) or partially (with the hydrophilic ends either crossing the membrane or remaining in the cytosol) [109]. The ER-bound proteins are translocated by a passive channel that associates with three different types of partners to drive the force for translocation: with ribosomes (mostly used for secretory proteins and most membrane proteins) for co-translational translocation, with the Sec62/Sec63 complex (a tetrameric membrane-protein complex) or with the luminal chaperone binding immunoglobulin protein (BiP) for post-translational translocation [44, 109, 110]. While in the translocon, the signal peptidases are cleaved off [96], the nascent chain is glycosylated by oligosaccharyltransferase (OST) and co-translational folding is conducted. Assembly of multicomponent complexes may also occur co-translationally in some cases [64]. After release from the ribosome/translocon complex and having almost reached the final conformation, oligomers are assembled in three stages under the presence of three folding enzymes (BOX 2) [30]. Misfolded proteins are identified in a three-stage process featuring molecular folding sensors and chaperones [30] and possibly retrotranslocated into the cytosol via the translocon for degradation in the cytosol by the cytoplasmic proteasome [74, 133].

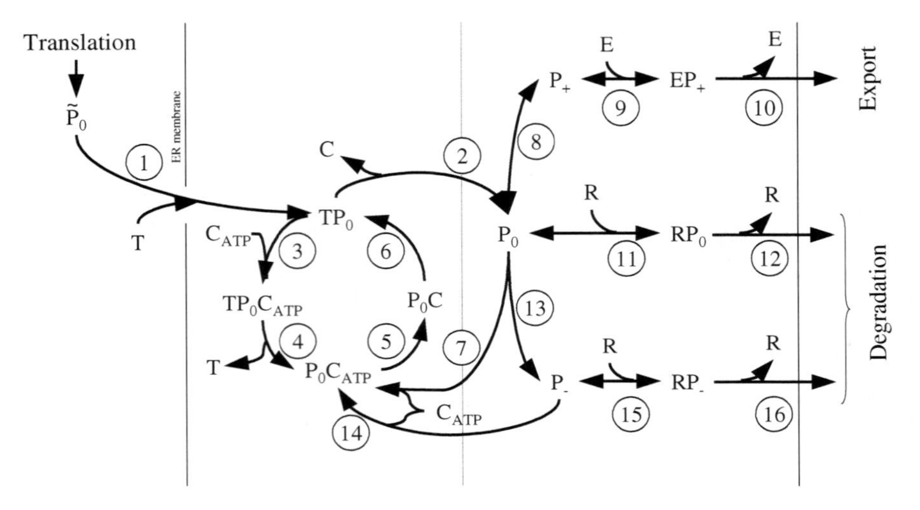

Fig. 4 Overview of protein export model according to [141]. See text and Table 1 for explanations and details

7.2 FoldEx: ER Protein Export Model

A non-compartmented protein trafficking model called "FoldEx" describing the folding-end export of proteins from the ER has been published in [141]. It provides a quantitative description of the dynamics of the competing pathways that are reponsible for ER import of the unfolded proteins facilitated by the translocon; high-affinity (ATP-bound) and low-affinity binding to chaperones; recognition of correct, incomplete or uncorrect folding; re-folding procedures; and association to export or retranslocation pathways. An overview of the model is given in Fig. 4 and Table 1. The pathways are implemented utilizing Michaelis–Menten kinetics.

The authors conclude that the export efficiency of fast-folding (small) proteins mainly depends on the relative activities of the export pathways and ERAD (ER-associated degradation pathway), i.e. the corresponding concentrations ratio c_E/c_R, while being largely constant with varying k_8 at wide ranges. On the other hand, the export efficiency of slow-folding proteins is more sensitive to folding kinetics; thus even stable proteins may be degraded to a considerable extent when they fold slowly (i.e. transmembrane proteins). Misfolded proteins may re-enter the ER-assisted folding (ERAF) pathway to some extent, increasing the export efficiency of proteins prone to misfolding.

Table 1 Desciption of pathways and rate constants in the protein export model according to [141], as outlined in Fig. 4

No.	Description	c	Val.	Unit
1	Cotranslocational insertion of unfolded cytosolic protein ($\tilde{P}_0$) into ER facilitated by translocon (T)	k_1	0.03	s^{-1}
2	Direct release ofP_0into ER when protein translocation is complete	k_2	0.03	s^{-1}
3	P_0 bound by high affinity (ATP-bound) chaperone(s) $\Rightarrow TP_0C_{ATP}$	k_3	10^6	$M^{-1}s^{-1}$
4	Release from T	k_4	0.03	s^{-1}
5	Turnover to low-bound state via ATP hydrolysis $\Rightarrow P_0C$	k_5	1	s^{-1}
6	Release of protein, releasing chaperone(s). Chaperone switches back to high-affinity state via ATP binding	k_6	100	s^{-1}
7	If folding fails $\Rightarrow$ rebinding to chaperone(s) possible	k_7	10^6	$M^{-1}s^{-1}$
8	Correct folding $\Rightarrow P_+$	k_8	var.	s^{-1}
		k_8'	var.	s^{-1}
9	Reversible association to export machinery (E)	k_9	10^6	$M^{-1}s^{-1}$
		k_9'	0.1	s^{-1}
10	Export to Golgi	k_{10}	1	s^{-1}
11	Incomplete folding $\Rightarrow$ reversible association to retranslocation machinery (R)	k_{11}	10^8	$M^{-1}s^{-1}$
		k_{11}'	0.1	s^{-1}
12	Export to cyotosol for proteolysis	k_{12}	1	s^{-1}
13	Misfolding $\Rightarrow P_-$	k_{13}	var.	s^{-1}
14	Rebinding to chaperone(s) possible for re-folding	k_{14}	10^6	$M^{-1}s^{-1}$
15	Reversible association to retranslocation machinery (R)	k_{15}	10^6	$M^{-1}s^{-1}$
		k_{15}'	0.1	s^{-1}
16	Export to cytosol for proteolysis	k_{16}	1	s^{-1}

Notation: $\tilde{P}_0$ denotes unfolded conformational ensemble in cytosol (before ER import), P_0 during and after import, T interaction with translocon, C_{ATP} chaperone(s) in high-affinity (ATP-bound) state, C in low-affinity state, E denotes export machinery, at a nominal concentration of c_E; c_R for retranslocation machinery, and c_C for chaperones. *ERAF* ER-assisted folding (ERAF) pathways, *ERAD* ER-associated degradation pathway

7.3 Protein Targeting

Although the prediction of protein localization from the amino acid sequence is beyond the scope of this overview, some published methods are worth mentioning. A mainly empirical method originating from experimental and computational observations, called PSORT, has been developed since the beginning of the 1990s [99] and is available at http://psort.ims.u-tokyo.ac.jp. More mechanistic approaches, i.e. simulating the mechanism of cellular sorting, are MultiLoc [56] (available at (http://www-bs.informatik.uni-tuebingen.de/Services/MultiLoc), LOCtree [98] (available at http://cubic.bioc.columbia.edu/services/loctree), and PLOC [105] (available at http://www.genome.jp/SIT/plocdir).

8 Exchange of Model Data

In the last two decades, with the rapidly increasing number and complexity of publicly available models, efforts have been made to introduce standards for the flexible exchange and re-use of biological model descriptions between different simulation packages. These efforts try to minimize the need for re-formulation of model implementations when switching between different simulation and/or analysis tools. They involve both the development of multi-purpose unified model description languages (described below), as well as the need for the definition of standards for the model architecture and implementation, such as the necessity for a single reference description, the traceability of the choice of parameters, initial conditions etc. [60]. A widely accepted standard is *Minimum information requested in the annotation of biochemical models* (MIRIAM) [102]. Additionally, published model implementations in public databases such as CellML are often subject to increased requests for annotations and modification history documentation. Also, equations and identifiers are unified or re-annotated if possible [81] according to ontological frameworks [140].

The following sections will focus on the most important standard model exchange formats. For further overviews of this topic, the reader is referred to dedicated reviews [125]. A variety of modeling software packages will be summarized in Sect. 9.

8.1 SBML

The systems biology markup language (SBML) was introduced in 2001 and extended to SBML level 2 in 2002 [58], the most recent version 4 having been published in 2008 [59]. It is based on XML (extensible markup language), a widespread general hierarchical document description format. It provides a generalized description of biochemical reaction systems, e.g. cell signaling pathways, metabolic pathways, biochemical reactions, gene regulation etc.

Since SBML is strictly standardized and commonly used, there is a variety of helpful applications available in the web, e.g. an automated SBML validator that checks for syntax and consistency (http://sbml.org/Facilities/Validator), a converter into human-readable documentation formats (PDF, LaTeX etc., http://www.ra.cs.uni-tuebingen.de/software/SBML2LaTeX/index.htm) [28], automated creation of kinetic equations (http://www.ra.cs.uni-tuebingen.de/software/SBMLsqueezer) [27] and others.

An increasing number (over 450 by August 2010) of peer-reviewed, quantitative models of biochemical and cellular systems is collected in the BioModels database (http://www.ebi.ac.uk/biomodels-main) [101]. To be accepted in BioModels, the model must comply with MIRIAM [102]; additionally, the model components are annotated according to standard databases (UniProt, KEGG, Reactome etc).

8.2 CellML

Another popular model description language is CellML [80], which has been developed starting in 1998. The first specification was published in 2001 [49]. Like SBML, it is XML-based; however, it is more general in that it is not restricted to biological systems.

The CellML repository (http://www.cellml.org/models) currently holds over 480 models (as of August 2010), mostly derived from peer-reviewed literature, covering a wide range of biological processes such as signal transduction pathways, metabolic pathways, electrophysiology, immunology, cell cycle, muscle contraction, mechanical models and constitutive laws etc. [81].

As with SBML, a variety of helper tools is available for CellML, e.g. converters to SBML: CellML2SBML [117] etc. A review of CellML-associated software can be found in [37].

9 Cell Modeling Program Packages

A variety of simulation software packages is available for the facilitation of whole-cell simulations or for solving specific biological questions without the need for extensive programming and mathematical and numerical optimization efforts. This overview concentrates on six different packages that are applicable to simulation of mesoscale intracellular transport processes: VirtualCell, MCell, Bio-SPICE, StochSim, COPASI and BioDrive. Not included in this overview are simulators for multicellular systems [55], pure genomic network-based simulation environments such as ECell [131]/ ECell2 [127] and CellDesigner [35], or tools that are dedicated for neural simulation, e.g. NEURON [54] or GENESIS [8]. Also, simulation suites for molecular dynamics simulations, such as CHARMM [9], AMBER [14] and GROMACS [53], which are to some extent suitable for simulation of transport processes at the nanoscale level [5, 68], are beyond the scope of this overview.

9.1 Virtual Cell

The Virtual Cell, or VCell, project, which was first published in [116], is developed at the National Institutes of Health (NIH). It is available from http://vcell.org. It provides a framework for modeling biochemical, electrophysiological and transport phenomena and allows the user to build complex models with a

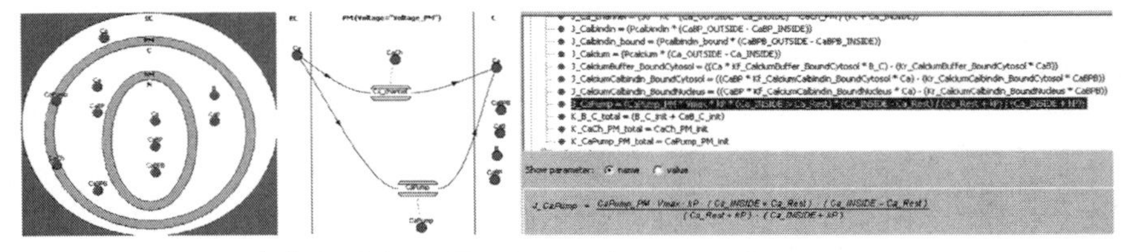

(a) Physiology: Topology and reactions/fluxes.

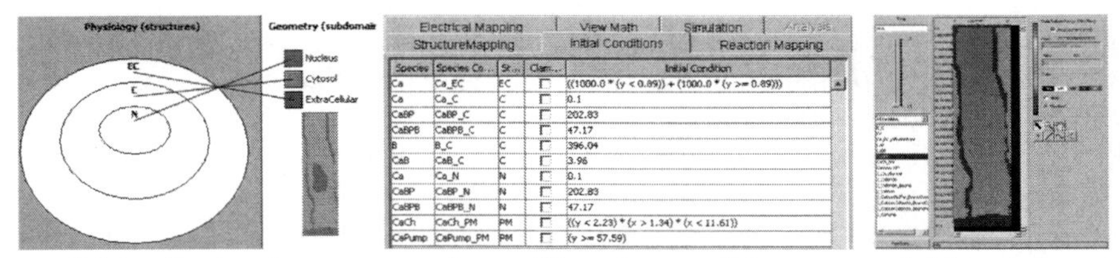

(b) Application: Structure mapping and Boundary conditions. **(c)** Simulation.

Fig. 5 General VCell workflow: from general physiology, over specific structures and applications, to simulation. **a** Physiology: topology and reactions/fluxes. **b** Application: structure mapping and boundary conditions. **c** Simulation

Java-based interface. Arbitrary (but temporally constant) compartment topologies and geometries are supported. The computation is performed on-line on computing clusters dedicated to VCell at the National Resource for Cell Analysis and Modeling (NRCAM).

VCell biological models are composed of three components—a physiological model containing the mechanistic hypothesis (i.e. reactions, fluxes, electrical currents on membranes), an application with experimental conditions, geometry and modeling approximations, and one or more simulations (Fig. 5). The underlying mathematics and physics as well as the working principles are described in [94, 121]. For improved visualization of simulation results on physiological data, a 3D interactive visualization tool is integrated.

9.2 MCell

MCell, available at http://www.mcell.cnl.salk.edu, is a general Monte Carlo simulator of cellular microphysiology. It enables simulation of (intra-)cellular signalling in and around cells. The framework utilizes Monte Carlo random walk and chemical reaction algorithms based on pseudo-random-number generation and is thus capable of tracking the stochastic behavior of discrete molecules in space. A prominent feature of the software is its ability to stop the simulation at arbitrary time positions, change parameters and morphologies on-line and continue with the simulation, which in principle allows for modeling of dynamic cell shape variations. The software utilizes a special model description language (MDL) and currently does not provide an interactive interface.

Although originally having been developed for miniature endplate current generation in the vertebrate neuromuscular junction [7], this software suite is meanwhile generalized enough to be used for different biological questions, such as Ca^{2+} dynamics in dendritic spines [33] and modulation of impulse propagation in nodes of Ranvier depending on Na^{+}channel distribution [124]. The simulations can be performed on distributed computing environments—the "grid" [32]—to increase computation speed [13].

9.3 Bio-SPICE

Bio-SPICE is an open-source project (http://biospice.sourceforge.net), originally developed at the Defense Advanced Research Projects Agency (DARPA) in 2002. It is intended for modeling and simulation of spatiotemporal processes in living cells. The toolbox combines different software from various vendors by integrating them into the "Dashboard", the Bio-SPICE core application that provides a consistent workflow for modeling, analysis and simulation. It is closely related to the Systems Biology Workbench (SBW), a software framework that allows heterogeneous application components written in multiple programming languages on different platforms [115].

9.4 StochSim

StochSim, a discrete stochastic simulator for chemical and biochemical reactions [31], available at http://www.pdn.cam.ac.uk/groups/comp-cell/StochSim.html, with a graphical user interface at http://www.ebi.ac.uk/lenov/stochsim.html, has been developed as part of a study of bacterial chemotaxis [95]. This algorithm considers chemical reaction partners as individual interacting objects. The StochSim algorithm is in most cases more efficient than the older Gillespie [38, 39] algorithm, especially when multi-state molecules are considered [92].

9.5 GEPASI and COPASI

For simulation and analysis of biochemical reaction networks with support of recent model exchange standards such as SBML and CellML (Sect. 8), the COmplex PAthway SImulator (COPASI) software suite has been developed [57, 91] as a successor to the GEneral PAthway SImulator (GEPASI) [90] that was first published in the early 1990s. It is available for multiple platforms under http://www.copasi.org. The software considers sensitivity and metabolic control analysis methods as well as various optimization algorithms such as evolutionary programming, Nelder–Mead, particle swarm, simulated annealing and others [91].

9.6 BioDrive

The BioDrive biochemical reaction and gene expression simulation software suite has been developed by the group of Kyoda et al. since the late 1990s. The authors emphasize its easy usability, or "biologist-friendliness", as well as its ability to cope with multi-cellular organisms and extra-cellular processes. It is based on ordinary differential equations and incorporates diffusion and spatiotemporal patterning [75].

10 Conclusions

In the past two decades, significant efforts have been made to deliver more in-depth insights into the dynamics of mesoscale intracellular transport processes in eukaryotic cells and their impact on cell function. A large variety of general-use biochemical simulation software is available, though only a subset (VCell, MCell, Bio-SPICE) considers spatially resolved models. Active transport via, e.g., the actin filament or microtubule network is not yet explicitly incorporated. The influence of such processes has increasingly moved into the scope of research, hence numerous more or less ad-hoc Monte Carlo methods have been implemented and published. Unfortunately, the interchangeability of such models is limited compared to standardized model description languages. The unification of different models covering specialized aspects to more holistic approaches with a perspective towards whole-cell simulations is therefore still a very cumbersome process; however, the foundations for the necessary infrastructure are laid (e.g. MIRIAM) and are being further extended.

References

1. Agre P, Bonhivers M, Borgnia MJ (1998) The aquaporins, blueprints for cellular plumbing systems. J Biol Chem 273(24):14,659–14,662
2. Alieva IB, Borisy GG, Vorobjev IA (2008) The spatial organization of centrosome-attached and free microtubules in 3T3 fibroblasts. Tsitologiia 50(11):936–946
3. Alvarez J, Torres J (1985) Slow axoplasmic transport: a fiction?. J Theor Biol 112(3):627–651

4. Anderson RGW, Jacobson K (2002) A role for lipid shells in targeting proteins to caveolae, rafts, and other lipid domains. Science 296(5574):1821–1825
5. Bahar I, Lezon TR, Bakan A, Shrivastava IH (2010) Normal mode analysis of biomolecular structures: functional mechanisms of membrane proteins. Chem Rev 110(3):1463–1497
6. Ballarini P, Guido R, Mazza T, Prandi D (2009) Taming the complexity of biological pathways through parallel computing. Brief Bioinform 10(3):278–288
7. Bartol TM Jr, Land BR, Salpeter EE, Salpeter MM (1991) Monte Carlo simulation of miniature endplate current generation in the vertebrate neuromuscular junction. Biophys J 59(6):1290–1307
8. Bower JM, Beeman D (1998) The book of GENESIS (2nd ed): exploring realistic neural models with the general neural simulation system. Springer-Verlag , New York Inc
9. Brooks BR, Bruccoleri RE, Olafson BD, States DJ, Swaminathan S, Karplus M (1983) CHARMM: a program for macromolecular energy, minimization, and dynamics calculations. J Comput Chem 4(2):187–217
10. Brown PH, Schuck P (2006) Macromolecular size-and-shape distributions by sedimentation velocity analytical ultracentrifugation. Biophys J 90(12):4651–4661
11. Brune D, Kim S (1993) Predicting protein diffusion coefficients. Proc Nat Acad Sci U S A 90(9):3835–3839
12. Carbrey JM, Agre P (2009) Discovery of the aquaporins and development of the field. Handb Exp Pharmacol 190:3–28
13. Casanova H, Berman F, Bartol T, Gokcay E, Sejnowski T, Birnbaum A, Dongarra J, Miller M, Ellisman M, Faerman M, Obertelli G, Wolski R, Pomerantz S, Stiles J (2004) The virtual instrument: Support for grid-enabled MCell simulations. Int J High Perf Comput Appl 18:3–17
14. Case DA, Cheatham TE III, Darden TA, Gohlke H, Luo R, Merz KM Jr, Onufriev A, Simmerling CL, Wang B, Woods RJ (2005) The amber biomolecular simulation programs. J Comput Chem 26(16):1668–1688
15. Chakraborty P, Wang Y, Wei JH, van Deursen J, Yu H, Malureanu L, Dasso M, Forbes DJ, Levy DE, Seemann J, Fontoura BM (2008) Nucleoporin levels regulate cell cycle progression and phase-specific gene expression. Dev Cell 15(5):657–667
16. Conde C, Cáceres A (2009) Microtubule assembly, organization and dynamics in axons and dendrites. Nat Rev Neurosci 10(5):319–332
17. Conlon I, Raff M (2003) Differences in the way a mammalian cell and yeast cells coordinate cell growth and cell-cycle progression. J Biol 2(1):7
18. Csikáasz-Nagy A, Novàk B, Tyson JJ (2008) Reverse engineering models of cell cycle regulation. Adv Exp Med Biol 641:88–97
19. Dauty E, Verkman AS (2005) Actin cytoskeleton as the principal determinant of size-dependent DNA mobility in cytoplasm. J Biol Chem 280(9):7823–7828
20. Day CA, Kenworthy AK (2009) Tracking microdomain dynamics in cell membranes. Biochim Biophys Acta 1788(1):245–253
21. Dayel MJ, Hom EFY, Verkman AS (1999) Diffusion of green fluorescent protein in the aqueous-phase lumen of endoplasmic reticulum. Biophys J 76(5):2843–2851
22. Dinh AT, Pangarkar C, Theofanous T, Mitragotri S (2006) Theory of spatial patterns of intracellular organelles. Biophys J 90(10):L67–L69
23. Dinh AT, Pangarkar C, Theofanous T, Mitragotri S (2007) Understanding intracellular transport processes pertinent to synthetic gene delivery via stochastic simulations and sensitivity analyses. Biophys J 92(3):831–846
24. Dinh AT, Theofanous T, Mitragotri S (2005) A model for intracellular trafficking of adenoviral vectors. Biophys J 89(3):1574–1588
25. Dolznig H, Grebien F, Sauer T, Beug H, Müllner EW (2004) Evidence for a size-sensing mechanism in animal cells. Nat Cell Biol 6(9):899–905
26. Donnert G, Keller J, Wurm CA, Rizzoli SO, Westphal V, Schönle A, Jahn R, Jakobs S, Eggeling C, Hell SW (2007) Two-color far-field fluorescence nanoscopy. Biophys J 92(8):L67–L69

27. Dräger A, Hassis N, Supper J, Schröder A, Zell A (2008) SBMLsqueezer: a CellDesigner plug-in to generate kinetic rate equations for biochemical networks. BMC Syst Biol 2(1):39
28. Dräger A, Planatscher H, Wouamba D, Schröder A, Hucka M, Endler L, Golebiewski M, Müller W, Zell A (2009) SBML2LaTeX: conversion of SBML files into human-readable reports. Bioinformatics 25(11):1455–1456
29. Economou A, Christie PJ, Fernandez RC, Palmer T, Plano GV, Pugsley AP (2006) Secretion by numbers: protein traffic in prokaryotes. Mol Microbiol 62(2):308–319
30. Ellgaard L, Helenius A (2003) Quality control in the endoplasmic reticulum. Nat Rev Mol Cell Biol 4(3):181–191
31. Firth C, Bray D (2004) Stochastic simulation of cell signaling pathways. In: Bower JM, Bolouri H (eds) Computational modeling of genetic and biochemical networks. MIT Press, Cambridge, MA, pp 263–286
32. Foster I, Kesselman C, Tuecke S (2001) The anatomy of the grid: enabling scalable virtual organizations. Int J High Perf Comput Appl 15:200–222
33. Franks KM, Bartol TM, Sejnowski TJ (2001) An MCell model of calcium dynamics and frequencydependence of calmodulin activation in dendritic spines. Neurocomputing 38-40, 9–16
34. Friedman A, Craciun G (2005) A model of intracellular transport of particles in an axon. J Math Biol 51:217–246
35. Funahashi A, Morohashi M, Kitano H, Tanimura N (2003) CellDesigner: a process diagram editor for gene-regulatory and biochemical networks. Biosilico 1(5):159–162
36. Garciadela Torre J, Bloomfield V (1981) Hydrodynamic properties of complex, rigid, biological macromolecules: theory and applications. Q Rev Biophys 14(1):81–139
37. Garny A, Nickerson DP, Cooper J, dos Santos RW, Miller AK, McKeever S, Nielsen PM, Hunter PJ (2008) CellML and associated tools and techniques. Phil Trans R Soc 366(1878):3017–3043
38. Gibson MA, Bruck J (2000) Efficient exact stochastic simulation of chemical systems with many species and many channels. J Phys Chem 104(9):1876–1889
39. Gillespie DT (1977) Exact stochastic simulation of coupled chemical reactions. J Phys Chem 81(25):2340–2361
40. Goranov AI, Amon A (2010) Growth and division—not a one-way road. Curr Opin Cell Biol 22:1–6
41. Grewal SS, Edgar BA (2003) Controlling cell division in yeast and animals: Does size matter?. J Biol 2(1):5
42. Gunkel M, Erdel F, Rippe K, Lemmer P, Kaufmann R, Hörmann C, Amberger R, Cremer C (2009) Dual color localization microscopy of cellular nanostructures. Biotechnol J 4:927–938
43. Gutiérrez R, del Valle EMM, Galán MA (2007) Theoretical model to predict the diffusion coefficients of enzymes on adsorption processes based on hard-core two-Yukawa potential. Ind Eng Chem Res 46(23):7410–7416
44. Haigh NG, Johnson AE (2002) Protein sorting at the membrane of the endoplasmic reticulum. In: Dalbey RE, von Heijne G (eds) Protein targeting, transport translocation. Academic Press, London, pp 74–106
45. Hama S, Akita H, Ito R, Mizuguchi H, Hayakawa T, Harashima H (2006) Quantitative comparison of intracellular trafficking and nuclear transcription between adenoviral and lipoplex systems. Mol Ther 13(4):786–794
46. Hancock JF (2006) Lipid rafts: contentious only from simplistic standpoints. Nat Rev Mol Cell Biol 7(6):456–462
47. Hannon R, Richards E, Gould H (1986) Facilitated diffusion of a DNA binding protein on chromatin. EMBO J 5(12):3313–3319
48. Hediger MA, Romero MF, Peng JB, Rolfs A, Takanaga H, Bruford EA (2004) The ABCs of solute carriers: physiological, pathological and therapeutic implications of human membrane transport proteins. Pflugers Arch 447(5):465–468

49. Hedley WJ, Nelson MR, Bellivant DP, Nielsen PF (2001) A short introduction to CellML. Phil Trans R Soc 359(1783):1073–1089
50. Hell SW (2003) Towards fluorescence nanoscopy. Nat Biotechnol 21(11):1347–1355
51. Helms JB, Zurzolo C (2004) Lipids as targeting signals: lipid rafts and intracellular trafficking. Traffic 5(4):247–254
52. Herbert DN, Molinari M (2007) In and out of the ER: protein folding, quality control, degradation, and related human diseases. Physiol Rev 87(4):1377–1408
53. Hess B, Kutzner C, van der Spoel D, Lindahl E (2008) GROMACS 4: algorithms for highly efficient, load-balanced, and scalable molecular simulation. J Chem Theor Comput 4:435–447
54. Hines ML, Carnevale NT (1997) The NEURON simulation environment. Neural Comput 9(7):1179–1209
55. Hoehme S, Drasdo D (2010) A cell-based simulation software for multicellular systems. Bioinformatics 26(20):2641–2642
56. Höglund A, Dönnes P, Blum T, Adolph HW, Kohlbacher O (2006) Multiloc: prediction of protein subcellular localization using N-terminal targeting sequences, sequence motifs and amino acid composition. Bioinformatics 22(10):1158–1165
57. Hoops S, Sahle S, Gauges R, Lee C, Pahle J, Simus N, Singhal M, Xu L, Mendes P, Kummer U (2006) COPASI – a COmplex PAthway SImulator. Bioinformatics 22(24):3067–3074
58. Hucka M, Finney A, Sauro HM, Bolouri H, Doyle JC, Kitano H, Arkin AP, Bornstein BJ, Bray D, Cornish-Bowden A, Cuellar AA, Dronov S, Gilles ED, Ginkel M, Gor V, Goryanin II, Hedley WJ, Hodgman TC, Hofmeyr JH, Hunter PJ, Juty NS, Kasberger JL, Kremling A, Kummer U, Novre NL, Loew LM, Lucio D, Mendes P, Minch E, Mjolsness ED, Nakayama Y, Nelson MR, Nielsen PF, Sakurada T, Schaff JC, Shapiro BE, Shimizu TS, Spence HD, Stelling J, Takahashi K, Tomita M, Wagner J, Wang J (2003) The systems biology markup language (SBML): a medium for representation and exchange of biochemical network models. Bioinformatics 19(4):524–531
59. Hucka M, Hoops S, Keating SM, Novére NL, Sahle S, Wilkinson DJ (2008) Systems biology markup language (SBML) level 2: structures and facilities for model definitions. Nature Precedings. URL:http://www.sbml.org/specifications/sbml-level-2/version-4/release-1/sbml-level-2-version-4-rel-1.pd. URL:http://www. sbml.org/specifications/sbml-level-2/version-4/release-1/sbml-level-2-version-4-rel-1.pdf
60. Hucka M, Schaff J (2009) Trends and tools for modeling in modern biology. In: Photosynthesis in silico. Laisk A., Nedbal L., Govindjee (eds.) Springer Netherlands pp 3–15
61. Invitrogen - Molecular Probes: BCECF - Introduction (2006). URL http://probes. invitrogen.com/media/pis/mp01150.pdf. URL:http://probes.invitrogen.com/media/pis/mp01150.pdf
62. Jacobson K, Mouritsen OG, Anderson RGW (2007) Lipid rafts: at a crossroad between cell biology and physics. Nat Cell Biol 9(1):7–14
63. Jandt U, Shao S, Wirth M, Zeng AP (2011) Spatiotemporal modeling and analysis of transient gene delivery. Biotechnol Bioeng 108(9):2205–2217
64. Johnson AE, van Waes MA (1999) The translocon: a dynamic gateway at the ER membrane. Annu Rev Cell Dev Biol 15:799–842
65. Johnson AW, Lund E, Dahlberg J (2002) Nuclear export of ribosomal subunits. Trends Biochem Sci 27(11):580–585
66. Kampmann M (2005) Facilitated diffusion in chromatin lattices: mechanistic diversity and regulatory potential. Mol Microbiol 57(4):889–899
67. Kelly RB (1985) Pathways of protein secretion in eukaryotes. Science 230(4721):25–32
68. Khalili-Araghi F, Gumbart J, Wen PC, Sotomayor M, Tajkhorshid E, Schulten K (2009) Molecular dynamics simulations of membrane channels and transporters. Curr Opin Struct Biol 19(2):128–137
69. Khanna-Gupta A, Ware VC (1989) Nucleocytoplasmic transport of ribosomes in a eukaryotic system: is there a facilitated transport process?. Proc Natl Acad Sci U S A 86(6):1791–1795

70. Khelashvili G, Weinstein H, Harries D (2008) Protein diffusion on charged membranes: a dynamic mean-field model describes time evolution and lipid reorganization. Biophys J 94(7):2580–2597
71. Kholodenko BN (2006) Cell-signalling dynamics in time and space. Nat Rev Mol Cell Biol 7(3):165–176
72. Komarova YA, Vorobjev IA, Borisy GG (2002) Life cycle of MTs: persistent growth in the cell interior, asymmetric transition frequencies and effects of the cell boundary. J Cell Sci 115:3527–3539
73. Koshland DE Jr, Hamadani K (2002) Proteomics and models for enzyme cooperativity. J Biol Chem 277(49):46,841–46,844
74. Kowalski JM, Parekh RN, Mao J, Wittrup KD (1998) Protein folding stability can determine the efficiency of escape from endoplasmic reticulum quality control. J Biol Chem 273(31):19,453–19,458
75. Kyoda KM, Muraki M, Kitano H (2000) Construction of a generalized simulator for multi-cellular organisms and its application to SMAD signal transduction. Pac Symp Biocomput 5:314–325
76. Lemmer P, Gunkel M, Baddeley D, Kaufmann R, Urich A, Weiland Y, Reymann J, Müller P, Hausmann M, Cremer C (2008) SPDM: light microscopy with single-molecule resolution at the nanoscale. Appl Phys B 93(1):1–12
77. Levi SK, Glick BS (2007) GRASPing unconventional secretion. Cell 130(3):407–409
78. Lingwood D, Kaiser HJ, Levental I, Simons K (2009) Lipid rafts as functional heterogeneity in cell membranes. Biochem Soc Trans 37(Pt 5):955–960
79. Liu YH, Bi JX, Zeng AP, Yuan JQ (2007) A population balance model describing the cell cycle dynamics of myeloma cell cultivation. Biotechnol Progr 23(5):1198–1209
80. Lloyd CM, Halstead MDB, Nielsen PF (2004) CellML: its future, present and past. Prog Biophys Mol Biol 85(2–3):433–450
81. Lloyd CM, Lawson JR, Hunter PJ, Nielsen PF (2008) The CellML model repository. Bioinformatics 24(18):2,122–2,123
82. Lüders J, Stearns T (2007) Microtubule-organizing centres: a re-evaluation. Nat Rev Mol Cell Biol 8(2):161–167
83. Lukacs GL, Haggie P, Seksek O, Lechardeur D, Freedman N, Verkman AS (2000) Size-dependent DNA mobility in cytoplasm and nucleus. J Biol Chem 275(3):1625–1629
84. Makhnevych T, Lusk C, Anderson AM, Aitchison JD, Wozniak RW (2003) Cell cycle regulated transport controlled by alterations in the nuclear pore complex. Cell 115(7):813–823
85. Männistö M, Reinisalo M, Ruponen M, Honkakoski P, Tammi M, Urtti A (2007) Polyplex-mediated gene transfer and cell cycle: effect of carrier on cellular uptake and intracellular kinetics, and significance of glycosaminoglycans. J Gene Med 9(6):479–487
86. Markovitch O, Chen H, Izvekov S, Paesani F, Voth GA, Agmon N (2008) Special pair dance and partner selection: elementary steps in proton transport in liquid water. J Phys Chem B 112(31):9456–9466
87. Mayor S, Rao M (2004) Rafts: scale-dependent, active lipid organization at the cell surface. Traffic 5(4):231–240
88. McCarthy MR, Vandegriff KD, Winslow RM (2001) The role of facilitated diffusion in oxygen transport by cell-free hemoglobins: implications for the design of hemoglobin-based oxygen carriers. Biophys Chem 92:103–117
89. van Meer G, Voelker DR, Feigenson GW (2008) Membrane lipids: where they are and how they behave. Nat Rev Mol Cell Biol 9(2):112–124
90. Mendes P (1993) GEPASI: a software package for modelling the dynamics, steady states and control of biochemical and other systems. Bioinformatics 9(5):563–571
91. Mendes P, Hoops S, Sahle S, Gauges R, Dada J, Kummer U (2009) Computational modeling of biochemical networks using COPASI. Methods Mol Biol 500:17–59
92. Meng TC, Somani S, Dhar P (2004) Modeling and simulation of biological systems with stochasticity. In Silico Biol 4(3):293–309

93. Mohr PJ, Taylor BN, Newell DB (2008) CODATA recommended values of the fundamental physical constants: 2006. Rev Mod Phys 80(2):633–730
94. Moraru II, Schaff JC, Slepchenko BM, Blinov M, Morgan F, Lakshminarayana A, Gao F, Li Y, Loew LM (2008) The virtual cell modeling and simulation software environment. IET Syst Biol 2(5):352–362
95. Morton-Firth CJ, Shimizu TS, Bray D (1999) A free-energy-based stochastic simulation of the tar receptor complex. J Mol Biol 286(4):1059–1074
96. Mothes W, Prehn S, Rapoport TA (1994) Systematic probing of the environment of a translocating secretory protein during translocation through the ER membrane. EMBO J 13(17):3973–3982
97. Mukherjee S, Maxfield FR (2004) Membrane domains. Annu Rev Cell Dev Biol 20:839–866
98. Nair R, Rost B (2005) Mimicking cellular sorting improves prediction of subcellular localization. J Mol Biol 348(1):85–100
99. Nakai K, Horton P (1999) PSORT: a program for detecting sorting signals in proteins and predicting their subcellular localization. Trends Biochem Sci 24(1):34–36
100. Nogales E (2000) Structural insights into microtubule function. Annu Rev Biochem 69:277–302
101. Novére NL, Bornstein B, Broicher A, Courtot M, Donizelli M, Dharuri H, Li L, Sauro H, Schilstra M, Shapiro B, Snoep JL, Hucka M (2006) BioModels database: a free, centralized database of curated, published, quantitative kinetic models of biochemical and cellular systems. Nucleic Acids Res 34:D689–D691
102. Novére NL, Finney A, Hucka M, Bhalla US, Campagne F, Collado-Vides J, Crampin EJ, Halstead M, Klipp E, Mendes P, Nielsen P, Sauro H, Shapiro B, Snoep JL, Spence HD, Wanner BL (2005) Minimum information requested in the annotation of biochemical models (MIRIAM). Nat Biotechnol 23(12):1509–1515
103. Pangarkar C, Dinh AT, Mitragotri S (2005) Dynamics and spatial organization of endosomes in mammalian cells. Phys Rev Lett 95(15):158101
104. Papadopoulos MC, Saadoun S, Verkman AS (2008) Aquaporins and cell migration. Pflugers Arch 454:693–700
105. Park KJ, Kanehisa M (2003) Prediction of protein subcellular locations by support vector machines using compositions of amino acids and amino acid pairs. Bioinformatics 19(13):1656–1663
106. Partikian A, Olveczky B, Swaminathan R, Li Y, Verkman A (1998) Rapid diffusion of green fluorescent protein in the mitochondrial matrix. J Cell Biol 140(4):821–829
107. Pfeffer SR, Rothman JE (1987) Biosynthetic protein transport and sorting by the endoplasmic reticulum and Golgi. Annu Rev Biochem 56:829–852
108. Ploeger L, Dullens H, Huisman A, van Diest P (2008) Fluorescent stains for quantification of DNA by confocal laser scanning microscopy in 3-D. Biotech Histochem 83(2):63–69
109. Rapoport TA (2007) Protein translocation across the eukaryotic endoplasmic reticulum and bacterial plasma membranes. Nature 450(7170):663–669
110. Rapoport TA, Jungnickel B, Kutay U (2007) Protein transport across the eukaryotic endoplasmic reticulum and bacterial inner membranes. Annu Rev Biochem 65:271–303
111. Ritchie K, Iino R, Fujiwara T, Murase K, Kusumi A (2003) The fence and picket structure of the plasma membrane of live cells as revealed by single molecule techniques. Mol Membr Biol 20(1):13–18
112. van Royen ME, Farla P, Mattern KA, Geverts B, Trapman J, Houtsmuller AB (2009) Fluorescence recovery after photobleaching (FRAP) to study nuclear protein dynamics in living cells. Methods Mol Biol 464:363–385
113. Sander G, Pardee AB (1972) Transport changes in synchronously growing CHO and L cells. J Cell Physiol 80(2):267–271
114. Saparov SM, Kozono D, Rothe U, Agre P, Pohl P (2001) Water and ion permeation of aquaporin-1 in planar lipid bilayers. major differences in structural determinants and stoichiometry. J Biol Chem 276(34):31, 515–31, 520

115. Sauro HM, Hucka M, Finney A, Wellock C, Bolouri H, Doyle J, Kitano H (2003) Next generation simulation tools: the systems biology workbench and BioSPICE integration. OMICS 7(4):355–372
116. Schaff J, Fink CC, Slepchenko B, Carson JH, Loew LM (1997) A general computational framework for modeling cellular structure and function. Biophys J 73(3):1135–1146
117. Schilstra MJ, Li L, Matthews J, Finney A, Hucka M, Novére NL (2006) CellML2SBML: conversion of CellML into SBML. Bioinformatics 22(8):1018–1020
118. Schmidt R, Wurm CA, Punge A, Egner A, Jakobs S, Hell SW (2009) Mitochondrial cristae revealed with focused light. Nano Lett 9(6):2508–2510
119. Schwehm M (2002) Parallel stochastic simulation of whole-cell models. In: Proceedings of international conference on architecture of computing systems. ARCS 2002, pp 333–341
120. Simons K, Ikonen E (1997) Functional rafts in cell membranes. Nature 387:569–572
121. Slepchenko BM, Schaff JC, Macara I, Loew LM (2003) Quantitative cell biology with the virtual cell. Trends Cell Biol 13(11):570–576
122. Slepchenko BM, Semenova I, Zaliapin I, Rodionov V (2007) Switching of membrane organelles between cytoskeletal transport systems is determined by regulation of the microtubule-based transport. J Cell Biol 179(4):635–641
123. Smith DA, Simmons RM (2001) Models of motor-assisted transport of intracellular particles. Biophys J 80(1):45–68
124. Sosinsky GE, Deerinck TJ, Greco R, Buitenhuys CH, Bartol TM, Ellisman MH (2005) Development of a model for microphysiological simulations: small nodes of Ranvier from peripheral nerves of mice reconstructed by electron tomography. Neuroinformatics 3(2):133–162
125. Strömbäck L, Hall D, Lambrix P (2007) A review of standards for data exchange within systems biology. Proteomics 7:857–867
126. Verkman AS (2002) Solute and macromolecule diffusion in cellular aqueous compartments. Trends Biochem Sci 27(1):27–33
127. Takahashi K, Ishikawa N, Sadamoto Y, Sasamoto H, Ohta S, Shiozawa A, Miyoshi F, Naito Y, Nakayama Y, Tomita M (2003) E-Cell 2: multi-platform E-Cell simulation system. Bioinformatics 19(13):1727–1729
128. Takenaka T, Gotoh H (1984) Simulation of axoplasmic transport. J Theor Biol 107(4):579–601
129. Tolentino TP, Wu J, Zarnitsyna VI, Fang Y, Dustin ML, Zhu C (2008) Measuring diffusion and binding kinetics by contact area FRAP. Biophys J 95(2):920–930
130. Tomita M (2001) Whole-cell simulation: a grand challenge of the 21st century. Trends Biotechnol 19(6):205–210
131. Tomita M, Hashimoto K, Takahashi K, Shimizu T, Matsuzaki Y, Miyoshi F, Saito K, Tanida S, Yugi K, Venter J, Hutchison 3rd C (1999) E-CELL: software environment for whole-cell simulation. Bioinformatics 15(1):72–84
132. Tran EJ, Wente SR (2006) Dynamic nuclear pore complexes: life on the edge. Cell 125(6):1041–1053
133. Tsai B, Ye Y, Rapoport TA (2002) Retro-translocation of proteins from the endoplasmic reticulum into the cytosol. Nat Rev Mol Cell Biol 3(4):246–255
134. Tsai J, Sun E, Gao Y, Hone JC, Kam LC (2008) Non-brownian diffusion of membrane molecules in nanopatterned supported lipid bilayers. Nano Lett 8(2):425–430
135. Verkman AS (2005) More than just water channels: unexpected cellular roles of aquaporins. J Cell Sci 118(Pt 15):3225–3232
136. Vorobjev IA, Svitkina TM, Borisy GG (1997) Cytoplasmic assembly of microtubules in cultured cells. J Cell Sci 110(Pt 21):2635–2645
137. Walter A, Gutknecht J (1986) Permeability of small nonelectrolytes through lipid bilayer membranes. J Membr Biol 90(3):207–217
138. Weis K (2003) Regulating access to the genome: nucleocytoplasmic transport throughout the cell cycle. Cell 112:441–451

139. Weiss DG, Keller F, Gulden J, Maile W (1986) Towards a new classification of intracellular particle movements based on quantitative analyses. Cell Motil Cytoskeleton 6(2):128–135
140. Wimalaratne SM, Halstead MDB, Lloyd CM, Crampin EJ, Nielsen PF (2009) Biophysical annotation and representation of CellML models. Bioinformatics 25(17):2263–2270
141. Wiseman RL, Powers ET, Buxbaum JN, Kelly JW, Balch WE (2007) An adaptable standard for protein export from the endoplasmic reticulum. Cell 131(4):809–821
142. Wittenberg JB (2007) On optima: the case of myoglobin-facilitated oxygen diffusion. Gene 398:156–161
143. Wyman J (1966) Facilitated diffusion and the possible role of myoglobin as a transport mechanism. J Biol Chem 241(1):115–121
144. Yvon AMC, Wadsworth P (1997) Non-centrosomal microtubule formation and measurement of minus end microtubule dynamics in A498 cells. J Cell Sci 110:2391–2401
145. Zeidel ML, Nielsen S, Smith BL, Ambudkar SV, Maunsbach AB, Agre P (1994) Ultrastructure, pharmacologic inhibition, and transport selectivity of aquaporin channel-forming integral protein in proteoliposomes. Biochemistry 33(6):1606–1615

Adv Biochem Engin/Biotechnol (2012) 127: 251–284
DOI: 10.1007/10_2011_117

Published Online: 9 November 2011

Genetic Aspects of Cell Line Development from a Synthetic Biology Perspective

L. Botezatu, S. Sievers, L. Gama-Norton, R. Schucht, H. Hauser and D. Wirth

Abstract Animal cells can be regarded as factories for the production of relevant proteins. The advances described in this chapter towards the development of cell lines with higher productivity capacities, certain metabolic and proliferation properties, reduced apoptosis and other features must be regarded in an integrative perspective. The systematic application of systems biology approaches in combination with a synthetic arsenal for targeted modification of endogenous networks are proposed to lead towards the achievement of a predictable and technologically advanced cell system with high biotechnological impact.

Keywords Synthetic biology · Mammalian cells · Controlled genetic engineering · Synthetic expression cassettes · Controlled intervention

Contents

L. Botezatu · S. Sievers · L. Gama-Norton · R. Schucht · H. Hauser · D. Wirth (✉)
Helmholtz Centre for Infection Research, Braunschweig Germany
e-mail: dagmar.wirth@helmholtz-hzi.de

1 Introduction

Mammalian cell lines have been used for decades for the production of biopharmaceuticals. To date, the development of producer cells mainly focuses on the optimization of the features of the recombinant expression cassettes that encode the biopharmaceutical proteins of interest. Highly potent promoter elements that allow high levels of expression have been identified. In addition, protocols for amplification of these cassettes have been developed. Recent approaches also acknowledge the impact of the chromosomal integration site on the performance of the recombinant expression cassettes.

However, due to the great complexity of mammalian cells and the multifactorial facets of cell productivity, it has so far been difficult to systematically circumvent limitations that still affect the production of recombinant proteins. There is growing evidence that high-producer cell clones must have not only optimal design but also an appropriate chromosomal integration site of the expression cassette. Cells often accumulate random, uncontrolled mutations that render them better producer systems. On the other hand, considerable efforts have been made to increase genomic stability and productivity by manipulating cellular properties such as cell growth or apoptosis.

While the first example of a pure synthetic prokaryotic organism was given recently [58], such an endeavor does not seem to be feasible for mammalian cells in the near future. Due to the greater complexity of the mammalian cell, synthetic approaches focus on modulation of selected pathways or networks. For this purpose, strategies are required that allow us to specifically combine cellular networks or to interfere with them. Examples of such approaches concern cellular metabolic networks and cellular proliferation. While much data and information on relevant pathways have been assembled, a deep systems-biology-based analysis of these networks is not yet available. This is the focus of current work in the field. A systems-biology-driven rational engineering of cells is expected to increase the productivity of cell systems directly or indirectly.

To date, approaches for development of cell systems for production have not been guided by systems biology analysis. They have been based on a patchy knowledge of cellular processes. A consequent systems biology analysis as it is carried out in this rapidly growing field will permit the expansion of these efforts to new targets. In the first part of this review, the state of the art of methods which allow controlled expression of genes is summarized. Genetic engineering of cells is a prerequisite for rationally coupling or interfering with distinct cellular pathways. This section is followed by one that describes selected approaches to inducing specific expression phenotypes. Finally, approaches that have been initiated to alter certain cellular phenotypes are described. The methods presented here are perfect tools with which to intervene in all types of cellular processes and are thus key for challenging or confirming new systems-biology-driven networks.

2 Approaches to Tuning Gene Expression in Mammalian Cells

To control expression of genes in mammalian cells, regulatory circuits that are largely independent of the cellular metabolism have been developed. These comprise both transcriptionally regulated modules for tunable expression of transgenes, and post-transcriptionally regulated elements such as the ones that control protein stability. In addition, siRNA-based approaches are used to down-regulate cellular genes. The basic principles of these approaches together with examples will be discussed in the following paragraphs.

2.1 Transcriptional Control

In the last two decades, synthetic systems for controlling transcription in mammalian cells have been developed [50, 119, 191]. These are based on the principles that regulate transcription of synthetic prokaryotic gene-networks. Generally, these systems rely on prokaryotic protein moieties that bind to their cognate DNA operator sequences, an event that is necessarily modulated by external signals (i.e. inducer molecules). In order to adapt these systems for transcriptional control in mammalian cells, prokaryotic DNA binding moieties are fused to eukaryotic protein domains, constituting a chimeric transactivating protein with the capacity to activate or repress the activity of promoters. Further, synthetic regulatable promoters are designed by combining a minimal eukaryotic promoter with prokaryotic DNA elements (operators) that allow binding of the cognate protein moiety. The transcriptional activator or repressor domain can thereby be recruited to the synthetic promoter (Fig. 1a). For these complex systems to be successfully employed in mammalian cells, the following requirements have to be met:

- Binding of the transactivating protein to the promoter tunable in a dose-dependent manner;
- Fast reversibility upon withdrawal of the inducer;
- No or low basal expression in the absence of the inducer;
- No or minimal interference with the host cell system and the endogenous gene network.

In order to achieve controlled expression of a protein, initial approaches relied on temperature-sensitive (ts) protein mutants displaying altered activities upon shifting temperature. Today, different kinds of small-molecular-weight inducers are used to achieve controlled expression. These include antibiotics, steroid hormones, *quorum*-sensing molecules, urea, immunosuppressive and anti-diabetic drugs, phloretin, biotin, L-arginine and volatile compounds like acetaldehyde. In addition, light-induced systems have been employed in mammals [100, 203, 209]. For a comprehensive list of inducer molecules used in transgene expression

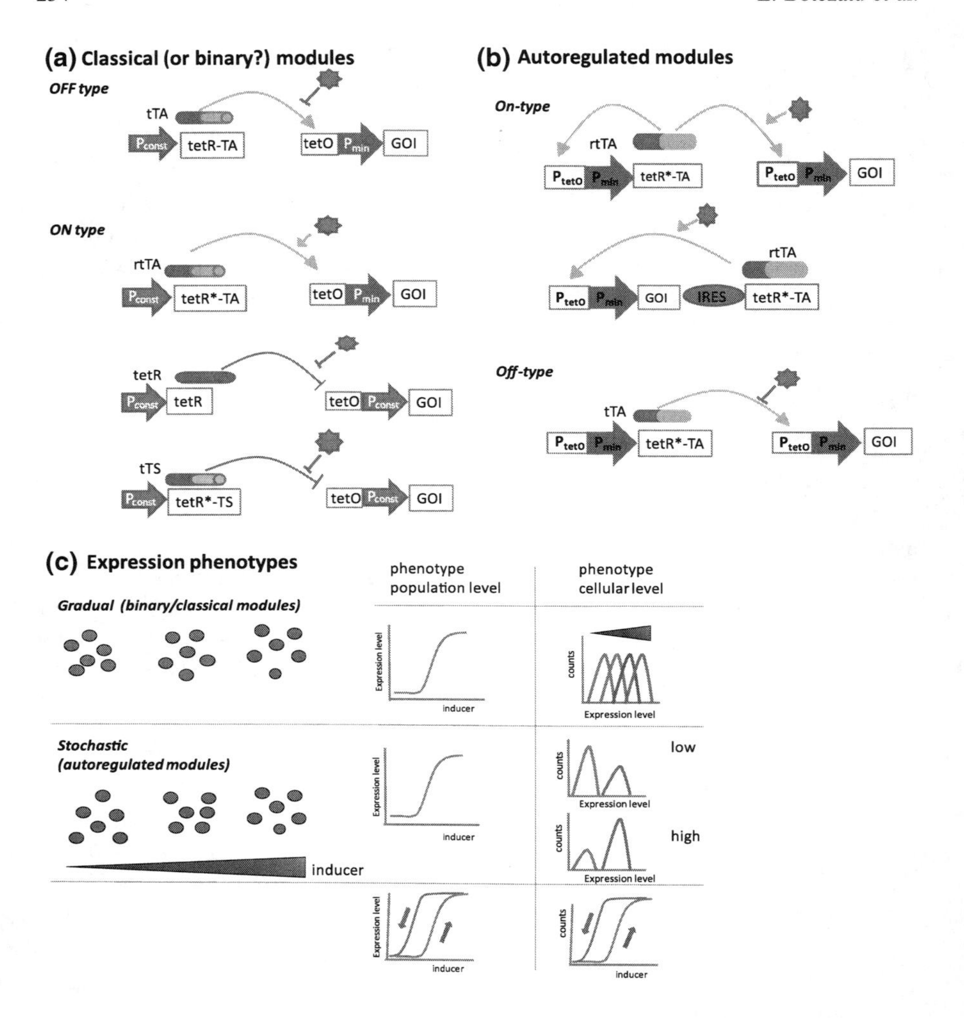

regulatable systems, see Table 1. In the following, the principle of regulatory synthetic modules will be exemplified by the "Tet-system". This system has been shown to allow tight control of gene expression in many different biological contexts, from diverse primary cells to animal models.

2.1.1 The Tet-System

The tetracycline regulation response in *E.coli* relies on the binding of a protein (tetR) to its cognate operator sequences (tetO) located in the *E.coli* tetracycline operon. Gossen and Bujard developed a fusion protein consisting of the tetR binding domain and the transactivating domain of herpes simplex virus VP16 [63].

◄**Fig. 1** **a** *Binary expression modules*. Binary Tet-dependent expression modules consist of two independent expression units: a constitutive promoter (Pconst) that drives expression of a transactivator (tTA, rtTA) or a transrepressor (tetR, tTS) and a synthetic, inducible promoter driving the gene of interest (GOI). The transacting proteins consist of a DNA binding moiety derived from the bacterial tetR (tR) fused to a transactivating (TA) or silencing (TS) domain. Binding of the transacting molecules to the tetO sequence in synthetic promoters is achieved by the tetR binding domain tR or mutated variants tR*. The transactivating or repressing domains are thereby directed to the synthetic promoters. Doxycycline (Dox) or tetracycline (depicted as a *star*) binds to the transacting molecules and induces a conformational change that modulates binding to the tetO sequence. While Dox inhibits binding of proteins harbouring the tR domain (tTA, tTS and tetR), it is required for binding in the rtTAs that incorporate mutant tR* domains. In Off-type expression cassettes, transcription is switched off on addition of Tet, while On-type expression cassettes require Tet to be induced. **b** *Autoregulated modules*. In these modules, the transacting protein is controlled by a Tet-responsive promoter, thereby generating a positive feedback loop. This can be realized in two independent expression units for the transacting proteins and the GOI, respectively. Expression of these two proteins can be linked in a bicistronic message. Alternatively, bidirectional Tet-responsive promoters can be exploited for this purpose (not depicted). Both On-type and Off-type designs can be realized. **c** Expression phenotypes generated with binary modules result in gradual increase of expression upon administration of increasing concentrations C of the inducer. This is observed on both the population level and the cellular level. In contrast, autoregulated modules result in a stochastic activation of gene expression. In this case, in individual cells expression is either ‘On’ or ‘Off’. Accordingly, two distinct expression states can be maintained which are visible on the single cell level only. Increasing concentrations of the inducer increase the probability that cells express the transgene. This is exemplified for low concentrations of the inducer (c_{low}) and high concentrations (c_{high}). Stochastic gene activation results in hysteresis [90, 120]: while cells need a certain concentration of inducer to switch gene expression on (*blue arrow*), lower concentration is needed to maintain expression (*red arrow*)

This fusion protein, called tTA, can transactivate synthetic promoters in which the Tet operator sequences have been fused to a minimal eukaryotic promoter. Transcription is thereby activated, leading to expression of the gene of interest (GOI) (see Fig. 1a and b). The binding of tetracycline [or its derivatives such as doxycycline (Dox)] to the tetR binding domain induces a conformational change that leads to the release of the protein complex from the promoter, abrogating transcription (Tet-off system). Reverse transactivators (rtTAs) have been developed in which mutations in the prokaryotic domain reverse its binding properties. In these mutants, the binding to the promoter is only achieved in presence of the inducer Dox [64]. In order to increase the stringency of the expression system, these mutants have been significantly improved [10, 183]. Moreover, new transactivator variants have been developed that are optimized for human codon usage [114]. It was shown that HSV VP16 can outcompete binding of transcription factors, thereby inhibiting transcription of cellular genes, an effect called transcriptional squelching [59]. To overcome this side effect, a panel of different transactivating domains from cellular activator proteins such as p65 and E2F4 have been functionally fused to the Tet-binding domain [3, 62, 183, reviewed in 14].

Since the performance of inducible systems is limited by basal expression levels, efforts to modify the Tet-dependent promoter have been undertaken.

Table 1 Transcription regulatable systems

Inducers	References
Antibiotics	
Coumermycin	Zhao et al. [215]
Macrolides (e.g. Erythromycin)	Weber et al. [192]
Streptogramines (e.g. Pristinamycin)	Fussenegger et al. [50]
Tetracycline (and derivatives)	Gossen and Bujard [63]; Gossen et al. [64]; Urlinger et al. [183]
Steroids	
Mifepristone	Wang et al. [187]
Estrogen	Braselmann et al. [24]
Ecdysone/Muristerone A	No et al. [134]; Yao et al. [207]
Physiological molecules	
NADH	Weber et al. [192]
Urea	Kemmer et al. [83]
Acetaldehyde	Weber et al. [192]
Nicotine	Malphettes et al. [111]
Vitamin H (Biotin)	Weber et al. [192]; Weber et al. [195]
Arginine	Hartenbach et al. [68]; Weber et al. [196]; Weber et al. [192]
Environmental signals	
Temperature	Boorsma et al. [20, 21]; Siddiqui et al. [163]; Weber et al. [193]
Light	Wu et al. [203]; Yazawa et al. [209]; Levskaya et al. [100]
Hypoxia	Binley et al. [16]
Radiation	Mezhir et al. [123]
Metal ions (Zinc)	Searle et al. [159]
Quorum sensing signals	
Butyrolactone	Weber et al. [194]
Acylated homoserine lactone	Neddermann et al. [130]
Others	
IPTG	Hu and Davidson [73]
Cumate	Mullick et al. [129]
Rosiglitazone	Tascou et al. [176]
Electricity	Weber et al. [197]
Phloretin	Gitzinger et al. [60]

An increase of stringency in transgene regulatable expression has been achieved by optimizing the Tet-responsive promoter through modifications of heptameric tetO sequences [1] as well as by the use of other minimal promoter elements, such as the ones from MMTV and HIV [31, 70, 166].

With these improvements, tight regulation of a plethora of transgenes has been realized in various cell types. Moreover, simultaneous expression of several transgenes was achieved on coupling these genes in multicistronic expression units or with the help of a bidirectional promoter [9].

Tet-induced repression of promoter activity. A reverse approach to the activation of a minimal promoter consists in the repression of gene expression by binding of a repressor moiety to the regulatable promoter. For this purpose, a chimeric protein consisting of the Krueppel Associated Box (KRAB) domain of the mammalian *Kox1* gene and the TetR binding domain was generated [34, 112]. In the absence of the inducer, this protein binds to the tetO sequence of the regulatable promoter and exerts its epigenetic silencing activity by inducing the formation of heterochromatin. On administration of the inducer, the repressor dissociates from the tetO site leaving it available for its activation. This strategy was successfully employed for reduction of the basal activity of the Tet-inducible promoter. Forster et al. combined the rtTA-dependent activation of the Tet-on system with a specific repression of the uninduced state [45]. For this purpose, a repressor is co-expressed that is inversely activated by the inducer. To avoid heterodimerization of repressor and activator proteins, heterologous elements for dimerization and DNA recognition are employed in the transactivator and repressor proteins, respectively. Basal expression of the system could thereby be reduced and tighter regulation was achieved. Freundlieb et al. employed the epigenetically modulating KRAB domain derived from the human kidney protein Kid-1 for a fusion with the DNA binding moiety of the tetR binding domain to establish a Tet-dependent silencer tTS [49]. This fusion represses basal expression from the Tet-responsive promoter leading to a strong reduction of basal activity, also in vivo [216]. Apart from these artificial fusion proteins that direct repressor domains to synthetic promoters, the prokaryotic tetR protein itself can directly impair transcription of a mammalian promoter [207].

Two-vector versus one-vector systems. Most synthetic transcriptional regulation systems rely on two components: one ensures the expression of the transactivator or repressor; the other contains the inducible cassette with the promoter and the GOI. In most cases, these two elements are delivered by separate vectors (two-vector system). As an alternative, one-vector systems were developed: in this set-up all necessary components are assembled in one vector. The two components might be oriented in the same or in opposite directions with regard to the transcriptional activity. The major advantages of having both elements in one vector are the single gene transfer step and the possibility of packaging in viral vectors. However, the promoters can influence each other's activity due to the close proximity of the regulatory elements, which complicates the vector design.

Autoregulation. One elegant way to overcome promoter interactions is the use of autoregulated systems in which the expression of both the transactivator and the GOI are controlled by a single ligand-responsive promoter. This can be achieved in multicistronic cassette designs or by the use of bidirectional promoters. Such modules can be transduced by retroviral and lentiviral vectors [114, 121, 182]. Autoregulated systems depend on a low level of basal expression that is required for inducibility. It is important to note that this basal expression is low and does not necessarily induce a biological effect [116]. Autoregulated expression cassettes have been shown to result in a stochastic, i.e. bimodal, type of expression. This is in contrast to the classical systems which rely on constitutive transactivator expression and which give rise to gradual expression response [120]. For a comprehensive distinction between the characteristics associated with autoregulated and constitutive expression systems, the reader is referred to Fig. 1b. In addition, it has been shown that autoregulated expression circuits are characterized by increased expression stability and reduced sensitivity to noise-inducing fluctuations [13].

2.1.2 Steroid-Inducible Systems

Another important group of transcription inducible systems are those relying on steroids. In the human proteome, steroid hormone receptors constitute the largest group of transcription factors. Activation of these cytosolic receptors depends on binding to their cognate ligands, the steroid hormones, which easily cross plasma membranes or epithelial borders. Upon binding, a conformational change is induced that results in release of heat shock proteins and facilitates translocation of the receptor to the nucleus where gene expression is induced by binding of the receptors to their cognate promoters (see [135] for review).

Drawbacks are associated with this kind of system. The respective inducers and/or repressors of GOI expression may modulate not only the transgene, but also endogenous gene expression. Expression of endogenous genes is avoided by using synthetic steroids that do not interact with the cellular machinery [26]. A number of agonists and antagonists of synthetic progesterone receptors were identified and their function as inducers was characterized in more detail [132].

2.2 Post-transcriptional Control Systems

2.2.1 Gene Silencing Via RNA Interference

A new regulatory mechanism of post-transcriptional gene silencing (PTGS) which is based on RNA interference (RNAi) was discovered and explored in the last decade (reviewed in [162]). In mammalian cells, this mechanism is endogenously realized by so-called miRNAs, small RNAs that show homology to endogenous mRNAs. Upon binding to cognate mRNA targets, they can specifically down-regulate expression. miRNAs are encoded in hairpin structures of many cellular mRNAs. This hairpin structure is processed by cellular enzymes Drosha/DGCR8 to pre-miRNAs and further cleaved by Dicer. The resulting short RNA molecule of 21–23 nucleotides (nt) can then bind to homologous sequences in their cognate mRNA in a complex with Argonate proteins; this complex is called RISC, the RNA-induced silencing complex. Depending on the extent of homology, the consequence of this binding is either translational repression or degradation of the mRNA, resulting in a specific gene silencing. The discovery and subsequent exploitation of this gene-silencing mechanism via RNAi has opened new avenues towards specific manipulation of genes.

To down-regulate target genes, three different options can be followed. Long-term down-regulation is achieved (a) by Pol II-dependent transcription of miRNA encoding mRNAs, thereby exploiting the cellular pathway of processing pre-miRNAs down to the short 21–23 nt entities, (b) by transducing Pol III-dependent expression units to transcribe small hairpin RNAs that rely on cleavage by Dicer to generate the single-stranded RNAs that bind to their cognate target sequences or (c) by transient down-regulation through direct transfection of synthetically produced short (21–23 nt) RNAs, so called siRNAs.

2.2.2 Steroid Receptor Fusion Proteins

Several studies have shown that the regulatory domains of steroid receptors constitute a reversible molecular switch for the post-translational regulation of a wide variety of cytoplasmic and nuclear proteins [140]. Control of gene expression is achieved by exploiting steroid receptors such as the estrogen receptor and the progesterone receptor. Here, the pre-existing protein is inactive and becomes activated by the addition of the steroid. Indeed, it has been shown that upon fusing the ligand-binding domain of steroid receptors to the GOI, nuclear translocation can be controlled. Various examples have been reported. These include cell cycle regulation proteins such as myc, fos and interferon regulatory factor-1 (IRF-1) [38, 85, 172]. This principle has also been successfully employed to regulate recombinases such as Cre, Flp and PhiC31 that allow site-specific DNA modifications including targeted integrations [42, 74, 107, 161].

Mutations in the hormone-binding domain allowed restriction of activation of the fusion proteins to synthetic hormone analogues such as tamoxifen or mifepristone [82, 185]. Cross-activation by endogenous hormones is thereby overcome and also allows application of this principle in vivo. It should be mentioned that these compounds might still influence cell physiology: e.g., tamoxifen acts as an estradiol antagonist.

2.2.3 Control of Protein Stability

Recently, a novel post-translational strategy for controlling the protein level has been introduced [8]. Fusing the protein of interest to a specific destabilization domain leads to rapid degradation of the protein. This degradation can be blocked by administration of drugs that bind to the destabilization domain and thereby shield the protein from degradation. This activity has been shown to be dose-dependent. Meanwhile, a number of applications to various proteins highlight the flexibility of this strategy [8, 61, 142].

A reverse approach was taken by Nishimura et al. [133]. Fusion proteins of the GOI with *Arabidopsis thaliana* auxin/IAA transcription repressor 17 (Aid) are ubiquitinated and readily degraded by co-expressed TIR-1 in an auxin-dependent manner. In this system, TIR-1 interacts with the mammalian ubiquitination complex SCF. In the presence of auxin, a phytohormone, degradation is induced.

Apart from the obvious advantages for controlling expression of proteins that negatively affect the cells, these strategies will also be of interest for perturbing cells for a limited period of time.

3 Stable and Long-Term Expression of Synthetic Cassettes in Mammalian Cells

Gene transfer is a prerequisite for modification of cells. For the implementation of synthetic cassettes into mammalian cells, standard physicochemical transfection methods can be used. Biotechnologically relevant cell lines such as BHK, Chinese hamster ovary (CHO) and HEK 293 cells can also be robustly and reliably manipulated using standard protocols for liposome-based transfer and electroporation-based protocols [138, 141, 198]. Other cell types and also many primary cells require more efficient transduction systems. In particular, viral transduction methods have been proven to efficiently transduce cassettes. This includes adenoviral vectors [186], adenovirus-associated viral vectors [25], and retroviral and lentiviral vectors [11, 17, 115].

While most gene transfer procedures give rise to a transient expression period, stable gene expression is less frequent. In most cases, it is achieved by integration of the transgene into the chromosomal DNA of the host. Stable expression normally allows the cloning of transgenic cells and thus enables studies with homogenous cell populations that can be further characterized. The following section is focused on stable gene transfer and highlights gamma-retroviral and lentiviral transduction as well as transposon-based methods for efficient and stable genome engineering.

3.1 Gamma-Retroviral and Lentiviral Vectors

Gene transduction mediated by gamma-retroviral and lentiviral vectors is a highly efficient method for stably modifying cells of different species [17, 165]. For production of these vectors, safe packaging systems that rely on helper cell lines or transient packaging systems have been established [57, 124, 157, 189]. For both systems, so-called self-inactivating (SIN) vectors are available, in which the viral promoter/enhancer activity is eliminated upon infection [94, 153, 214]. SIN vectors allow the transduction of cassettes without potentially interfering viral regulatory sequences; accordingly, these are the vector types of choice for transduction of synthetic expression modules.

Gamma-retroviral and lentiviral vectors can accommodate all elements necessary for Tet-dependent expression. In particular, those encoding modules for graded expression [70, 72, 105] have been described. Apart from this, autoregulated modules have also been successfully transduced by gamma-retroviral [96, 181] and lentiviral vectors [113, 114, 120]. Appropriate vector systems have also been developed for the coumermycin system, a system that is similar to the Tet system (see Table 1) [215]. Moreover by lentiviral transduction autoregulated cassettes bimodal expression characteristics have been achieved [120].

3.2 Transposon-Mediated Integration of DNA

In recent years, transposon-mediated integration of transgene cassettes has emerged as a highly efficient method of achieving long-term expression [78]. Various transposon-based systems have been described [48, 77, 88, 125]. Transposable elements derived from natural transposons are non-viral gene delivery vehicles capable of efficient genomic insertion. Briefly, they rely on a transposase that efficiently recombines specific inverted repeat sequences that flank a plasmid-based cassette, resulting in its integration into chromosomes. The transposase integrates plasmid-based DNA into the chromosomal DNA through a precise, recombinase-mediated mechanism. Long-term expression of the GOI is thereby achieved. The transposase can be provided on the same plasmid (*cis*) or in another plasmid (*trans*), but also can be transferred as mRNA or as protein. Different methods can be employed to transfer the transposon plasmids to the cells. The enzymatic efficiency of transposases has been recently improved, now reaching the efficiency of viral transduction [78]. Recent reports include the successful application of this method for the chromosomal integration of Tet-dependent expression modules [70].

3.3 Episomal Vectors

As an alternative to genomic integration, extra-chromosomal maintenance of replicating vectors can provide long-term gene expression, even in proliferating cells. Such vectors are derived from viral entities such as Epstein–Barr virus but can also be based on non-viral episomally replicating vectors (reviewed in [108]). These episomal vectors can persist in the nucleus without integrating into the cellular genome and are not thereby affected by position effects. Episomal vectors comprising synthetic modules have been reported and shown to be efficiently replicated with tightly controlled expression proprieties [7, 22]. In this regard, they are an attractive tool for genetic engineering of cells, including therapeutic applications [37, 202].

3.4 Strategies for Predictable and Stable Expression of Synthetic Modules in Mammalian Cells: Targeted Integration

Most gene transfer protocols—including the above-mentioned protocols employing viruses and transposons—rely on the stable integration of the modules into the genome of the host cell via undirected, i.e. sequence-independent, illegitimate recombination, which is a largely random process. Accordingly, the sites of integration are mostly spread all over the genome [204]. As a consequence, individual cell clones are characterized by specific integration site(s) of the module(s).

Once integrated into the cellular DNA, transgenic modules are affected by neighboring chromosomal elements that modulate promoters to a large extent (see [199] for a recent review). Enhancers and silencers directly affect the synthetic promoters of individual modules in *cis*. Besides this, chromatin-modeling elements such as locus control regions and S/MARs significantly influence the transgene expression level [18, 101, 199]. Finally, evidence has been provided that proximal promoter elements can also interact with incoming promoters (promoter crosstalk) resulting in their down-regulation (e.g. by promoter occlusion) or potentiation [66]. These interactions of the synthetic modules with the chromosomal flanking regions can affect both the basal activity of a synthetic module and its regulation capacity, a phenomenon described as "position effect". Thus, on random integration, individual cell clones display a highly heterogeneous expression pattern of the synthetic module and have to be screened for appropriate, i.e. tightly controlled, expression. Accordingly, the reproducibility is often difficult on random integration of expression cassettes. On the other hand, the heterogeneity might be of advantage if infrequent phenotypes in individual cells have to be studied.

Homologous recombination is used in stem cells for targeting transgenes to specific loci. This method allows the optimization of the performance of a Tet-dependent cassette in a characterized integration site [167]. However, in differentiated cells, homologous recombination is a very infrequent event. For these cells, recombinase-based methods such as targeted integration or recombinase-mediated cassette exchange (RMCE) (reviewed in [54, 200]) represent potent tools for targeting individual chromosomal sites for various applications in order to overcome the limitations of random integrations. Indeed, such approaches have been successfully performed for reliable modification of the mouse genome [136, 149]. In addition, it has been exploited for the generation of production cells [55, 106, 131].

As an alternative to site-specific recombinases, strategies have been followed to increase the frequency of homologous recombination by specific induction of double-strand breaks. For this purpose, zinc (Zn) finger-based nucleases are designed to induce DNA double-strand breaks at specific sites of the genome [87, 184]. Triggered by this DNA break, co-transfected expression cassettes are integrated through cellular repair enzymes. Targeted integration of cassettes into virtually any chromosomal site can thereby be achieved.

4 Expression Characteristics Created by Synthetic Modules

While pioneering work in prokaryotes and lower eukaryotes has provided proof of concept for implementation of even complex synthetic systems [40], synthetic networks have also been designed to create new functionalities in mammalian cells. Strategies for establishing complex artificial circuits in mammalian cells have been summarized in a number of excellent reviews [6, 65, 191, 192]. It is anticipated that new functionalities that mimic endogenous functions can be generated with the help of complex synthetic circuits. One example concerns the engineering of signal transduction pathways. The challenges in specific engineering of signal transduction pathways by synthetic tools have been highlighted [84]. These concern (a) the fact that signalling transduction pathways are much faster than transcriptional responses generated with synthetic circuits, as recently reviewed [143], (b) the impact of subcellular localization of signalling proteins, and (c) the inherent noise of genetic circuits in contrast to signalling molecules whose actions are less stochastic due to the large number of molecules involved.

In Table 2, synthetic approaches are exemplified that allow the establishment of defined expression characteristics. Here, we focus on several elegant studies.

One of the first synthetic circuits created was a synthetic three-step transcription cascade. This was realized by interconnecting the regulatory modules responsive to tetracyclines, erythromycin and streptogramin in a sequential manner [91]. This cascade was shown to be controlled individually at the various levels and provided four defined levels of expression in response to the three antibiotics. Furthermore, functional coupling of endogenous signals and synthetic cassettes was recently realized. In this cascade, hypoxic conditions induced translocation of endogenous HIF-1alpha to the nucleus, which resulted in activation of an artificial multi-step regulatory module initiated by a HIF-1-responsive artificial module [93].

By combining two antibiotic-dependent Off-type switches in a mutually repressing manner, a biological toggle switch could be created [92]. In this study, the authors used two independently controlled synthetic modules to express mutual repressor proteins that switch off the other expression unit. For this purpose they relied on the KRAB transrepressor that epigenetically silences the neighbouring DNA region [127] (for details please consult Sect. 2). Binding of the KRAB domain to the respective promoter was realized upon fusion of the KRAB domain to the erythromycin- or pristinamycin-regulated protein domains, thereby directing the KRAB domain to the respective synthetic promoter. This binding can be impaired by addition of the respective small molecule drugs pristinamycin and erythromycin. Upon releasing one of the repressors by addition of the antibiotic, expression of the second regulatory module is initiated. This expression is maintained even in presence of the repressor, thus providing a toggle-switch-like expression phenotype. Furthermore, with this setting a bimodal (On/Off) expression phenotype was created.

In the bimodal, "bistable" or "switch-like" expression pattern, the concentration of the inducer increases the probability that a gene will be expressed. Once it

Table 2 Switches in mammalian cells realized by synthetic cassettes

Synthetic switches	References
Boolean gates: AND, OR, NOR,…	Kramer et al. [92]; Rinaudo et al. [144]
Synthetic transcription cascade	Kramer et al. [91]
Semi-synthetic transcription cascade	Kramer et al. [93]
Coupled transcription/ translation cascades	Malphettes and Fussenegger [110]; Deans et al. [33]; Greber and Fussenegger [65]
Toggle switch	Kramer et al. [92]; Greber and Fussenegger [65]
Hysteresis	May et al. [120]; Kramer et al. [93]
Semi-synthetic oscillator	Toettcher [179]
Synthetic oscillator	Swinburne et al. [173]; Tigges et al. [177]; Tigges et al. [178]
Epigenetic memory device	Kramer et al. [92]; Greber and Fussenegger [65]
Time delay expression	Weber et al. [192]

is turned on, expression is always maximal. Natural examples for this switch include E2F activation of Rb [208] and the response of mouse embryonic stem cells to leukemia inhibitory factor (LIF) [32]. Bimodally expressed genes in the mouse genome were identified based on microarray data [41]. In an engineered system, bimodal expression has been realized by implementing autoregulated switches [90, 120].

Bimodal expression can result in stochastic activation of gene expression. Moreover, such expression patterns have been shown to result in hysteresis. In natural systems, hysteretic switches are considered to make gene expression more robust and largely independent of noise. This is due to the fact that subtle changes in the concentration of the inducers do not affect the expression phenotype. Indeed, such an expression pattern is also predicted by modelling approaches based on stochastic mathematical tools [120].

More complex expression phenotypes are represented by oscillators. Oscillating gene expression can be achieved by interconnecting positive and negative feedback loops. Periodic gene expression results in the proper timing of events in periodic processes. The first oscillating gene network in mammalian cells has recently been introduced [173]. This approach employed a tetR-based negative feedback loop which was shown to be modulated by the length of the intronic region preceding the tetR coding sequence. By increasing the intron size up to 16 kb, higher oscillation periods were created. A more recent study [177] combined a positive feedback loop with a time-delayed negative feedback loop and could thereby generate tunable and sustained oscillation as exemplified by expression of fluorescent proteins. By adjusting the concentration of the inducer, periods could be manipulated from 140 to 330 min. Significantly slower oscillation periods were achieved by combining a tTA-dependent positive feedback loop with a delayed negative feedback loop. This was realized by implementation of an intronically encoded siRNA that negatively regulates expression of an autoregulated tTA loop.

Interestingly, in contrast to the above-mentioned rapid oscillation in this system, prolonged periods of about 25 h were realized.

While the oscillatory examples are highly promising, to date they could only be realized in transient expression systems. It seems that only a fraction of the transduced cells react in the desired way because the composition of the individual elements must match a certain stoichiometry. This has so far limited the application.

5 Applications of Synthetic Modules for Alteration of Cellular Function

To increase cellular productivity, a number of studies have been performed that aim at the modification of certain cellular features that directly or indirectly regulate production of relevant biomolecules. This includes specific modulation of cellular pathways such as metabolic pathways, proliferation or survival. Here, we focus on strategies for controlling cell proliferation of immortalized cell lines as well as novel approaches for controlled expansion by conditional immortalization. In addition, we discuss approaches towards controlled product secretion.

5.1 Control of Cell Proliferation

Control of cell proliferation is of central interest in mammalian cell biotechnology. The motivation for controlling cell cycle progression and cell proliferation emerged for two different reasons. Firstly, fed-batch production schemes are limited by the fact that ongoing cell proliferation and accompanying increase in cell mass eventually leads to a cellular/system collapse and thus defines the end of the production phase. Therefore, based on the fact that cell proliferation consumes a significant portion of cellular energy, it was hypothesized that control of cell proliferation would allow redirection of this energy to production of the desired product, thereby increasing productivity.

Initial cell engineering strategies developed to increase cell-specific recombinant protein production rate by manipulation of cell cycle progression remained below expectations. The first evidence that recombinant protein production might be uncoupled from cell proliferation was obtained in CHO cells and was described 20 years ago [69]. Since then, the development of strategies that lead to an increase of cellular productivity by manipulating cell cycle progression has become a major area in the metabolic engineering field. These first results opened avenues for improvement of strategies that aim at a more balanced and controlled cell growth in a biotechnological perspective, being an active area of cell engineering research nowadays.

5.1.1 Controlled Growth of Immortalized Cells

Proliferation control of immortalized cell lines has been accomplished by at least three different main strategies. The first involves cell growth arrest based on the engineering of molecular networks that control progression of the mammalian cell cycle [51, 52, 56, 122]. The second is based on cultivation at sub-physiological temperatures (i.e. below 37° C) [19, 46, 47, 81, 154, 210–213]. The third approach consists in the retardation of proliferation by chemical induction of cell cycle arrest [43, 71, 103].

The following paragraph describes approaches towards increased productivity of recombinant proteins by engineering cellular pathways that control the cell cycle progression of immortalized mammalian cell lines such as CHO or HEK293.

Early studies on the mammalian cell cycle and cancer research uncovered a multitude of proteins whose function relies on negative effects on cell cycle progression. The control of the cell cycle of immortalized cell lines predominantly dealt with arrest at the G1 phase, considered to be the ideal time period in which the productivity of the protein of interest can reach its maximal levels ([15, 28, 75, 170, 180, 210, 211] and [170] for a recent review). Nevertheless, other reports described the production of recombinant proteins when the cell cycle is arrested in S phase [47] or even irrespective of cell cycle phase [104].

One of the first studies aimed at controlling cell growth by genetic engineering implicated the over-expression of IRF-1. Besides the pleiotropic effects of IRF-1 in activating several cellular genes (reviewed in [95]) ranging from IFN-β to IFN-stimulated genes, it was shown that IRF-1 also acts as a tumor suppressor [174, 175] and, importantly, acts as a negative regulator of cell proliferation [85, 145]. It was shown that the downstream effects of IRF-1 converge, at least partially, in the induction of anti-proliferative genes, such as *p21* [137, 175] leading to a G1 cell cycle-specific arrest. Thus, the control of cell cycle progression by expressing IRF-1 can be regarded as a multigenic approach towards an increase of recombinant protein production by modulating the cellular transcriptome.

The activation of IRF-1 function was achieved by transcriptional activation of the IRF-1 transgene and by post-translational activation. Pioneering studies on growth manipulation of recombinant mammalian cells were performed by establishing a murine C243-based cellular system in which proliferation was controlled by inducible expression and activation of IRF-1 protein [89]. In this proof-of-principle study, the transcription of IRF-1 is driven by a Tet-repressible promoter and thus IRF-1 is expressed only in the absence of the inducer. Growth inhibition could be demonstrated by omitting the inducer from the growth medium. The second approach employed a fusion of IRF-1 to the estrogen receptor. In this case, the function of the constitutively expressed IRF-1 fusion protein is only activated by the addition of β-estradiol. The biotechnological relevance of this achievement was later demonstrated by modulating the IRF-1-responsive cellular system towards production of a relevant model protein [56]. The authors showed that proliferation control of a BHK-derived cell line in a perfused cell-culture process was effectively regulated for over 7 weeks. Moreover, heterodimeric IgG antibody productivity increased up to sixfold during growth arrest triggered by inducible expression of IRF-1. The drawback of efficient and rapid cell growth inhibition by IRF-1 over-expression in BHK-derived cells is related to a decrease in the cell viability over time. In an attempt to keep cell viability at satisfactory levels, it was shown that manipulation of cycles by addition and removal of estradiol modulates IRF-1 activity, thus overcoming the loss in cell viability of IRF-1 growth-arrested BHK cells [27].

Other studies were performed by over-expressing proteins that cause interventions in the cell cycle control. Such a "one gene metabolic engineering" approach was first performed and described by Bailey et al. [51], over-expressing inhibitors of cell cycle progression transiently transfected into CHO-derived cells. The authors achieved an enhanced productivity in CHO-derived cells upon transient expression of *p21*, *p27* or the *p53*-derived mutant *p53-175P*, (mutant showing specific loss of programmed cell death function, [146]). Importantly, the expression of these genes was coordinated with expression of a model protein, secreted alkaline phosphatase (SEAP), and expression of both genes was driven by a Tet-repressible promoter. Upon triggering expression of each of these inhibitors of cell cycle progression, growth arrest was achieved with concomitant increase in recombinant protein production. Moreover, the growth capacity of the cell lines was rescued when the expression of cell cycle inhibitor proteins was repressed. The concept was further extended by modulating the "one gene metabolic engineering" by constitutive expression of cytostatic genes [122]. The authors found that conditional expression of these genes in stable clones does not entirely recapitulate what was observed in transient expression settings [51]. Stable integration of *p21* did not lead to generation of clones with growth arrest capacity and *p53-175P* expression was related to induction of apoptosis in growth-arrested cells. Nevertheless, the authors succeeded in generating CHO-derived clones expressing *p27* with cell growth arrest properties in a regulatable manner and with a concomitant 10–15-fold increase in relevant protein production per cell.

Cell cycle control is an intricate network of cellular proteins that act in an orchestrated manner in regulatory circuits. This includes some intervening molecules that can be overcome by the function of other proteins. The redundancy of such pathways in controlling cell cycle progression was believed to account for the impairment in establishing stable integrated mutants with cell arrest properties on inducible over-expression of certain cytostatic proteins such as *p21* [122]. In an attempt to circumvent this limitation, Fussenegger et al. described a "one-step multigene metabolic engineering technology" in CHO cells [53] as a means to regulate the multifactorial process of cell cycle progression. In order to increase the levels of active *p21* in the cells, the authors developed a strategy in which *p21* was coordinatedly expressed with the transcription factor C/EBPα (a protein that increases *p21* production and protein half-life). Taking advantage of Tet-controlled tricistronic expression vectors [52], the regulated and coordinated expression of *p21*, *C/EPBα* and the model protein with concomitant cell cycle arrest in stable cell clones were achieved [53]. The authors showed that regulation of cell cycle progression could be achieved in a multifactorial manner, confirming that an individual cell is more productive in a cell-cycle-arrested state.

The concept of cell growth arrest by regulatable over-expression of cytostatic proteins was further extended towards increase of recombinant protein production. It proved to be applicable not only to other cell lines, and different regulation systems but to expression of proteins with high potential application value [75, 190]. The authors established a low-density batch culture of NS0-derived cell line expressing *p21* in an inducible manner. One of the major achievements in this work was the use of a regulatable promoter whose expression is triggered by the addition of an inducer (On-system). The authors describe the G1-phase arrest in response to high levels of *p21* expression and concomitant increase of IgG4 antibody by more than fourfold in the arrested state.

In another study, the approach described above was extended to a high-cell-density continuous perfusion culture [75]. A multigenic manipulation of the cell cycle was evaluated by expression of the *p21* gene and the anti-apoptotic protein *Bcl-2*. Although the authors could confirm the role of *p21* in the arrest of cell cycle progression and increased productivity, *Bcl-2* expression had no effect on cell viability of the arrested cells. The coordinated expression of anti-apoptotic proteins (such as *Bcl-2* and *BclX*$_1$), cytostatic genes (*p27*, *p21*) and a GOI was earlier described in CHO-derived cell lines [53]. The authors found that only with coordinated expression of *BclX*$_1$ and *p27* cell cycle arrest was possible in G1 phase. Stable cell clones with G1 growth arrest proprieties were generated and no signs of apoptosis during growth-arrested periods were observed (for a further detailed description of cellular engineering strategies that aim at the prolongation of cell viability by over-expression of anti-apoptotic proteins, the reader is referred to the Sect. 5.2).

The positive relationship between cell cycle arrest in the G1 phase and the increase in recombinant protein productivity was found in different cell lines [51, 53, 56, 75, 122, 190]. Importantly, the physiological basis of this effect was revealed by different groups [15, 28, 104]. Bi et al. were able to show that in CHO-derived cells expressing inducible *p21*, the increase in productivity of the model protein (GOI) is related to an increased cell size, mitochondrial mass and activity, and ribosome biogenesis [15]. Furthermore, they showed that these alterations were uncoupled from the cell cycle, i.e. these physiological changes occur in the cell despite the abolition of cell cycle progression and division. Characterization of cellular metabolism of a CHO cell line over-expressing the cytostatic *p27* protein showed the same line of evidence [27]. The cell cycle arrest in G1 phase was found to be related to an increase in cell protein content and concomitantly with a higher expenditure of cell energy.

5.1.2 Conditional Cell Expansion

Many immortal cell lines have been either isolated from tumor tissue or have resulted from spontaneous immortalization. In addition, cell lines have been

Table 3 Conditional and reversible immortalization approaches

Immortalized cell lines	Types of immortalizations	Types of switches	References
Murine			
Embryonic fibroblasts	Conditional	Tet-system	May et al. [116]
Ear fibroblasts (adult)	Conditional	Tet-system	May et al. [118]
Lung microvascular endothelial cells	Conditional	Tet-system	May et al. [121]
Myogenic clonal cell lines	Conditional	Temperature	Morgan et al. [128]
Hepatocytes	Conditional	Temperature	Yanai et al. [205]
Tissue specific microvascular endothelial cells	Conditional	Temperature	Langley et al. [98]
Astrocytes	Conditional	Temperature	Langley et al. [99]
Cardiomyocytes	Reversible	Cre-recombinase	Rybkin et al. [147]
Hematopoietic cells	Reversible	Estrogen	Ito et al. [76]; Wang et al. [188]
Human			
Umbilical vein endothelial cells (HUVEC)	Conditional	Tet-system	May et al. [121]
Liver endothelial cells	Reversible	Cre-recombinase	Salmon et al. [148]

successfully established by introduction of immortalizing genes into primary cells. However, in many cases, such constitutively immortalized cells have limited benefit. They often do not reflect the properties of the primary cells they have been derived from. One issue concerns the expression of the immortalization gene per se, which has a significant impact on the cellular phenotype. In recent years, strategies have been followed aiming at a systematic construction of new cell lines from primary cells. The emphasis of these approaches is to restrict expression of the immortalizing gene(s) to the period of cell expansion, in other words controlled expansion of primary cells. This is achieved by reverting or switching off the immortalizing genes (see Table 3 for an overview of recent approaches). With such approaches, current limitations of immortalized cells which are related to prolonged constitutive expression of the immortalizing gene(s) causing changes in cellular properties might be overcome. The resulting cells may be used for analytic issues, the production of recombinant proteins, and gene and cell therapies.

Conditional cell expansion is achieved by conditional expression of specific genes. One example concerns the gene encoding the simian virus 40 large T antigen (TAg). TAg is a powerful oncogene able to inactivate cellular key tumor suppressors like p53 and retinoblastoma proteins, and is therefore frequently used in generation of immortalized cell lines. There is evidence that the transforming potential of TAg is also attributed to an anti-apoptotic activity which is not related to the activation of p53 [4]. In a pioneering study aimed at generating conditionally immortalized cell lines, Jat et al. created a transgenic mouse called ImmortoMouse for which they developed a thermo-labile mutant of the SV40 TAg [79].

The thermo-labile mutant of SV40 TAg is inactive at the physiological body temperature and permissive at 33 °C. In the ImmortoMouse the thermo-labile mutant of SV40 TAg is under the control of the mouse major histocompatibility complex H-2K^b promoter which is expressed in a broad range of tissues and is induced by interferon. On isolation of cells from the ImmortoMouse, several conditionally immortalized cell lines have been established, including myogenic cell lines [128], hepatocyte cell lines [205], tissue-specific micro-vascular endothelial cells [98] and, more recently, an astrocyte cell line [99]. In these cell lines, permissive conditions (33 °C in the presence of medium containing interferon-γ) turn on the expression and activity of SV40 TAg leading to exponential proliferation of the cells. Importantly, in this proliferative state the established cell lines retained most of their tissue-specific morphological and biochemical properties. When the cells were transferred to non-permissive temperatures (37 °C or higher) they either ceased proliferation or started to differentiate.

While the cell lines developed from the ImmortoMouse gave first proof-of-concept for de-novo establishment of growth-controlled cell lines, the approach is limited to the temperature-labile mutant of SV40 TAg. Obviously, such an approach is not generic or easily translatable to other immortalizing genes. Thus, another approach was employed that makes use of specific excision of the immortalizing gene(s) mediated by DNA recombinases such as Cre or Flp, resulting in reversible expression of the immortalizing gene. Salmon et al. showed that human liver endothelial cells were immortalized through lentiviral-mediated gene transfer of SV40 large T and telomerase. Cre-mediated excision of the immortalizing genes resulted in complete growth arrest within 2 days [148].

In a similar study, Rybkin et al. used TAg to create cardiomyocyte cell lines capable of proliferating and reversibly withdrawing from the cell cycle [147]. Cre recombinase was used to switch the cells from a proliferative to a quiescent state. Although the reversible Cre/loxP approach enables conditional immortalization, this method is challenging since high transduction efficiency of the recombinase is hard to achieve. In addition to this, the selection schemes must be very strict to ensure a homogenously reverted cell line in which the non-transduced cells are not overgrown [121].

However, recent developments in control of gene expression (see Sect. 2) have allowed the principle of conditional expression to be employed in more generic approaches. This was shown both for transcriptionally and post-transcriptionally controlled approaches.

The estrogen receptor was used to construct a selective amplifier gene for controllable expansion of genetically modified hematopoietic cells. Specifically, the fusion of a steroid receptor hormone-binding domain to the growth factor G-CSF exerts a reversible activation in a steroid-dependent manner [76]. Therefore, the growth signal is active only upon estrogen treatment of the transduced murine hematopoietic stem cells, and most of the cells expand even if some of them enter the differentiation pathway. This strategy is applicable to the in vivo expansion of genetically modified hematopoietic stem cells [97]. In another study, the principle of estrogen receptor-mediated control was successfully applied to hematopoietic progenitor cells using tamoxifen-controlled expression of HoxB8 [188], giving rise to expandable stem cells that are still susceptible to differentiation.

More recent studies have shown that cell proliferation control can be achieved by an autoregulated, Dox-dependent transcriptional immortalization strategy [116]. Autoregulation imposes a bimodal decision on the cells. Thus, only fully induced cells receive the proliferation signal while non-induced cells remain un-induced and are overgrown by the proliferating cells. The installation of the positive feedback loop requires a low basal expression of the transactivator and is initiated through administration of Dox. In the presence of Dox, the system is turned on and the immortalizing gene is expressed, leading to the exponential proliferation of cells [158]. In the absence of Dox, the induced effects are reverted and the cells stop proliferating.

Through lentiviral transduction of the autoregulated expression cassette, a broad range of cell types can be transduced. Lentiviral transduction is much more efficient than plasmid transfer, thereby dramatically increasing the immortalization efficiency up to 1,000-fold [121].

A series of such lentiviral vectors were constructed that can be distinguished by the expression of the proliferator gene such as *SV40 TAg*, *PymT*, *hTert* and *c-Myc*. This allows the simultaneous transduction of cells with different proliferator genes. Indeed, the combination of such genes [86, 206] led to the expansion of cell types that were not previously accessible (Table 1). Examples include lung micro-vascular-derived and HUVEC-derived endothelial cells from mouse and human origin [121]. Importantly, these cells expressed cell-type-specific markers and, in the case of endothelial cells, showed de-novo vessel formation in vitro and in vivo. Controlled proliferation was achieved due to strict co-regulation of these cells which harbor several proliferator genes. Importantly, the conditionally immortalized cells did not induce oncogenic transformation [118].

Proliferation-controlled fibroblast cells were used to express recombinant erythropoietin (EPO) without any signs of degradation, while expression was higher in the growth-arrested state [117].

5.2 Apoptosis

Apoptosis is a form of regulated or programmed cell death controlled by signaling pathways in mitochondria, endoplasmic reticulum (ER), or cellular surface death receptor(s). A series of cysteine proteases (caspases) execute the cell death program [5, 109]. Apoptosis is an inherited property of cells derived from their original function in the multi-cellular organism which is mostly non-beneficial in technological cell culture. Many mammalian cell lines are susceptible to apoptosis under conditions of typical bioreactor growth cultivation [139, 164], thereby decreasing the overall productivity achievable. Inhibiting or controlling apoptosis is thus a major target of cell engineering, e.g. by blocking pro-apoptotic mechanisms or by over-expressing apoptosis-counteracting proteins.

Apoptosis can be initiated by two general mechanisms, which are characterized by extrinsic or intrinsic pathways. Extrinsically induced apoptosis is initiated through the binding of ligands (e.g. FasL) to receptors of the tumor necrosis factor (TNF) family. FasL binding induces the formation of the death-inducing signaling complex (DISC) which contains the Fas-associated death domain (Fadd), caspase-8 and caspase-10. The intrinsic mechanism of apoptosis initiation is the answer to cellular stress. DNA damage, for instance, causes the activation of *p53* which induces the expression of pro-apoptotic members of the *Bcl-2* family (*Bax*, *Bak*, *Bok*, *Bik*, *Bad*, and *Bid*). These factors are responsible for the release of cytochrome c from the mitochondria which binds to the protease activation factor 1 (Apaf-1). Apaf-1 in turn is now able to activate procaspase-9.

Among the intrinsic factors leading to apoptosis, the ER stress-induced death pathway is certainly of great significance in high-expression bioprocesses. It is caused by an unfolded protein response (UPR), hypoxia, changes in intracellular calcium levels or lack of glucose. Enhanced cell survival in bioreactors has been achieved by the supplementation of chemicals (e.g. the polysulfated naphthylurea Suramin) and certain recombinant proteins. Examples are the expression of the insulin-like growth factor receptor along with its ligand IGF-I and transferrin added to the medium [171], and elevated concentrations of amino acids which have shown to protect cells from environmental stress [35].

Genetic strategies for preventing apoptosis in cell culture can be divided in the over-expression of anti-apoptotic factors and the knockdown of pro-apoptotic factors. The over-expression of anti-apoptotic genes like *bcl-2* and $bcl\text{-}x_L$ in NS0, CHO, BHK or hybridoma cells was shown to improve cell viability when exposed to cellular stress such as starvation or toxins (see [5] for review). Caspases-9, -3 and -7 are targets of the X-linked inhibitor of apoptosis (XIAP) which can be used for over-expression [151]. Cho et al. demonstrated that the over-expression of the Ca^{2+}-dependent enzyme transglutaminase 2 (TG2) inhibits apoptosis through suppression of caspase-3 and -9 activities [29].

Since apoptosis is a complex biological program with redundancy, a combination of several factors involved in different stages of apoptosis is more effective than the over-expression of a single one. This was demonstrated by the co-expression of *Bcl-XL* and *XIAPΔ* in CHO and myeloma cell lines [150–152]. Dorai et al. have systematically tested the combination of a set of anti-apoptotic genes in a CHO cell line. The best results in terms of viability and productivity could be achieved by over-expression of either single anti-apoptotic genes (*Bcl-2Δ* or *Bcl-XL*), or a combination of two or three anti-apoptotic genes (*E1B-19K*, *Aven*, and *XIAPΔ*) [36].

The knockdown of pro-apoptotic factors represents another genetic engineering strategy to circumvent apoptosis. Silencing *Bax* and *Bak* in CHO cells using shRNA vectors was reported to increase cell viability and improve the production of recombinant interferon-γ in producer CHO cell clones [102]. Silencing caspases-3 and -7 expression, but not caspase-3 alone, can improve cell viability and recombinant thrombopoietin production in CHO cells following treatment with sodium butyrate [168, 169]. Silencing the apoptosis-linked gene 2 (*Alg-2*) and

the Zn-finger protein transcriptional factor *Requiem* in CHO cells was also reported to improve cell viability and recombinant protein production [201]. The disadvantage of these knockdown approaches is that usually less than 100% of the target
mRNA is silenced, leading to a remaining pro-apoptotic function. Using Zn-finger nucleases, Cost et al. established a double knockout of both genes *Bak* and *Bax* in a CHO cell line and could improve the production of IgG up to fivefold under starvation conditions and higher cell densities under normal conditions compared to wild-type cells.

5.3 Secretion Engineering

The majority of therapeutic proteins which are produced by mammalian cells in bioreactors are secreted into the medium. Consequently, efforts have been made to improve cellular productivity by the introduction of biological modules which enhance protein transport and post-translational modifications. All secreted proteins are co-translationally targeted to the endoplasmic reticulum (ER) and then translocated across the membrane of the ER to the Golgi apparatus. The proteins are further processed within the trans-Golgi network before packed into secretory vesicles.

The transport and modification of heterologous proteins through the secretory machinery can be seen as a bottle-neck of protein production. In the ER, cellular quality control of proteins is performed and only "perfect" products pass this barrier [39]. Misfolded, unfolded or unassembled proteins will be degraded and mechanisms are activated to lower the biosynthetic burden of the ER and to protect it. As mentioned above, this unfolded protein response can induce apoptosis; consequently, strategies have been developed to engineer cell lines by altering processes in the ER [156].

Florin et al. generated stable CHO cell lines expressing heterologous ceramide transfer protein (CERT) [44]. CERT mediates ATP-dependent ceramide transport from the ER to the Golgi complex [67]. These cells showed significantly higher specific productivities of the protein HSA and enhanced monoclonal antibody secretion. The expression of the transcription factor X-box binding protein-1 (XBP-1) was shown to increase the ER content of a therapeutic antibody. This, in turn, led to a 40% higher productivity of a CHO cell line [12]. XBP-1 regulates this process by binding to the ER stress-responsive elements within the promoters of a wide spectrum of secretory pathway genes, resulting in enhanced total protein synthesis [160].

Chaperones have been linked to many ER functions such as protein translocation, folding, and oligomerization [2, 155]. The lack of co-chaperones seems to be rate-limiting. However, engineering chaperone systems by over-expression of a single component of the ER secretion machinery has yielded mixed results regarding productivity [80, 126]. It was possible to increase antibody productivity in a CHO line by over-expression of PDI, but this failed to increase thrombopoietin secretion. PDI is an ER enzyme that catalyses the formation and breakage of disulfide bonds between thiol groups of cysteine residues using the substrate glutathione. It operates as a chaperone to inhibit the aggregation of misfolded proteins. Another group was able to demonstrate that the over-expression of BiP, a member of the hsp70 family, decreased the secretion of a recombinant antibody in CHO cells [23].

In contrast, over-expression of the calreticulin and calnexin chaperones was found to nearly double the specific productivity of thrombopoietin in recombinant CHO cultures [30]. The example of PDI over-expression demonstrates that the engineering of complex systems such as the chaperone system may need more sophisticated engineering to improve secretion rates.

References

1. Agha-Mohammadi S, O'Malley M, Etemad A, Wang Z, Xiao X, Lotze MT (2004) Second-generation tetracycline-regulatable promoter: repositioned tet operator elements optimize transactivator synergy while shorter minimal promoter offers tight basal leakiness. J Gene Med 6:817–828
2. Ailor E, Betenbaugh MJ (1998) Overexpression of a cytosolic chaperone to improve solubility and secretion of a recombinant IgG protein in insect cells. Biotechnol Bioeng 58:196–203
3. Akagi K, Kanai M, Saya H, Kozu T, Berns A (2001) A novel tetracycline-dependent transactivator with E2F4 transcriptional activation domain. Nucleic Acids Res 29:E23
4. Ali SH, DeCaprio JA (2001) Cellular transformation by SV40 large T antigen: interaction with host proteins. Semin Cancer Biol 11:15–23
5. Arden N, Betenbaugh MJ (2004) Life and death in mammalian cell culture: strategies for apoptosis inhibition. Trends Biotechnol 22:174–180
6. Aubel D, Fussenegger M (2010) Mammalian synthetic biology—from tools to therapies. BioEssays 32:332–345
7. Bach M, Grigat S, Pawlik B, Fork C, Utermöhlen O, Pal S, Banczyk D, Lazar A, Schömig E, Gründemann D (2007) Fast set-up of doxycycline-inducible protein expression in human cell lines with a single plasmid based on Epstein-Barr virus replication and the simple tetracycline repressor. FEBS J 274:783–790
8. Banaszynski LA, Sellmyer MA, Contag CH, Wandless TJ, Thorne SH (2008) Chemical control of protein stability and function in living mice. Nat Med 14:1123–1127
9. Baron U, Freundlieb S, Gossen M, Bujard H (1995) Co-regulation of two gene activities by tetracycline via a bidirectional promoter. Nucleic Acids Res 23:3605–3606
10. Baron U, Gossen M, Bujard H (1997) Tetracycline-controlled transcription in eukaryotes: novel transactivators with graded transactivation potential. Nucleic Acids Res 25:2723–2729
11. Barquinero J, Eixarch H, Pérez-Melgosa M (2004) Retroviral vectors: new applications for an old tool. Gene Ther 11:S3–S9
12. Becker E, Florin L, Pfizenmaier K, Kaufmann H (2008) An XBP-1 dependent bottle-neck in production of IgG subtype antibodies in chemically defined serum-free Chinese hamster ovary (CHO) fed-batch processes. J Biotechnol 135:217–223
13. Becskei A, Serrano L (2000) Engineering stability in gene networks by autoregulation. Nature 405:590–593

14. Berens C, Hillen W (2003) Gene regulation by tetracyclines: constraints of resistance regulation in bacteria shape TetR for application in eukaryotes. Eur J Biochem 270:3109–3121
15. Bi JX, Shuttleworth J, Al-Rubeai M (2004) Uncoupling of cell growth and proliferation results in enhancement of productivity in p21CIP1-arrested CHO cells. Biotechnol Bioeng 85:741–749
16. Binley K, Iqball S, Kingsman A, Kingsman S, Naylor S (1999) An adenoviral vector regulated by hypoxia for the treatment of ischaemic disease and cancer. Gene Ther 6:1721–1727
17. Blesch A (2004) Lentiviral and MLV based retroviral vectors for ex vivo and in vivo gene transfer. Methods 33:164–172
18. Bode J, Benham C, Knopp A, Mielke C (2000) Transcriptional augmentation: modulation of gene expression by scaffold/matrix-attached regions (S/MAR elements). Crit Rev Eukaryot Gene Expr 10:73–90
19. Bollati-Fogolin M, Forno G, Nimtz M, Conradt HS, Etcheverrigaray M, Kratje R (2005) Temperature reduction in cultures of hGM-CSF-expressing CHO cells: effect on productivity and product quality. Biotechnol Prog 21:17–21
20. Boorsma M, Nieba L, Koller D, Bachmann MF, Bailey JE, Renner WA (2000) A temperature-regulated replicon-based DNA expression system. Nat Biotechnol 18:429–432
21. Boorsma M, Hoenke S, Marrero A, Fischer R, Bailey JE, Renner WA, Bachmann MF (2002) Bioprocess applications of a Sindbis virus-based temperature-inducible expression system. Biotechnol Bioeng 79:602–609
22. Bornkamm GW, Berens C, Kuklik-Roos C, Bechet JM, Laux G, Bachl J, Korndoerfer M, Schlee M, Hölzel M, Malamoussi A, Chapman RD, Nimmerjahn F, Mautner J, Hillen W, Bujard H, Feuillard J (2005) Stringent doxycycline-dependent control of gene activities using an episomal one-vector system. Nucleic Acids Res 33:1–11
23. Borth N, Mattanovich D, Kunert R, Katinger H (2005) Effect of increased expression of protein disulfide isomerase and heavy chain binding protein on antibody secretion in a recombinant CHO cell line. Biotechnol Prog 21:106–111
24. Braselmann S, Graninger P, Busslinger M (1993) A selective transcriptional induction system for mammalian cells based on Gal4-estrogen receptor fusion proteins. Proc Natl Acad Sci U S A 90:1657–1661
25. Büning H, Nicklin SA, Perabo L, Hallek M, Baker AH (2003) AAV-based gene transfer. Curr Opin Mol Ther 5:367–375
26. Burcin MM, O'Malley BW, Tsai SY (1998) A regulatory system for target gene expression. Front Biosci J Virtual Library 3:c1–c7
27. Carvalhal AV, Moreira JL, Carrondo MJ (2001) Strategies to modulate BHK cell proliferation by the regulation of IRF-1 expression. J Biotechnol 92:47–59
28. Carvalhal AV, Marcelino I, Carrondo MJ (2003) Metabolic changes during cell growth inhibition by p27 overexpression. Appl Microbiol Biotechnol 63:164–173
29. Cho SY, Lee JH, Bae HD, Jeong EM, Jang GY, Kim CW, Shin DM, Jeon JH, Kim IG (2010) Transglutaminase 2 inhibits apoptosis induced by calcium overload through down-regulation of Bax. Exp Mol Med 42:639–650
30. Chung JY, Lim SW, Hong YJ, Hwang SO, Lee GM (2004) Effect of doxycycline-regulated calnexin and calreticulin expression on specific thrombopoietin productivity of recombinant Chinese hamster ovary cells. Biotechnol Bioeng 85:539–546
31. Danke C, Grunz X, Wittmann J, Schmidt A, Agha-Mohammadi S, Kutsch O, Jäck HM, Hillen W, Berens C (2010) Adjusting transgene expression levels in lymphocytes with a set of inducible promoters. J Gene Med 12:501–515
32. Davey RE, Onishi K, Mahdavi A, Zandstra PW (2007) LIF-mediated control of embryonic stem cell self-renewal emerges due to an autoregulatory loop. FASEB J 21:2020–2032
33. Deans TL, Cantor CR, Collins JJ (2007) A tunable genetic switch based on RNAi and repressor proteins for regulating gene expression in mammalian cells. Cell 130:363–372
34. Deuschle U, Meyer WKH, Thiesen HJ (1995) Tetracycline-reversible silencing of eukaryotic promoters. Mol Cell Biol 15:1907–1914

35. deZengotita VM, Schmelzer AE, Miller WM (2002) Characterization of hybridoma cell responses to elevated pCO(2) and osmolality: intracellular pH, cell size, apoptosis, and metabolism. Biotechnol Bioeng 77:369–380
36. Dorai H, Ellis D, Keung YS, Campbell M, Zhuang M, Lin C, Betenbaugh MJ (2010) Combining high-throughput screening of caspase activity with anti-apoptosis genes for development of robust CHO production cell lines. Biotechnol Prog 26:1367–1381
37. Ehrhardt A, Haase R, Schepers A, Deutsch MJ, Lipps HJ, Baiker A (2008) Episomal vectors for gene therapy. Curr Gene Ther 8:147–161
38. Eilers M, Picard D, Yamamoto KR, Bishop JM (1989) Chimaeras of myc oncoprotein and steroid receptors cause hormone-dependent transformation of cells. Nature 340:66–68
39. Ellgaard L, Helenius A (2003) Quality control in the endoplasmic reticulum. Nat Rev Mol Cell Biol 4:181–191
40. Elowitz MB, Leibler S (2000) A synthetic oscillatory network of transcriptional regulators. Nature 403:335–338
41. Ertel A, Tozeren A (2008) Human and mouse switch-like genes share common transcriptional regulatory mechanisms for bimodality. BMC Genomics 9:628
42. Feil R, Wagner J, Metzger D, Chambon P (1997) Regulation of Cre recombinase activity by mutated estrogen receptor ligand-binding domains. Biochem Biophys Res Commun 237: 752–757
43. Fiore M, Zanier R, Degrassi F (2002) Reversible G(1) arrest by dimethyl sulfoxide as a new method to synchronize Chinese hamster cells. Mutagenesis 17:419–424
44. Florin L, Pegel A, Becker E, Hausser A, Olayioye MA, Kaufmann H (2009) Heterologous expression of the lipid transfer protein CERT increases therapeutic protein productivity of mammalian cells. J Biotechnol 141:84–90
45. Forster K, Helbl V, Lederer T, Urlinger S, Wittenburg N, Hillen W (1999) Tetracycline-inducible expression systems with reduced basal activity in mammalian cells. Nucleic Acids Res 27:708–710
46. Fox SR, Patel UA, Yap MG, Wang DI (2004) Maximizing interferon-gamma production by Chinese hamster ovary cells through temperature shift optimization: experimental and modeling. Biotechnol Bioeng 85:177–184
47. Fox SR, Tan HK, Tan MC, Wong SC, Yap MG, Wang DI (2005) A detailed understanding of the enhanced hypothermic productivity of interferon-gamma by Chinese-hamster ovary cells. Biotechnol Appl Biochem 41:255–264
48. Fraser MJ, Ciszczon T, Elick T, Bauser C (1996) Precise excision of TTAA-specific lepidopteran transposons piggyBac (IFP2) and tagalong (TFP3) from the baculovirus genome in cell lines from two species of *Lepidoptera*. Insect Mol Biol 5:141–151
49. Freundlieb S, Schirra-Müller C, Bujard H (1999) A tetracycline controlled activation/repression system with increased potential for gene transfer into mammalian cells. J Gene Med 1:4–12
50. Fussenegger M (2001) The impact of mammalian gene regulation concepts on functional genomic research, metabolic engineering, and advanced gene therapies. Biotechnol Prog 17:1–51
51. Fussenegger M, Mazur X, Bailey JE (1997) A novel cytostatic process enhances the productivity of Chinese hamster ovary cells. Biotechnol Bioeng 55:927–939
52. Fussenegger M, Mazur X, Bailey JE (1998a) pTRIDENT, a novel vector family for tricistronic gene expression in mammalian cells. Biotechnol Bioeng 57:1–10
53. Fussenegger M, Schlatter S, Datwyler D, Mazur X, Bailey JE (1998b) Controlled proliferation by multigene metabolic engineering enhances the productivity of Chinese hamster ovary cells. Nat Biotechnol 16:468–472
54. Gama-Norton L, Riemer P, Sandhu U, Nehlsen K, Schucht R, Hauser H, Wirth D (2009) Defeating randomness—targeted integration as a boost for biotechnology. In: Al-Rubeai M (ed) Cell engineering. Springer, London
55. Gama-Norton L, Herrmann S, Schucht R, Coroadinha AS, Low R, Alves PM, Bartholomae CC, Schmidt M, Baum C, Schambach A, Hauser H, Wirth D (2010) Retroviral vector performance in defined chromosomal loci of modular packaging cell lines. Hum Gene Ther 21:979–991

56. Geserick C, Bonarius HP, Kongerslev L, Hauser H, Mueller PP (2000) Enhanced productivity during controlled proliferation of BHK cells in continuously perfused bioreactors. Biotechnol Bioeng 69:266–274
57. Ghani K, Cottin S, Kamen A, Caruso M (2007) Generation of a high-titer packaging cell line for the production of retroviral vectors in suspension and serum-free media. Gene Ther 14:1705–1711
58. Gibson DG, Glass JI, Lartigue C, Noskov VN, Chuang RY, Algire MA, Benders GA, Montague MG, Ma L, Moodie MM, Merryman C, Vashee S, Krishnakumar R, Assad-Garcia N, Andrews-Pfannkoch C, Denisova EA, Young L, Qi ZQ, Segall-Shapiro TH, Calvey CH, Parmar PP, Hutchison CA III, Smith HO, Venter JC (2010) Creation of a bacterial cell controlled by a chemically synthesized genome. Science 329:52–56
59. Gill G, Ptashne M (1988) Negative effect of the transcriptional activator GAL4. Nature 334:721–724
60. Gitzinger M, Kemmer C, El-Baba MD, Weber W, Fussenegger M (2009) Controlling transgene expression in subcutaneous implants using a skin lotion containing the apple metabolite phloretin. Proc Natl Acad Sci U S A 106:10638–10643
61. Glaß M, Busche A, Wagner K, Messerle M, Borst EM (2009) Conditional and reversible disruption of essential herpesvirus proteins. Nat Methods 6:577–579
62. Go WY, Ho SN (2002) Optimization and direct comparison of the dimerizer and reverse tet transcriptional control systems. J Gene Med 4:258–270
63. Gossen M, Bujard H (1992) Tight control of gene expression in mammalian cells by tetracycline- responsive promoters. Proc Natl Acad Sci U S A 89:5547–5551
64. Gossen M, Freundlieb S, Bender G, Muller G, Hillen W, Bujard H (1995) Transcriptional activation by tetracyclines in mammalian cells. Science 268:1766–1769
65. Greber D, Fussenegger M (2007) Mammalian synthetic biology: engineering of sophisticated gene networks. J Biotechnol 130:329–345
66. Hampf M, Gossen M (2007) Promoter crosstalk effects on gene expression. J Mol Biol 365:911–920
67. Hanada K, Kumagai K, Yasuda S, Miura Y, Kawano M, Fukasawa M, Nishijima M (2003) Molecular machinery for non-vesicular trafficking of ceramide. Nature 426:803–809
68. Hartenbach S, Daoud-El Baba M, Weber W, Fussenegger M (2007) An engineered L-arginine sensor of Chlamydia pneumoniae enables arginine-adjustable transcription control in mammalian cells and mice. Nucleic Acids Res 35:e136
69. Hayter PM, Curling EM, Baines AJ, Jenkins N, Salmon I, Strange PG, Bull AT (1991) Chinese hamster ovary cell growth and interferon production kinetics in stirred batch culture. Appl Microbiol Biotechnol 34:559–564
70. Heinz N, Schambach A, Galla M, Maetzig T, Baum C, Loew R, Schiedlmeier B (2010) Retroviral and transposon-based tet-regulated all-in-one vectors with reduced background expression and improved dynamic range. Hum Gene Ther 22:166–176
71. Hendrick V, Winnepenninckx P, Abdelkafi C, Vandeputte O, Cherlet M, Marique T, Renemann G, Loa A, Kretzmer G, Werenne J (2001) Increased productivity of recombinant tissular plasminogen activator (t-PA) by butyrate and shift of temperature: a cell cycle phases analysis. Cytotechnology 36:71–83
72. Hofmann A, Nolan GP, Blau HM (1996) Rapid retroviral delivery of tetracycline-inducible genes in a single autoregulatory cassette. Proc Natl Acad Sci U S A 93:5185–5190
73. Hu MC, Davidson N (1987) The inducible lac operator-repressor system is functional in mammalian cells. Cell 48:555–566
74. Hunter NL, Awatramani RB, Farley FW, Dymecki SM (2005) Ligand-activated Flpe for temporally regulated gene modifications. Genesis 41:99–109
75. Ibarra N, Watanabe S, Bi JX, Shuttleworth J, Al-Rubeai M (2003) Modulation of cell cycle for enhancement of antibody productivity in perfusion culture of NS0 cells. Biotechnol Prog 19:224–228

76. Ito K, Ueda Y, Kokubun M, Urabe M, Inaba T, Mano H, Hamada H, Kitamura T, Mizoguchi H, Sakata T, Hasegawa M, Ozawa K (1997) Development of a novel selective amplifier gene for controllable expansion of transduced hematopoietic cells. Blood 90:3884–3892
77. Ivics Z, Hackett PB, Plasterk RH, Izsvák Z (1997) Molecular reconstruction of sleeping beauty, a Tc1-like transposon from fish, and its transposition in human cells. Cell 91:501–510
78. Izsvák Z, Chuah MKL, VandenDriessche T, Ivics Z (2009) Efficient stable gene transfer into human cells by the Sleeping Beauty transposon vectors. Methods 49:287–297
79. Jat PS, Noble MD, Ataliotis P, Tanaka Y, Yannoutsos N, Larsen L, Kioussis D (1991) Direct derivation of conditionally immortal cell lines from an H-2Kb-tsA58 transgenic mouse. Proc Natl Acad Sci U S A 88:5096–5100
80. Jenkins N, Meleady P, Tyther R, Murphy L (2009) Strategies for analysing and improving the expression and quality of recombinant proteins made in mammalian cells. Biotechnol Appl Biochem 53:73–83
81. Kaufmann H, Mazur X, Fussenegger M, Bailey JE (1999) Influence of low temperature on productivity, proteome and protein phosphorylation of CHO cells. Biotechnol Bioeng 63:573–582
82. Kellendonk C, Tronche F, Monaghan AP, Angrand PO, Stewart F, Schutz G (1996) Regulation of Cre recombinase activity by the synthetic steroid RU 486. Nucleic Acids Res 24:1404–1411
83. Kemmer C, Gitzinger M, Daoud-El Baba M, Djonov V, Stelling J, Fussenegger M (2010) Self-sufficient control of urate homeostasis in mice by a synthetic circuit. Nat Biotechnol 28:355–360
84. Kiel C, Yus E, Serrano L (2010) Engineering signal transduction pathways. Cell 140:33–47
85. Kirchhoff S, Schaper F, Hauser H (1993) Interferon regulatory factor 1 (IRF-1) mediates cell growth inhibition by transactivation of downstream target genes. Nucleic Acids Res 21:2881–2889
86. Kiyono T, Foster SA, Koop JI, McDougall JK, Galloway DA, Klingelhutz AJ (1998) Both Rb/p16INK4a inactivation and telomerase activity are required to immortalize human epithelial cells. Nature 396:84–88
87. Klug A. (2010) The discovery of zinc fingers and their applications in gene regulation and genome manipulation. Annu Rev Biochem 79:213–231
88. Koga A, Suzuki M, Inagaki H, Bessho Y, Hori H (1996) Transposable element in fish [11]. Nature 383:30
89. Köster M, Kirchhoff S, Schaper F, Hauser H (1995) Proliferation control of mammalian cells by the tumor suppressor IRF-1. Cytotechnology18:67–75
90. Kramer BP, Fussenegger M (2005) Hysteresis in a synthetic mammalian gene network. Proc Natl Acad Sci U S A 102:9517–9522
91. Kramer BP, Weber W, Fussenegger M (2003) Artificial regulatory networks and cascades for discrete multilevel transgene control in mammalian cells. Biotechnol Bioeng 83:810–820
92. Kramer BP, Viretta AU, Baba MDE, Aubel D, Weber W, Fussenegger M (2004) An engineered epigenetic transgene switch in mammalian cells. Nat Biotechnol 22:867–870
93. Kramer BP, Fischer M, Fussenegger M (2005) Semi-synthetic mammalian gene regulatory networks. Metab Eng 7:241–250
94. Kraunus J, Schaumann DH, Meyer J, Modlich U, Fehse B, Brandenburg G, von Laer D, Klump H, Schambach A, Bohne J, Baum C (2004) Self-inactivating retroviral vectors with improved RNA processing. Gene Ther 11:1568–1578
95. Kroger A, Koster M, Schroeder K, Hauser H, Mueller PP (2002) Activities of IRF-1. J Interferon Cytokine Res 22:5–14
96. Kühnel F, Fritsch C, Krause S, Mundt B, Wirth T, Paul Y, Malek NP, Zender L, Manns MP, Kubicka S (2004) Doxycycline regulation in a single retroviral vector by an autoregulatory loop facilitates controlled gene expression in liver cells. Nucleic Acids Res 32:e30
97. Kume A, Koremoto M, Xu R, Okada T, Mizukami H, Hanazono Y, Hasegawa M, Ozawa K (2003) In vivo expansion of transduced murine hematopoietic cells with a selective amplifier gene. J Gene Med 5:175–181

98. Langley RR, Ramirez KM, Tsan RZ, Van Arsdall M, Nilsson MB, Fidler IJ (2003) Tissue-specific microvascular endothelial cell lines from H-2K(b)-tsA58 mice for studies of angiogenesis and metastasis. Cancer Res 63:2971–2976
99. Langley RR, Fan D, Guo L, Zhang C, Lin Q, Brantley EC, McCarty JH, Fidler IJ (2009) Generation of an immortalized astrocyte cell line from H-2Kb-tsA58 mice to study the role of astrocytes in brain metastasis. Int J Oncol 35:665–672
100. Levskaya A, Weiner OD, Lim WA, Voigt CA (2009) Spatiotemporal control of cell signalling using a light-switchable protein interaction. Nature 461:997–1001
101. Li Q, Peterson KR, Fang X, Stamatoyannopoulos G (2002) Locus control regions. Blood 100:3077–3086
102. Lim SF, Chuan KH, Liu S, Loh SO, Chung BY, Ong CC, Song Z (2006) RNAi suppression of Bax and Bak enhances viability in fed-batch cultures of CHO cells. Metab Eng 8: 509–522
103. Liu C, Chu I, Hwang S (2001) Pentanoic acid, a novel protein synthesis stimulant for Chinese hamster ovary (CHO) cells. J Biosci Bioeng 91:71–75
104. Lloyd DR, Holmes P, Jackson LP, Emery AN, Al-Rubeai M (2000) Relationship between cell size, cell cycle and specific recombinant protein productivity. Cytotechnology 34:59–70
105. Loew R, Vigna E, Lindemann D, Naldini L, Bujard H (2006) Retroviral vectors containing Tet-controlled bidirectional transcription units for simultaneous regulation of two gene activities. J Mol Genet Med 2:107–118
106. Loew R, Meyer Y, Kuehlcke K, Gama-Norton L, Wirth D, Hauser H, Stein S, Grez M, Thornhill S, Thrasher A, Baum C, Schambach A (2010) A new PG13-based packaging cell line for stable production of clinical-grade self-inactivating gamma-retroviral vectors using targeted integration. Gene Ther 17:272–280
107. Logie C, Stewart AF (1995) Ligand-regulated site-specific recombination. Proc Natl Acad Sci U S A 92:5940–5944
108. Lufino MMP, Edser PAH, Wade-Martins R (2008) Advances in high-capacity extrachromosomal vector technology: episomal maintenance, vector delivery, and transgene expression. Mol Ther 16:1525–1538
109. Major MB, Camp ND, Berndt JD, Yi X, Goldenberg SJ, Hubbert C, Biechele TL, Gingras AC, Zheng N, Maccoss MJ, Angers S, Moon RT (2007) Wilms tumor suppressor WTX negatively regulates WNT/beta-catenin signaling. Science 316:1043–1046
110. Malphettes L, Fussenegger M (2006) Improved transgene expression fine-tuning in mammalian cells using a novel transcription-translation network. J Biotechnol 124:732–746
111. Malphettes L, Weber CC, El-Baba MD, Schoenmakers RG, Aubel D, Weber W, Fussenegger M (2005) A novel mammalian expression system derived from components coordinating nicotine degradation in arthrobacter nicotinovorans pAO1. Nucleic Acids Res 33:e107
112. Margolin JF, Friedman JR, Meyer WKH, Vissing H, Thiesen HJ, Rauscher Iii FJ (1994) Kruppel-associated boxes are potent transcriptional repression domains. Proc Natl Acad Sci U S A 91:4509–4513
113. Markusic D, Seppen J (2010) Doxycycline regulated lentiviral vectors. Meth Mol Biol 614:69–76
114. Markusic D, Oude-Elferink R, Das AT, Berkhout B, Seppen J (2005) Comparison of single regulated lentiviral vectors with rtTA expression driven by an autoregulatory loop or a constitutive promoter. Nucleic Acids Res 33:e63
115. Mátrai J, Chuah MK, Vandendriessche T (2010) Recent advances in lentiviral vector development and applications. Mol Ther 18:477–490
116. May T, Hauser H, Wirth D (2004a) Transcriptional control of SV40 T-antigen expression allows a complete reversion of immortalization. Nucleic Acids Res 32:5529–5538
117. May T, Lindenmaier W, Wirth D, Mueller PP (2004b) Application of a reversible immortalization system for the generation of proliferation-controlled cell lines. Cytotechnology 46:69–78

118. May T, Mueller PP, Weich H, Froese N, Deutsch U, Wirth D, Kroger A, Hauser H (2005) Establishment of murine cell lines by constitutive and conditional immortalization. J Biotechnol 120:99–110
119. May T, Hauser H, Wirth D (2006) Current status of transcriptional regulation systems. Cytotechnology 50:109–119
120. May T, Eccleston L, Herrmann S, Hauser H, Goncalves J, Wirth D (2008) Bimodal and hysteretic expression in mammalian cells from a synthetic gene circuit. PLoS One 3:e2372
121. May T, Butueva M, Bantner S, Marcusic D, Seppen J, Weich H, Hauser H, Wirth D (2010) Synthetic gene regulation circuits for control of cell expansion. Tissue Eng Part A 16: 441–452
122. Mazur X, Fussenegger M, Renner WA, Bailey JE (1998) Higher productivity of growth-arrested Chinese hamster ovary cells expressing the cyclin-dependent kinase inhibitor p27. Biotechnol Prog 14:705–713
123. Mezhir JJ, Smith KD, Posner MC, Senzer N, Yamini B, Kufe DW, Weichselbaum RR (2006) Ionizing radiation: a genetic switch for cancer therapy. Cancer Gene Ther 13:1–6
124. Miller AD (2001) Production of retroviral vectors. Curr Protoc Hum Genet Chap 12: Unit 12–15
125. Miskey C, Izsvák Z, Plasterk RH, Ivics Z (2003) The frog prince: a reconstructed transposon from *Rana pipiens* with high transpositional activity in vertebrate cells. Nucleic Acids Res 31:6873–6881
126. Mohan C, Park SH, Chung JY, Lee GM (2007) Effect of doxycycline-regulated protein disulfide isomerase expression on the specific productivity of recombinant CHO cells: thrombopoietin and antibody. Biotechnol Bioeng 98:611–615
127. Moosmann P, Georgiev O, Thiesen HJ, Hagmann M, Schaffner W (1997) Silencing of RNA polymerases II and III-dependent transcription by the KRAB protein domain of KOX1, a Kruppel-type zinc finger factor. Biol Chem 378:669–677
128. Morgan JE, Beauchamp JR, Pagel CN, Peckham M, Ataliotis P, Jat PS, Noble MD, Farmer K, Partridge TA (1994) Myogenic cell lines derived from transgenic mice carrying a thermolabile T antigen: a model system for the derivation of tissue-specific and mutation-specific cell lines. Dev Biol 162:486–498
129. Mullick A, Xu Y, Warren R, Koutroumanis M, Guilbault C, Broussau S, Malenfant F, Bourget L, Lamoureux L, Lo R, Caron AW, Pilotte A, Massie B (2006) The cumate gene-switch: a system for regulated expression in mammalian cells. BMC Biotechnol 6:43
130. Neddermann P, Gargioli C, Muraglia E, Sambucini S, Bonelli F, De Francesco R, Cortese R (2003) A novel, inducible, eukaryotic gene expression system based on the quorum-sensing transcription factor TraR. EMBO Rep 4:159–165
131. Nehlsen K, Schucht R, da Gama-Norton L, Kromer W, Baer A, Cayli A, Hauser H, Wirth D (2009) Recombinant protein expression by targeting pre-selected chromosomal loci. BMC Biotechnol 9:100
132. Ngan ESW, Schillinger K, DeMayo F, Tsai SY (2002) The mifepristone-inducible gene regulatory system in mouse models of disease and gene therapy. Semin Cell Dev Biol 13:143–149
133. Nishimura K, Fukagawa T, Takisawa H, Kakimoto T, Kanemaki M (2009) An auxin-based degron system for the rapid depletion of proteins in nonplant cells. Nat Methods 6:917–922
134. No D, Yao TP, Evans RM (1996) Ecdysone-inducible gene expression in mammalian cells and transgenic mice. Proc Natl Acad Sci U S A 93:3346–3351
135. Nordstrom JL (2002) Antiprogestin-controllable transgene regulation in vivo. Curr Opin Biotechnol 13:453–458
136. Palais G, Cat AND, Friedman H, Panek-Huet N, Millet A, Tronche F, Gellen B, Mercadier JJ, Peterson A, Jaisser F (2009) Targeted transgenesis at the HPRT locus: an efficient strategy to achieve tightly controlled in vivo conditional expression with the tet system. Physiol Genom 37:140–146

137. Pamment J, Ramsay E, Kelleher M, Dornan D, Ball KL (2002) Regulation of the IRF-1 tumour modifier during the response to genotoxic stress involves an ATM-dependent signalling pathway. Oncogene 21:7776–7785
138. Pathak A, Patnaik S, Gupta KC (2009) Recent trends in non-viral vector-mediated gene delivery. Biotechnol J 4:1559–1572
139. Perani A, Singh RP, Chauhan R, Al-Rubeai M (1998) Variable functions of bcl-2 in mediating bioreactor stress-induced apoptosis in hybridoma cells. Cytotechnology 28: 177–188
140. Picard D (1994) Regulation of protein function through expression of chimaeric proteins. Curr Opin Biotechnol 5:511–515
141. Pichon C, Billiet L, Midoux P (2010) Chemical vectors for gene delivery: uptake and intracellular trafficking. Curr Opin Biotechnol 21:640–645
142. Pruett-Miller SM, Connelly JP, Maeder ML, Joung JK, Porteus MH (2008) Comparison of zinc finger nucleases for use in gene targeting in mammalian cells. Mol Ther 16:707–717
143. Pryciak PM (2009) Designing new cellular signaling pathways. Chem Biol 16:249–254
144. Rinaudo K, Bleris L, Maddamsetti R, Subramanian S, Weiss R, Benenson Y (2007) A universal RNAi-based logic evaluator that operates in mammalian cells. Nat Biotechnol 25:795–801
145. Romeo G, Fiorucci G, Chiantore MV, Percario ZA, Vannucchi S, Affabris E (2002) IRF-1 as a negative regulator of cell proliferation. J Interferon Cytokine Res 22:39–47
146. Rowan S, Ludwig RL, Haupt Y, Bates S, Lu X, Oren M, Vousden KH (1996) Specific loss of apoptotic but not cell-cycle arrest function in a human tumor derived p53 mutant. Embo J 15:827–838
147. Rybkin II, Markham DW, Yan Z, Bassel-Duby R, Williams RS, Olson EN (2003) Conditional expression of SV40 T-antigen in mouse cardiomyocytes facilitates an inducible switch from proliferation to differentiation. J Biol Chem 278:15927–15934
148. Salmon P, Oberholzer J, Occhiodoro T, Morel P, Lou J, Trono D (2000) Reversible immortalization of human primary cells by lentivector-mediated transfer of specific genes. Mol Ther 2:404–414
149. Sandhu U, Cebula M, Behme S, Riemer P, Wodarczyk C, Metzger D, Reimann J, Schirmbeck R, Hauser H, Wirth D (2010) Strict control of transgene expression in a mouse model for sensitive biological applications based on RMCE compatible ES cells. Nucleic acids Res 39:e1
150. Sauerwald TM, Betenbaugh MJ, Oyler GA (2002) Inhibiting apoptosis in mammalian cell culture using the caspase inhibitor XIAP and deletion mutants. Biotechnol Bioeng 77:704–716
151. Sauerwald TM, Oyler GA, Betenbaugh MJ (2003) Study of caspase inhibitors for limiting death in mammalian cell culture. Biotechnol Bioeng 81:329–340
152. Sauerwald TM, Figueroa B Jr, Hardwick JM, Oyler GA, Betenbaugh MJ (2006) Combining caspase and mitochondrial dysfunction inhibitors of apoptosis to limit cell death in mammalian cell cultures. Biotechnol Bioeng 94:362–372
153. Schambach A, Galla M, Maetzig T, Loew R, Baum C (2007) Improving transcriptional termination of self-inactivating gamma-retroviral and lentiviral vectors. Mol Ther 15: 1167–1173
154. Schatz SM, Kerschbaumer RJ, Gerstenbauer G, Kral M, Dorner F, Scheiflinger F (2003) Higher expression of Fab antibody fragments in a CHO cell line at reduced temperature. Biotechnol Bioeng 84:433–438
155. Schroder M (2008) Engineering eukaryotic protein factories. Biotechnol Lett 30:187–196
156. Schroder M, Kaufman RJ (2005) The mammalian unfolded protein response. Annu Rev Biochem 74:739–789
157. Schucht R, Coroadinha AS, Zanta-Boussif MA, Verhoeyen E, Carrondo MJ, Hauser H, Wirth D (2006) A new generation of retroviral producer cells: predictable and stable virus production by Flp-mediated site-specific integration of retroviral vectors. Mol Ther 14: 285–292

158. Schucht R, Wirth D, May T (2009) Precise regulation of transgene expression level and control of cell physiology. Cell Biol Toxicol 26:29–42
159. Searle PF, Stuart GW, Palmiter RD (1985) Building a metal-responsive promoter with synthetic regulatory elements. Mol Cell Biol 5:1480–1489
160. Shaffer AL, Shapiro-Shelef M, Iwakoshi NN, Lee AH, Qian SB, Zhao H, Yu X, Yang L, Tan BK, Rosenwald A, Hurt EM, Petroulakis E, Sonenberg N, Yewdell JW, Calame K, Glimcher LH, Staudt LM (2004) XBP1, downstream of Blimp-1, expands the secretory apparatus and other organelles, and increases protein synthesis in plasma cell differentiation. Immunity 21:81–93
161. Sharma N, Moldt B, Dalsgaard T, Jensen TG, Mikkelsen JG (2008) Regulated gene insertion by steroid-induced PhiC31 integrase. Nucleic Acids Res 36:e67
162. Sibley CR, Seow Y, Wood MJ (2010) Novel RNA-based strategies for therapeutic gene silencing. Mol Ther 18:466–476
163. Siddiqui F, Li CY, Larue SM, Poulson JM, Avery PR, Pruitt AF, Zhang X, Ullrich RL, Thrall DE, Dewhirst MW, Hauck ML (2007) A phase I trial of hyperthermia-induced interleukin-12 gene therapy in spontaneously arising feline soft tissue sarcomas. Mol Cancer Ther 6:380–389
164. Singh RP, Al-Rubeai M, Gregory CD, Emery AN (1994) Cell death in bioreactors: a role for apoptosis. Biotechnol Bioeng 44:720–726
165. Sinn PL, Sauter SL, McCray PB Jr (2005) Gene therapy progress and prospects: development of improved lentiviral and retroviral vectors—design, biosafety, and production. Gene Ther 12:1089–1098
166. Sipo I, Hurtado Pico A, Wang X, Eberle J, Petersen I, Weger S, Poller W, Fechner H (2006) An improved Tet-On regulatable FasL-adenovirus vector system for lung cancer therapy. J Mol Med 84:215–225
167. Strathdee D, Ibbotson H, Grant SGN (2006) Expression of transgenes targeted to the Gt(ROSA)26Sor locus is orientation dependent. PLoS ONE 1:e4
168. Sung YH, Hwang SJ, Lee GM (2005) Influence of down-regulation of caspase-3 by siRNAs on sodium-butyrate-induced apoptotic cell death of Chinese hamster ovary cells producing thrombopoietin. Metab Eng 7:457–466
169. Sung YH, Lee JS, Park SH, Koo J, Lee GM (2007) Influence of co-down-regulation of caspase-3 and caspase-7 by siRNAs on sodium butyrate-induced apoptotic cell death of Chinese hamster ovary cells producing thrombopoietin. Metab Eng 9:452–464
170. Sunley K, Butler M (2010) Strategies for the enhancement of recombinant protein production from mammalian cells by growth arrest. Biotechnol Adv 28:385–394
171. Sunstrom NA, Gay RD, Wong DC, Kitchen NA, DeBoer L, Gray PP (2000) Insulin-like growth factor-I and transferrin mediate growth and survival of Chinese hamster ovary cells. Biotechnol Prog 16:698–702
172. Superti-Furga G, Bergers G, Picard D, Busslinger M (1991) Hormone-dependent transcriptional regulation and cellular transformation by Fos-steroid receptor fusion proteins. Proc Natl Acad Sci U S A 88:5114–5118
173. Swinburne IA, Miguez DG, Landgraf D, Silver PA (2008) Intron length increases oscillatory periods of gene expression in animal cells. Genes Dev 22:2342–2346
174. Tanaka N, Ishihara M, Kitagawa M, Harada H, Kimura T, Matsuyama T, Lamphier MS, Aizawa S, Mak TW, Taniguchi T (1994) Cellular commitment to oncogene-induced transformation or apoptosis is dependent on the transcription factor IRF-1. Cell 77:829–839
175. Tanaka N, Ishihara M, Lamphier MS, Nozawa H, Matsuyama T, Mak TW, Aizawa S, Tokino T, Oren M, Taniguchi T (1996) Cooperation of the tumour suppressors IRF-1 and p53 in response to DNA damage. Nature 382:816–818
176. Tascou S, Sorensen TK, Glenat V, Wang M, Lakich MM, Darteil R, Vigne E, Thuillier V (2004) Stringent rosiglitazone-dependent gene switch in muscle cells without effect on myogenic differentiation. Mol Ther 9:637–649
177. Tigges M, Marquez-Lago TT, Stelling J, Fussenegger M (2009) A tunable synthetic mammalian oscillator. Nature 457:309–312

178. Tigges M, Denervaud N, Greber D, Stelling J, Fussenegger M (2010) A synthetic low-frequency mammalian oscillator. Nucleic Acids Res 38:2702–2711
179. Toettcher JE, Mock C, Batchelor E, Loewer A, Lahav G (2010) A synthetic-natural hybrid oscillator in human cells. Proc Natl Acad Sci U S A 107:17047–17052
180. Trummer E, Fauland K, Seidinger S, Schriebl K, Lattenmayer C, Kunert R, Vorauer-Uhl K, Weik R, Borth N, Katinger H, Müller D (2006) Process parameter shifting: part II. Biphasic—a tool for enhancing the volumetric productivity of batch processes using Epo-Fc expressing CHO cells. Biotechnol Bioeng 94:1045–1052
181. Unsinger J, Kröger A, Hauser H, Wirth D (2001) Retroviral vectors for the transduction of autoregulated, bidirectional expression cassettes. Mol Ther 4:484–489
182. Unsinger J, Lindenmaier W, May T, Hauser H, Wirth D (2004) Stable and strictly controlled expression of LTR-flanked autoregulated expression cassettes upon adenoviral transfer. Biochem Biophys Res Commun 319: 879–887
183. Urlinger S, Baron U, Thellmann M, Hasan MT, Bujard H, Hillen W (2000) Exploring the sequence space for tetracycline-dependent transcriptional activators: novel mutations yield expanded range and sensitivity. Proc Natl Acad Sci U S A 97:7963–7968
184. Urnov FD, Rebar EJ, Holmes MC, Zhang HS, Gregory PD (2010) Genome editing with engineered zinc finger nucleases. Nat Rev Genet 11:636–646
185. Vegeto E, Allan GF, Schrader WT, Tsai MJ, McDonnell DP, O'Malley BW (1992) The mechanism of RU486 antagonism is dependent on the conformation of the carboxy-terminal tail of the human progesterone receptor. Cell 69:703–713
186. Volpers C, Kochanek S (2004) Adenoviral vectors for gene transfer and therapy. J Gene Med 6:S164–S171
187. Wang JD, Shi WL, Zhang GQ, Bai XM (1994) Tissue and serum levels of steroid hormones and RU 486 after administration of mifepristone. Contraception 49:245–253
188. Wang GG, Calvo KR, Pasillas MP, Sykes DB, Hacker H, Kamps MP (2006) Quantitative production of macrophages or neutrophils ex vivo using conditional Hoxb8. Nat Methods 3:287–293
189. Ward M, Sattler R, Grossman IR, Bell AJ Jr, Skerrett D, Baxi L, Bank A (2003) A stable murine-based RD114 retroviral packaging line efficiently transduces human hematopoietic cells. Mol Ther 8:804–812
190. Watanabe S, Shuttleworth J, Al-Rubeai M (2002) Regulation of cell cycle and productivity in NS0 cells by the over-expression of p21CIP1. Biotechnol Bioeng 77:1–7
191. Weber W, Fussenegger M (2010) Synthetic gene networks in mammalian cells. Curr Opin Biotechnol 21:690–696
192. Weber W, Bacchus W, Daoud-El Baba M, Fussenegger M (2007) Vitamin H-regulated transgene expression in mammalian cells. Nucleic Acids Res 35:e116
193. Weber W, Marty RR, Link N, Ehrbar M, Keller B, Weber CC, Zisch AH, Heinzen C, Djonov V, Fussenegger M (2003a) Conditional human VEGF-mediated vascularization in chicken embryos using a novel temperature-inducible gene regulation (TIGR) system. Nucleic Acids Res 31:e69
194. Weber W, Schoenmakers R, Spielmann M, El-Baba MD, Folcher M, Keller B, Weber CC, Link N, van de Wetering P, Heinzen C, Jolivet B, Sequin U, Aubel D, Thompson CJ, Fussenegger M (2003b) Streptomyces-derived quorum-sensing systems engineered for adjustable transgene expression in mammalian cells and mice. Nucleic Acids Res 31:e71
195. Weber W, Lienhart C, Baba MD, Fussenegger M (2009a) A biotin-triggered genetic switch in mammalian cells and mice. Metab Eng 11:117–124
196. Weber W, Schuetz M, Denervaud N, Fussenegger M (2009b) A synthetic metabolite-based mammalian inter-cell signaling system. Mol Biosyst 5:757–763
197. Weber W, Luzi S, Karlsson M, Sanchez-Bustamante CD, Frey U, Hierlemann A, Fussenegger M (2009c) A synthetic mammalian electro-genetic transcription circuit. Nucleic Acids Res 37:e33
198. Wells DJ (2010) Electroporation and ultrasound enhanced non-viral gene delivery in vitro and in vivo. Cell Biol Toxicol 26:21–28

199. West AG, Fraser P (2005) Remote control of gene transcription. Hum Mol Genet 14: R101–R111
200. Wirth D, Gama-Norton L, Riemer P, Sandhu U, Schucht R, Hauser H (2007) Road to precision: recombinase-based targeting technologies for genome engineering. Curr Opin Biotechnol 18:411–419
201. Wong DC, Wong KT, Nissom PM, Heng CK, Yap MG (2006) Targeting early apoptotic genes in batch and fed-batch CHO cell cultures. Biotechnol Bioeng 95:350–361
202. Wong SP, Argyros O, Coutelle C, Harbottle RP (2009) Strategies for the episomal modification of cells. Curr Opin Mol Ther 11:433–441
203. Wu YI, Frey D, Lungu OI, Jaehrig A, Schlichting I, Kuhlman B, Hahn KM (2009) A genetically encoded photoactivatable Rac controls the motility of living cells. Nature 461:104–108
204. Würtele H, Little KCE, Chartrand P (2003) Illegitimate DNA integration in mammalian cells. Gene Ther 10:1791–1799
205. Yanai N, Suzuki M, Obinata M (1991) Hepatocyte cell lines established from transgenic mice harboring temperature-sensitive simian virus 40 large T-antigen gene. Exp Cell Res 197:50–56
206. Yang G, Rosen DG, Mercado-Uribe I, Colacino JA, Mills GB, Bast RC Jr, Zhou C, Liu J (2007) Knockdown of p53 combined with expression of the catalytic subunit of telomerase is sufficient to immortalize primary human ovarian surface epithelial cells. Carcinogenesis 28:174–182
207. Yao F, Svensjo T, Winkler T, Lu M, Eriksson C, Eriksson E (1998) Tetracycline repressor, tetR, rather than the tetR-mammalian cell transcription factor fusion derivatives, regulates inducible gene expression in mammalian cells. Hum Gene Ther 9:1939–1950
208. Yao G, Lee TJ, Mori S, Nevins JR, You L (2008) A bistable Rb-E2F switch underlies the restriction point. Nat Cell Biol 10:476–482
209. Yazawa M, Sadaghiani AM, Hsueh B, Dolmetsch RE (2009) Induction of protein–protein interactions in live cells using light. Nat Biotechnol 27:941–945
210. Yoon SK, Kim SH, Lee GM (2003a) Effect of low culture temperature on specific productivity and transcription level of anti-4-1BB antibody in recombinant Chinese hamster ovary cells. Biotechnol Prog 19:1383–1386
211. Yoon SK, Song JY, Lee GM (2003b) Effect of low culture temperature on specific productivity, transcription level, and heterogeneity of erythropoietin in Chinese hamster ovary cells. Biotechnol Bioeng 82:289–298
212. Yoon SK, Hwang SO, Lee GM (2004) Enhancing effect of low culture temperature on specific antibody productivity of recombinant Chinese hamster ovary cells: clonal variation. Biotechnol Prog 20:1683–1688
213. Yoon SK, Hong JK, Choo SH, Song JY, Park HW, Lee GM (2006) Adaptation of Chinese hamster ovary cells to low culture temperature: cell growth and recombinant protein production. J Biotechnol 122:463–472
214. Yu SF, von Ruden T, Kantoff PW, Garber C, Seiberg M, Ruther U, Anderson WF, Wagner EF, Gilboa E (1986) Self-inactivating retroviral vectors designed for transfer of whole genes into mammalian cells. Proc Natl Acad Sci U S A 83:3194–3198
215. Zhao HF, Boyd J, Jolicoeur N, Shen SH (2003) A coumermycin/novobiocin-regulated gene expression system. Hum Gene Ther 14:1619–1629
216. Zhu Z, Ma B, Homer RJ, Zheng T, Elias JA (2001) Use of the tetracycline-controlled transcriptional silencer (tTS) to eliminate transgene leak in inducible overexpression transgenic mice. J Biol Chem 276:25222–25229

Adv Biochem Engin/Biotechnol (2012) 127: 285–288
DOI: 10.1007/978-3-642-28350-5

Index

L

M

N

O

P

R

PGSTL